Thorstein Veblen

and the Enrichment of Evolutionary Naturalism

Thorstein Veblen

and the Enrichment of Evolutionary Naturalism

Rick Tilman

University of Missouri Press
Columbia and London

Library of Congress Cataloging-in-Publication Data

Tilman, Rick.
  Thorstein Veblen and the enrichment of evolutionary naturalism /
Rick Tilman.
      p. cm.
  Summary: "Tilman argues that evolutionary naturalism provides
the philosophical foundations of Veblen's thought. He links evolutionary naturalism to Veblen's aesthetics, secular humanism, sociology of control, sociobiology, and sociology of knowledge, thereby
initiating observations regarding the relationship of Veblen's own
life to his thinking and his place as a cultural lag theorist"—Provided
by publisher.
  Includes bibliographical references and index.
  ISBN 978-0-8262-1714-1 (hard cover : alk. paper)
  1. Social sciences—Philosophy.   2. Naturalism.   3. Veblen,
Thorstein, 1857–1929.   I. Title.
  H61.15.T55 2007
  300.92—dc22   2006038994

Designer: *foleydesign*
Typesetter: The Composing Room of Michigan, Inc.
Printer and binder: The Maple-Vail Book Manufacturing Group
Typefaces: URW Antiqua and Palatino

To Jack Putnam,
      Don Lammers,
           Chuck Green,
      Pete Bateman,
Currin V. Shields,
      Bill McCormick,
           and my other teachers, 1944–1968

Contents

Preface

How is this book related to my earlier studies in which Thorstein Veblen is the key figure as well as to my larger research agenda which focuses on him? The first consists of three monographs on him, a three-volume anthology of writings about his life and work, an edited collection of his writings, and a coedited annotated bibliography. It also includes dozens of articles and chapters with Veblen as their central focus and a few reviews of books about him and related topics. My larger research agenda locates him in the context of modern intellectual history by comparing him with his contemporaries, analyzing the work of his critics, and interpreting his writing through textual exegesis.[1] I have also viewed him through the doctrinal lens of various thinkers, schools of thought, and theoretical perspectives. In a sense, the present study is a composite of much of what I have written and thought about Veblen since 1972 as well as recent archival work I have engaged in. He also is the central figure in *The Political Thought of Thorstein Veblen*, which I am coauthoring with Sidney Plotkin of Vassar College. However, Veblen is a minor figure in *The American Tradition of Empirical Collectivism*, the title of another book I am writing. In any case, *Thorstein Veblen and the Enrichment of Evolutionary Naturalism* is just another increment in an ongoing process and is best understood through evolutionary theory and philosophic naturalism.

For Veblen, the economy is not a self-regulating mechanism reflecting

I am indebted to Bill Dugger, Clare Eby, Edythe Miller, and two anonymous critics for several of the ideas expressed here.

1. As Edythe Miller recently pointed out, I affirm my affinity "With the growing group of Veblen scholars . . . who *reject* the long-held and formerly quite universally accepted interpretation advanced by Joseph Dorfman [of Veblen] as *culturally segregated, marginalized,* and a *philanderer*" [author's emphasis]. Miller, Review of Rick Tilman, "Thorstein Veblen, John Dewey, C. Wright Mills and the Generic Ends of Life," *Journal of Economic Issues* 39 (December 2005): 1088.

a stable human nature and a perpetual social order of landlords, laborers, and capitalists, each the social encapsulation of real elements of land, labor, and capital. Nor does the economic system by virtue of Say's law always tend toward an equilibrium of full employment. Veblen did not believe that conventional economics was free of class bias or value neutral. Instead, it was biased toward an individualistic ethos rooted in subjective preference that defines value simply as exchange value measured only by price. To Veblen, wealth is collectively generated, a result of a social process. The received economics of his day thus failed to locate the power and income inequality in the existing institutional fabric and thus sanctioned great inequalities of power and privilege.[2]

Veblen also held to an intellectual position that was nonpositivist, nonformalist, and nondualistic. On the whole, he interpreted inquiry as both value rooted and purposive. Yet he repudiated fixed truths and rigid ends and was committed to an egalitarian, participationist, communitarian ethos in which individual involvement was a valued ideal. He rejected the false dualisms provided by positivism between means and ends, thought and action, fact and value, and theory and practice. Veblen viewed the mind as an active agent rather than a passive recipient of sense data. Perhaps most important he functioned as a native critical theorist who sought a critique of critique by exposing the distinction between myths and facts, the ceremonial and the instrumental, ideology and reality in institutions and social structure. Social theory itself may be retarded by cultural lag—his own theorizing about consumption, for example, was an effort to overcome cultural lag at a time when few economists consciously theorized about it.

Veblen's more polemical writing is often interpreted as an idiosyncratic and purely individualist perspective. For example, his *Higher Learning in America* (1918) might be interpreted as evidence that he was not seriously interested in the university or its place within the larger society. In short, it is sometimes suggested that he had a vested interest in propagandizing some idealized version of the university so he could enjoy a life that pleased his intellectual interests. So, for example, he could teach a few well-prepared, highly motivated students, have abundant leisure to read Scandinavian folk tales, make fun of imbecile institutions, and so forth. The cynic might argue that Veblen believed that hierarchical, bureaucratic, businesslike arrangements were appropriate for most students and faculty But the few of us who have the intellect to engage in pursuits that we think appropriate should not have to work in the lesser colleges and universities and should not be accountable to those higher in the technocrat-

2. Ibid., 1087.

❧ Review Copy ☙

*Thorstein Veblen and the Enrichment
of Evolutionary Naturalism*

Rick Tilman

ISBN: 978-0-8262-1714-1
Price: $49.95s cloth, 368 pages
Publication Date: April 30, 2007

Please send our publicity department two copies of
your review when it appears.

University of Missouri Press
2910 LeMone Boulevard
Columbia, MO 65201
(573) 882-0180 ◆ Fax (573) 884-4498

ic ladder, or so it is claimed. Cynics of this ilk, who might place such an interpretation on Veblen, correspond to Oscar Wilde's definition of them as knowing "the price of everything and the value of nothing." They are found in abundance among the ranks of neoclassical economists, but they tell us little about Veblen's life or work that helps in understanding it.

The title of this book suggests that Veblen enriched—that is, enhanced, strengthened, and enlarged—the scope of evolutionary naturalism cumulatively and incrementally, and among other things changed it qualitatively. The study attempts to show how he did this. However, to accomplish this goal, it is essential to demonstrate how normative values both shape and mold evolutionary naturalism, how they are embedded in it, and how they can be extracted from it.

The critic may, of course, object that to do this would require overcoming or somehow not succumbing to the naturalistic fallacy of deriving ought from is. This assumes the validity of the epistemological distinction between facts and values as asserted by David Hume among others, and as the reader may know, Hume was Veblen's favorite philosopher among the British empiricists. In recent times, however, the philosopher John Searle has tried to show that the naturalistic fallacy is invalid. His argument is that if I "promise" to do something I "ought" to do it. Since promising to perform a certain action is conventionally understood to require honoring the promise, "ought" can, indeed, be derived from "is."[3]

Whether we accept this argument as valid or merely a clever play on words, it is evident that the naturalistic fallacy may no longer be the barrier it once was. This would not surprise Veblen, who mostly honored it in the breach anyway. While paying lip service at times to the positivist epistemological status of facts and values that sharply and definitively separated them from each other, he used massive barrages of satire and mockery to lampoon social values and practices he disliked.

To continue employing the naturalistic fallacy to analyze and indict Veblen is to aid in reducing his social critique to the status of the crotchets of the most bizarre and alienated of modern economists. Surely, he was much more than this. In any case, this is hardly the place for an extended discussion of a difficult subject about which most social scientists know even less than the author! Instead, the reader is referred to the more technical literature of philosophy and metatheory, some of which is listed in the bibliography.

3. See John Searle, "How to Derive 'Ought' from 'Is' in 'The Is/Ought Question,'" *A Collection of Papers on the Central Problem in Moral Philosophy*, ed. W. D. Hudson (London: Macmillan Press, 1969): 120–34.

It is interesting to note that two of the earliest studies of Veblen's thought, including one that focused on his philosophy, do not include "evolutionary naturalism" or even "naturalism" in their indexes. I am, of course, referring to Joseph Dorfman's *Thorstein Veblen and His America* (1934) and Stanley Daugert's *Philosophy of Thorstein Veblen* (1950). More recently, however, it has become more commonplace to attach these labels to his thought and, hopefully, as this study unfolds it will become evident as to why no other label than "evolutionary naturalism" so accurately fits his philosophy. Both the Dorfman and Daugert studies remain useful, if dated, analyses of Veblen's thought; but the recent articulation of "the progressive enrichment of evolutionary naturalism" by the philosopher Peter Hare is the source of the title of this book and a main source of inspiration for writing it.[4]

After analyzing Veblen's first published article, "Kant's Critique of Judgment" (1884), in the first chapter of his study of Veblen's philosophy, Daugert comments:

> When Veblen left Yale after receiving a Doctor's degree in philosophy in 1884 for a dissertation entitled "Ethical Grounds of a Doctrine of Retribution," he returned to his parents' home in the Mid-West where he lived the following seven years. He next entered academic life as a graduate student at Cornell in 1891. In the meantime he apparently had decided not to continue studying philosophy as such, but instead to concentrate his attention on political economy. The first article marking a changed philosophical orientation in his thinking was entitled "Some Neglected Points in the Theory of Socialism." Veblen's neo-Kantian metaphysics and epistemology now appeared to be developing into an empirico-genetic epistemology and social psychology.[5]

The last two sentences of this quotation are of particular value in indicating a "changed philosophical orientation" and movement away from neo-Kantian metaphysics and epistemology in the direction of "an empirico-genetic epistemology and social psychology," provided that the later is understood to be a form of evolutionary naturalism and its derivatives.

4. See Peter Hare, "The American Naturalist Tradition," 38; "Problems and Prospects in the Ethics of Belief," in *Pragmatic Naturalism and Realism*, ed. John R. Shook, 239–61; "Classical Pragmatism, Recent Naturalistic Theories of Representation, and Pragmatic Realism," in *The Role of Pragmatics in Contemporary Philosophy*, ed. P. Weingartner et al., 58–65; "Dewey, Analytic Epistemology, and Biology," in *Dewey, Pragmatism, and Economic Methodology*, ed. Ellias Khalil, 144–52; "The American Philosophical Tradition as Progressively Enriched Naturalism." Comments on Arnold Berleants, "Metapragmatism and the Future of American Philosophy" (Paper presented at meeting of SAAP, Bentley College, Waltham, MA, March 1995).

5. Stanley Daugert, *Philosophy of Thorstein Veblen*, 26.

Did Veblen have an early "non-naturalistic phase" as a young man before he became fully committed to naturalism? It has been suggested that his first published article in 1884 still contains at least the residues of his training in German Idealism. Certainly among the first influences in his childhood was his indoctrination in Protestant Christianity, that is, Norwegian Lutheranism. To what extent did these influences remain with him during his scholarly career? The position taken here is simply that Kant and Christ and company were secularized, the subjectivist and mystical elements in them sloughed off, and the elements congruent with evolutionary naturalism absorbed. If Veblen had an early non-naturalistic phase as a young man before he became fully committed to evolutionary naturalism, it had disappeared by 1892 when he began writing once again for scholarly publication.

At least two monographs whose titles contain the term "evolutionary naturalism" exist. The oldest one by several generations is the work of Roy Wood Sellars published in 1922. The other is by Michael Ruse written in 1995.[6] The view adopted here is that Veblen's evolutionary naturalism more closely resembles the earlier work. Sellars was younger than Veblen, but he wrote during Veblen's scholarly career and his views are more convergent with him. However, as I will later show, the title of this study, *Thorstein Veblen and the Enrichment of Evolutionary Naturalism,* also draws on the contemporary philosopher Peter Hare's contributions.

Veblen had a philosophy of history, to be sure, but it is often best revealed in his cultural and social anthropology, particularly in his four-stage evolutionary framework of social development, which is too well known to justify reiteration here. It is also evident in his belief in the likelihood of the persistence and resurgence of atavistic continuities, such as force, fraud, and superstition. More to the point, however, is where he falls on the continuum between idiographic and nomothetic views of history. As a philosophy of that discipline, Veblenism clearly does not rely on the narrow hyperfactualism of the idiographic view. Nor does it evolve grandiose theories of historical change as in the nomothetic view, despite occasional tugs in that direction. Long before C. Wright Mills rightly urged scholars to be culturally and historically specific and avoid the pit-

6. See Michael Ruse, *Evolutionary Naturalism Selected Essays* (London and New York: Routledge, 1995), and Roy Wood Sellars, *Evolutionary Naturalism* (Chicago: Open Court Publishing Company, 1922). The latter is more important for my purposes because it is contemporaneous with Veblen and exemplifies evolutionary naturalism with its crudities and flaws as it stood in his lifetime. Also, see Ernest W. Dewey and David L. Miller, "Veblen's Naturalism Versus Marxian Materialism," 165–74. What is novel here is my effort to link together his evolutionary naturalism and secular humanism and demonstrate that they are a logical and mostly coherent expression of his cosmology.

falls of both abstracted empiricism and grand systems theory, Veblen had uttered similar strictures, although in less rhetorical and strident tones than his disciple. But, more specifically, what was the precise nature of his view and did he practice what he preached? Although answering these two questions regarding his philosophy of history is not the central focus of this study, it provides information and analysis for those who wish to pursue these inquiries further.

The title of this study might suggest to the uninitiated that Veblen believed in social progress. He did accept the beneficence of higher rates of literacy, longer life expectancy, and higher disposable income when it consisted of socially useful goods and services. Certainly the decline in superstition and many religious practices and observances was pleasing to him when and where it occurred; the ebbing of nationalism and patriotism gratified him and its intensification appalled him. He did not specifically refer to his philosophy as evolutionary naturalism, nor did he endorse its progressive enrichment. But both of these resulted from his scholarly efforts and by way of explanation, they were latent as well as manifest, which in sociological jargon means unintended as well as intended. It is essential to recognize that Veblen succeeded in progressively enriching evolutionary naturalism by using it in realms of inquiry where it had often not been used before. That he deliberately and specifically set out to do this because he cherished utopian ideas about irreversible progress is doubtful.

Interestingly, an early Veblen scholar who detected the evolutionary naturalist qualities of Veblen was Lev Dobriansky (1957).[7] Although his work on Veblen is often marred by his Aristotelian and Thomist bias, it is only fair to credit him with identifying his secular humanism and locating its roots in American philosophic naturalism. Dobriansky is everything Veblen abhorred, a purveyor of archaic prejudices, indeed, a living exemplification of cultural lag and an apologist for the Roman Catholic Church—an institution Veblen strongly disliked (see chapter 5). But it was apparently these very deficiencies that alerted Dobriansky early on to the nature of Veblen's doctrines and their ideational affinities.

In any case, I am most concerned with asking and answering the following questions. What advantages accrued to Veblen as a theorist and critic from his evolutionary naturalist perspective and what biases did this foster in his work that matter? Closely linked to them is his role in the "progressive enrichment" of evolutionary naturalism. How and in what

7. See Lev Dobriansky, *The Values of Veblen* (Washington, DC: Public Affairs Press, 1957).

ways did he contribute to its development? In short, if Veblen was a leading American critical theorist, how is this revealed in his writings between 1884 and 1929 in his enhancement of evolutionary naturalism as it channeled his social theory and criticism?

Intellectuals and Intellectualism

The meaning of the term "intellectual" varies somewhat depending on who employs it and for what purposes. Probably three definitions of intellectual are most commonly used and most efficacious. The first has to do with the production, transmission, and dissemination of ideas. It focuses on books and the life of the mind. The second is more sociological and locates and identifies intellectuals as a specific group of related occupational categories such as academics, teachers, journalists, and so forth. The last sense in which "intellectual" is used is cultural and is permeated with the notion of cultural authority that is a claim to be recognized, even listened to on account of specialized training and knowledge that enables intellectuals of this sort to speak "authoritatively" or at least knowledgeably to more general audiences. Veblen was an "intellectual" in all three senses despite his lacerating, satirical, mocking way of expressing himself in his strenuous efforts to expose the source of the pathologies which afflict humankind.

But Veblen's intellectual pedigree and outlook need more definition and extrapolation. His general outlook included a skepticism toward the belief that moral and cultural progress were occurring, especially when a belief in progress was hidden in the language of special pleading such as "improvement," "melioration," and "providentiality." Perhaps no more skilled dissector of ideological verbiage appeared on the American scene during his lifetime than Veblen. Indeed, he attempted with some degree of success to rip away those parts of ideology that enmesh and inhibit the fulfillment of human potential. For he understood "ideology" in the pejorative sense as a distortive intrusion into and shaping of our thought. And, in fact, he held several, if not many, morally offensive cultural values—offensive, at least in the eyes of the leisure class and its political and doctrinal satellites. To make matters worse, he was astute in locating and criticizing the value-evidence relationship in the work of others. In general, it may be said that the idea of progress is the belief that things are getting better thanks to human effort and that they will continue to improve. But the version of progress Veblen might have accepted was one rooted in social indicators with empirical bias and import and with the proviso that atavistic continuities exist and are likely to reassert themselves and may undo progress that appears to have occurred.

Veblen's extensive formal training and reading in the discipline of philosophy that had become formally separated from psychology when he was a young man made him aware of the major fallacies other thinkers warned against. While warning his colleague economist John Cummings against logical fallacies such as the law of the excluded middle, he avoided philosophy's major fallacies—the "naturalistic," "genetic," and "misplaced" concreteness—as best he could. He did this not by direct confrontation, however, but by a mixture of satire, mockery, and avoidance, and added to this, by mostly not committing the fallacies themselves. He did not often confuse the *is* with the *ought,* the genesis of a theory with its validity, nor did he often mistake the abstract for the concrete. In fact, in the case of the latter he achieved some degree of notoriety for usually being historically and culturally specific despite his alleged predilections toward vague generalizations. Nevertheless, it is not always easy to ascertain how seriously to take his disclaimers nor is it easy to overlook entirely the instances in which he ignored his own strictures. In any case, he knew that understanding the genesis and development of a scientific theory did not prove its validity; no adequate understanding of theories such as Darwin's was possible without knowledge of its evolution within a certain historical and sociocultural context; the is/ought distinction could be a convenient device for avoiding critical analysis of existing institutions and the values undergirding them and the confusion between the abstract and the specific, likewise.

In fact, Veblen well recognized that the fact/value distinction was itself epistemologically value laden—the very belief that science does not exhibit human values was itself the result of a value-laden activity. Genuine science may contain no blatant impregnation by moral or ideological values. But, as Veblen knew, it is hardly value free or value neutral either. To illustrate, his work demonstrates that values effect the choice of topics, the use of language, and the selection of hypotheses to test and explore. Recognition of the ideological factors in morality and religion as impacting on "impartial" or "objective" inquiry was essential to scientific inquiry and sound scholarship, or so he would have us believe.

Veblen did not view religion, patriotism, and status emulation not to mention sports and gambling as things relative and functional only to a particular social system, that is, as mere expressions of the total structure of society. Although he was strongly opposed to the existing pockets of irrationality and inequity (ceremonialism), he doubted that these would soon disappear. Or even if they did temporarily diminish in size or perniciousness, they might reappear. In short, he was skeptical of the likelihood of the creation and maintenance of the best and most efficient final society in the calculable future unlike many contemporary Marxists. Also, unlike some Marxists Veblen did not demote to secondary status all questions of "inter-subjectivity of gender, race and ethnicity, of culture, psychology, and art, of ethics, human rights and democracy."[8] However important he may have regarded ownership and control of access to the means of production, he recognized that revolutionary changes in them were not ordinarily on the political agenda in advanced industrial communities.

Veblen recognized the tension that exists in human thought and its many cultures between the demands of objective knowledge of reality and the environment within which it emerges and often remains. But the question still persisted as to how to distinguish scientific enterprise and sound scholarship from mere myth-making. In short, Veblen was doubtful that there was such a thing as a totally disinterested participant in scientific inquiry. And his classic essay "The Place of Science in Modern Civilization" (1906) is a sensitive exploration of his skepticism in this regard.

Did Veblen believe that such things as good arguments and good evidence be essentially independent of a particular social context? Likewise, are moral or ethical choices and values ever independent of social context? Or did he simply believe in the limited relevance of relativism? As a transition figure in the shift from modernism to postmodernism, he neither endorsed the radical social construction of reality nor the complete relativism of values.

Yet Veblen was acutely aware of the claim that it was essential to distinguish between contributions to theoretical understanding whose factual validity does not depend on the social ideal to which a social scientist may subscribe. In short, that commitment to the dissemination or realization of certain social ideals may not be accepted by all social scientists. Is "real" science about objective, disinterested "external" reality—is it about the way that the world is rather than about the way that the world ought to be?

In response to such, Veblen apparently believed that Darwin and the new findings in evolutionary theory could be of aid in bridging the gap between the is and the ought, and that ethics in particular could be grounded in evolutionary thought. Was morality essentially cultural or

8. Lenny Moss, "Darwinism, Dualism, and Biological Agency," in *Darwinism and Philosophy*, ed. Vittorio Hosle and Christian Illies (Notre Dame: University of Notre Dame Press, 2005), 349.

was it fashioned by natural selection? Veblen seemed to believe that it was both, in part because the former was the product of the latter.

He realized that scientific truth is independent of one's own subjective wishes or dispositions; but a real world existed with its objective side and a subjective side, a reflection of the culture in which it was formed. But ultimately he reached no definitive position as to how to keep the two separate from each other, which may be a conclusion in itself.

Probably few knowledgeable Veblen scholars would deny that despite his eccentricities his work represents, if it does not exemplify, the naturalistic turn in the social sciences. But, unlike many of his critics and contemporaries, he knew only too well that demands for scientific rigor often concealed an insensitivity to the ethical and political aspects of inquiry. That is why it is necessary to contextualize his thought in the political-moral matrix made possible by evolutionary naturalism. Nor can the possibility be ignored that his own commitments to normatively substantive truths overrode his detachment and objectivity, thus opening a new paradigm for research. In any case, his thought encouraged unification of the social sciences in ways that enhanced critical reflection on the existing sociopolitical order. But did he seek to avoid institutional ossification and inertia through massive changes in our knowledge production processes? If so, he never made this clear. Nevertheless, Veblen was dubious regarding the suitability of the university as a site for critical social movements and inquiry. In fact, he had little confidence in the critical sensibility of academics; that academe could become a staging ground or a focal point for structural sociopolitical change probably struck him as unlikely.

Critics in the economics profession charged that Veblen's work was ideology disguised as economics. He rebutted them by pointing to the epistemological constraints normally operating on economists of which they were hardly aware. But Veblen's naturalistic epistemology made him increasingly skeptical that his critics could reconsider the validity of their beliefs because they were so enmeshed in the ceremonialism of their own discipline and the sociocultural network that enmeshed it. Although it is sometimes difficult to identify and locate the exact epistemological grounds for Veblen's claims, he clearly had reflected on the broader sociopolitical conditions and significance of these claims and the reasons his critics could not accept or even acknowledge them.

Veblen made no sharp distinction between the contexts of discovery and justification, a distinction important to many recent philosophers of science. Perhaps they would even argue that he had committed the "genetic fallacy." But then he knew that the intellectual history of the West was a succession of often-conflicting epistemologies. Consequently, he did not attempt to change its canonical arrangements so much as he sought to change the canons and their meanings themselves.

In any case, Veblen did not regard metatheory and its three components—epistemology, ontology, and axioms—as simply the reflection of a particular social formation such as classes or groups. Instead, he recognized the partial autonomy of ideational patterns as, for example, his development of a quasi-evolutionary model of economic change based on Darwin's rejection and denial of teleology or, for that matter, his subjection of epistemology to history, social class, and power considerations without submerging it in social formations or, at times, without being adequately explicit about what he was doing. The ontological pluralism that existed even in his day was exemplified by the fact that academic disciplines projected different ontologies, and Veblen made a scholarly reputation lampooning several of them. Was he, therefore, a metaphysical indeterminist? Hardly. He believed that social scientists had an ethical responsibility to develop an adequate social ontology regarding the structure of reality. Evaders, imitators, distorters, and saboteurs of this process were targets of his satire.

In what sense(s) can Veblen be said to have enriched evolutionary naturalism? First, are the ways in which he did not enrich it, or where his contributions were minimal at best. He did not contribute to neo-Darwinism because he was not a natural scientist, and his attempts to apply Darwinism to the social sciences were not an unqualified success. Indeed, as Geoff Hodgson and others have shown, he did not avoid the pitfalls and ambiguities of attempting to develop a fully evolutionary economics.[9] Secondly, in the technical philosophic sense, he did not refine American naturalism nor did he intend to. An adumbrated version of an evolutionary epistemology, ontology, and axiomatic inquiry exists in his work. But this is not where his real contribution lies. Rather, it resides in the historically social-theoretic and anthropologically rooted secular humanism that is an outgrowth of his evolutionary naturalism. Obviously, it is possible to adopt the policy stances of secular humanism on such topics as war, poverty, and various social pathologies without being a naturalist. But there is no denying the path Veblen took whether one agrees with the metatheoretical basis of his thought or not. In fact, his biases point ex-

9. See Hodgson's, *The Evolution of Institutional Economics Agency, Structure and Darwinism in American Institutionalism* (London and New York: Routledge, 2004); and "Generalizing Darwinism to Social Evolution: Some Early Attempts," *Journal of Economic Issues* 39 (December 2005): 899–914.

plicitly in the direction of the evolutionary naturalism on which his secular humanism is seated.

How does one explain the fact that not all evolutionary naturalists agreed with his indictment of basic social values and institutional practices? Evidently, his version of it did not always converge with the social theory and social criticism of other evolutionary naturalists. Was this because there was something distinctive or unique about it? This puzzle will be dealt with later.

Finally, how does evolutionary naturalism relate to secular humanism? It is difficult to see how secular humanism could avoid large-scale doctrinal convergence by incorporating evolutionary naturalism into its creed and vice versa. Secular humanism is, of course, capable of an eclectic absorption of doctrine from a multiplicity of sources. But there are metatheoretical limits to its eclecticism that will be articulated in the text.[10] It is to the scrutiny of these and other closely related questions that I now turn.

10. I am indebted to Steve Fuller, "The Case of Fuller vs. Kuhn," *Social Epistemology* 18, no. 1 (January–March 2004): 3–49, for questions raised and answers pondered in the last part of the Preface.

Acknowledgments

Paul Goldstene, Bob Farrell, Bob Baron, and several anonymous critics deserve thanks for reading earlier drafts of parts of the manuscript. Thanks are also due to the Tuesday morning coffee klatch which meets in the lounge of the History Department at Northern Arizona University. I am grateful to the philosopher Peter Hare, whose writings on evolutionary naturalism suggested the title of this book; as usual, I am indebted to Russell and Sylvia Bartley for their support. They and William Melton, Tom Veblen, Jonathan Larson, Eric Hilleman, Solidelle Wasser, Adil Mouhammed, Kyrill Thompson, Stephen Edgell, and Bill Waller contributed to solidification of my thinking on the subjects of Veblen's biography and his thought.

Some years ago historian John P. Diggins was generous enough to lend me copies of his correspondence and interviews with Veblen's stepdaughters, Becky Veblen Meyers and Ann Veblen Sims. I would like to thank him for lending me this material and also for his long-term help with my work in American intellectual history. On several occasions, his aid has been invaluable and it will not be forgotten.

I am heavily indebted to my friend and coauthor on other projects, Sidney Plotkin of Vasser College, for his aid, encouragement, and expertise as a political theorist. Also, my former student Michael Boyles contributed to the chapter on Veblen and sociobiology and is coauthor of it. My wife, Ruth Anne Porter Tilman, made a significant scholarly contribution to the section on gambling in chapter 6 when it was originally published in a chapter of a book edited by Doug Brown on the occasion of the centennial of *The Theory of the Leisure Class*. Finally, I thank Robert Griffin, Michael Hughey, and Cyrill Pasterk for their help with the chapter on Veblen's aesthetics and Carolyn Deibel for her typing skills and patience.

I would like to acknowledge the valuable and painstaking review of the manuscript for the University of Missouri Press by Clare Eby and William Dugger. As usual, the most penetrating critique of my work was by my fellow Veblen scholars, humanist Eby and economist Dugger, and I am in their debt.

The staff at Cline Library, Northern Arizona University, has been most helpful for many years. Dan Lee and Betty Lee, Sean Evans, Patty Shanholtzer, Gary Gustafson, Lee Gregory, John Doherty, Ian Hall, Brian Schuck, Bruce Palmer, Ann Eagan, Trey Rankine, and Penny Medina have all given me help before I asked for it.

Thorstein Veblen

and the Enrichment of Evolutionary Naturalism

Chapter 1

Introduction

Biography

There are two fundamental reasons for engaging in this study. First, Thorstein Veblen's ideas are still of use for contemporary social scientists. Second, his career and thought illuminates his time. As a historian of ideas, I have stressed the second over the first, although I am also convinced that his method of analysis and values are badly needed in our present circumstances.

Veblen is still one of the best-known and most influential social thinkers and critics America has produced. Indeed, the decline in the international influence of Karl Marx and classical Marxism signifies that Veblen and his radical institutionalism have a new opportunity to gain prominence in international intellectual circles. Furthermore, this is an auspicious time for the publication of a study on evolutionary naturalism and Veblen as a necessary antidote to the rise in anti-evolutionary thought and supernaturalism that are occurring in American life. Veblen's evolutionary naturalism, with its unflattering evaluation of America's self-selected special place in the international arena and its role in world affairs, may also reduce the impetus for intervention abroad and strengthen our convictions for the need of domestic social reform. At least this is a plausible reading of his potential long-term impact on U.S. foreign and domestic policy and that of like-minded thinkers such as John Dewey and C. Wright Mills.[1] "Globalization" is a popular buzzword as well as a growing political and social reality. But Veblen suggests that only the egomaniacal and the ideologically blind can imagine the United States as the sole social emulatory model for its realization.

---

1. See Rick Tilman, *Thorstein Veblen, John Dewey, C. Wright Mills, and the Generic Ends of Life,* for analysis of the similarities in their thought.

1

In any case, various aspects of his thought need further exploration if they are to be successfully linked with his contribution to American intellectual life and culture, that is, his progressive enrichment of evolutionary naturalism. However, a brief biographical sketch of his life is in order before a more detailed analysis of his thought is undertaken.

Born in 1857 in Wisconsin, Veblen was the fourth child of Norwegian immigrant farmers who raised a large family, most of whom received higher education. The Veblen children went to Carleton College in Northfield, Minnesota; there Thorstein obtained his training in economics and philosophy under the tutelage of John Bates Clark, who later became a prominent neoclassical economist. After receiving his bachelor's degree at Carleton, Veblen taught for a year and then enrolled at Johns Hopkins for graduate study. After a short stay, he transferred to Yale, where he obtained his Ph.D. in philosophy in 1884, while studying under such noted academicians as Noah Porter and William Graham Sumner.

Veblen was then idle for seven years, most of which was spent on the farms of relatives or in-laws in the Midwest. His agnosticism made him unemployable in schools with religious affiliations, and he had not yet established a reputation in economics. Finally, in 1891 he obtained a graduate position at Cornell University, where he once again became a doctoral student. The economist James Laurence Laughlin was impressed by him, and in 1892, when Laughlin moved to the newly founded University of Chicago, he took Veblen with him. Veblen soon became managing editor of the *Journal of Political Economy* and began publishing in the field of economics. In 1899 his most famous book, *The Theory of the Leisure Class,* appeared and achieved a notoriety all its own. But Veblen's personal idiosyncrasies and his failure to properly "advertise" the university offended the administration at Chicago, and he was forced to move. His next job was at Stanford where, in a few short years, he encountered similar difficulties that were exacerbated by his alleged "womanizing." He was compelled to move again, this time to the University of Missouri.

World War I found Veblen briefly in Washington as an employee of the U.S. Food Administration. After the war, he served for a short time as one of the editors of the *Dial,* a journal of literary and political opinion, and as a member of the faculty of the recently founded New School for Social Research in New York City. By then, even though his reputation as a scholar and publicist was at its peak, his academic career came to an end. Veblen retired and moved to Menlo Park, California, where he died in August 1929, shortly before the onset of the Depression.

Intent and Purpose

The social thought of Veblen is deeply rooted in evolutionary biology and American philosophical naturalism. In fact, it is impossible to adequately explain Veblen's institutional economics and radical sociology without understanding this. His epistemology, ontology, and axiology—that is, metatheory—while eclectic, can ultimately be traced to evolutionary naturalism and it is this perspective which underlies his social theory and criticism and even shapes his satire and mockery.

This is not the first attempt to relate his philosophy to his broader outlook. But it is partly novel in its selection of those aspects of his thought that it treats. These include his biography and its relationship to his thinking; cultural lag and its broader significance in his thought; and his analysis of sports, gambling, and religion. It also systematically links evolutionary naturalism to his aesthetics, secular humanism, sociology of control, sociobiology, and sociology of knowledge.

The study focuses on the intersection of social theory and social psychology, political economy and political theory, and modern philosophy and intellectual history in Veblen's thinking. It argues using textual exegesis of his published work, unpublished correspondence, related archival materials, and secondary sources that only evolutionary naturalism, or some similar perspective, could provide its philosophic foundations. The study also emphasizes his role in its "progressive enrichment," meaning his enhancement and embellishment of the social sciences and cultural studies and his insights into the processes of flux and change in the sociopolitical order of his time.

Although much is written about America's most influential heterodox economist, there is no unanimity among Veblen scholars left, right, or center, be they social scientists, humanists, or otherwise, about the central meaning of his work. Nor does agreement exist regarding his influence on public policy or American intellectual history. Given the politically, morally, and aesthetically fragmented nature of the social sciences and the humanities, along with their omnipresent doctrinal bias, perhaps it is chimerical to expect that any consensus should have emerged. To anticipate the crystallization of such a consensus is to assume that "objectivity" is somehow possible in the examination of the role of a historical figure such as Veblen. But, as Peter Novick in *That Noble Dream* and others have convincingly shown, the claim of "objectivity" in the writing of history is itself incoherent.[2] It is in such a spirit that this study is undertaken, for no

2. See Peter Novick, *That Noble Dream: The "Objectivity Question" and the American*

pretenses are made that it is value-neutral and thus lacking in political and ideological objectives, as if such a study of Veblen were really feasible. In any case, it is organized around themes found in Veblen's social theory and social criticism as viewed through the lens of his own evolutionary naturalism.

Veblen's "evolutionary naturalism" and its doctrinal assumptions provide much of the key to understanding and the guide to explaining his thought. This includes his aesthetics, history, anthropology, politics, economics, social theory, social criticism, and social psychology. His thought constitutes a coherent whole and it is evolutionary naturalism that binds the varied elements of his work together through its metatheoretical underpinnings. In short, his epistemology, ontology, axiology, and even his methodology are a variant fusion of American naturalism and Darwinian evolutionary theory, as is largely the case with his great contemporary, John Dewey.

Veblen worked within the American tradition of evolutionary naturalism and attempted with varying degrees of success to further develop a naturalistic theory of mind, a naturalized epistemology, naturalism in aesthetics and ethics, a naturalized methodology of the social sciences, and a naturalized study of economics. Perhaps, although he must have despaired at the prospects of achieving it, he wanted economists, whether evolutionary naturalists or not, to recognize that both economics and the economy as social products are, nevertheless, fully natural phenomena and nothing else.

Outline and Description of the Book

Recent interpretations of Veblen's theory of social change disagree about his role as a cultural lag theorist. Characteristically, he is interpreted as both a refiner and opponent of cultural lag whether it consisted of atavistic or predatory beliefs and practices or simply of institutional resistance to scientific and technological improvements. I argue both that he is a cultural lag theorist (with certain caveats) and that no other explanation of his larger theory of social change (of which cultural lag is a conceptual subset) is as serviceable.

Historical Profession, Intro. Also, see Alvin W. Gouldner, *For Sociology: Renewal and Critique in Sociology Today,* ch. 1. This chapter was originally published in article form as "Anti-minotaur: The Myth of a Value-Free Sociology" and is a valuable discussion of the problem of "objectivity" in sociology and the social sciences.

The cultural lag hypothesis in Veblen's critique of neoclassical economics is analyzed and compared with its utilization by John Dewey in philosophy and psychology, Frederick Jackson Turner in history, and Emile Durkheim in sociology. Moreover, the employment of cultural lag by Veblen and his contemporaries to critique and improve upon their own disciplines is the central focus of one chapter, both for their own sake and because they further illuminate his attack on the received political science and economics of his day.

For Veblen, sports, religion, and gambling are all part of the realm of make-believe, which he uses to explain their relationship to "matter-of-fact" social reality. He analyzes and satirizes all three for mostly similar reasons as products or byproducts of animism or its modern equivalents. Epistemologically, ontologically, and axiomatically they serve similar functions in the social order and are often indulged in by the same groups, classes, or subcultures. From his metatheoretical perspective, sportsmen, religionists, and gamblers all converge in their beliefs, a fact of which many Veblen scholars are aware, but have not adequately developed or used.

Elsewhere, I have articulated Veblen's theory of aesthetics, with its attack on wasteful decoration and self-indulgent display, and its emphasis on form following function and radical cleanness of design.[3] But inadequate focus still exists on art that is noninvidiously beautiful as far as the common person is concerned, and it is this latter proclivity in Veblen's aesthetics that needs explication and clarification.

Regardless of the fact that his aesthetics have been evaluated by aestheticians and social scientists of various ideological persuasions, Veblen's stature as an American critical theorist and the debate over the nature of his aesthetics justifies further attention to them. So one chapter is on Veblen's view of aesthetic experience, the nature of the noninvidiously beautiful in his aesthetics, his distinction between pure and applied art and its importance to the commoner as well as the social basis and artistic consequences of noninvidious art as exemplified in the painting of the American artist Winslow Homer (1836–1910). Problematic areas in Veblen's social theory and aesthetics are addressed and a noninvidious aesthetic of countervailence articulated, using Homer's work and that of Oscar Wilde to illustrate his view.

Was Veblen an "exceptionalist" whose view of American politics, cul-

---

3. See Rick Tilman et al., *The Intellectual Legacy of Thorstein Veblen: Unresolved Issues,* ch. 5, and "The Aesthetics of Thorstein Veblen Revisited," *Cultural Dynamics* 10 (November 1998): 325–40.

ture, and history was that these were essentially different from Western Europe and the rest of the world? In short, did he think that the United States was really a "city on a hill" waiting for the rest of the world to emulate it? Or did he believe that what the American people understood about it in terms of genesis and development were generically undifferentiated. These are not easy questions to answer, but they are certainly worth exploring if Veblen's view of his native land is to be better understood. Veblen sustained and nourished an empirical point of view and a naturalistic approach to values. He sought his standards in his own experience, to be sure, but these were rooted in the givenness of experience as well as what ought to be or what we might have had. "Is" was usually his guide to "ought" and both were a product of direct encounter with the natural order; and the latter included both the laws of nature and the data of science as it pertained to the social order. Veblen denied the efficacy of the mysterious and the providential for the purposes of men were not inscrutable, however difficult they might be to ascertain. In short, America was not a "city on a hill" to be admired and emulated, nor was its historical, cultural, and social experience so different from that of Europe to require that we view it as "exceptional."[4]

What was Veblen's sociology of control and how was it related to his theory of status emulation? Just as important, does this tie together the multifaceted strands of his thought into a broader form of social theory and criticism? This may be the linchpin of his social thought, but, whatever the answer to this question, much of the "law and the prophecy" as regards Veblen hang on it. In any case, I will explore it in some detail in this study.

No analysis of Veblen's evolutionary naturalism would be complete without a description of Darwin's influence on Veblen and the latter's use of his ideas in economics and sociology. This also provides an opportunity to evaluate the relationship between Veblen and his conservative Social Darwinist contemporaries, especially W. G. Sumner and, more important, sociobiologists such as the Harvard entomologist and social philosopher E. O. Wilson. The strong differences between the libertarian wing of evolutionary naturalism to which Sumner and, perhaps, Wilson belong and the collectivist wing to which Veblen belonged will be explored.

Veblen's sociology of knowledge is analyzed and other sociologists of knowledge are used to interpret it. Specifically, Max Weber, Karl Mann-

4. Although I have chosen to also use my analysis of Veblen and American exceptionalism in a forthcoming study of his political thought which Sidney Plotkin and I are now finishing, readers are referred to Rick Tilman, "Thorstein Veblen's Views on American Exceptionalism: An Interpretation," 177–204.

heim, Antonio Gramci, C. Wright Mills, Thomas Kuhn, Robert Merton, and Werner Stark all provide conceptual and theoretical lenses by which to better understand the social determinants of knowledge, including Veblen's. There is a substantial literature about his sociology of knowledge. But to my knowledge, important parts of the theoretical and conceptual work of these leading twentieth-century thinkers have never been applied to Veblen, and it is time to do so.

Veblen's use of the literature and data of anthropology has long interested social scientists, yet no systematic study of his broader employment of it exists. However, more to the point is the analysis of a salient, yet limited aspect of it, namely, criticism of his work using structural-functional theory. Talcott Parsons and three other sociologists trained or influenced by him—Robert Merton, Arthur Davis, and Bernard Rosenberg—used the theory in various ways to interpret Veblen. Interestingly, however, they failed to recognize his own selective use of structural-functionalism to attack the various social processes and practices they used the theory to defend.

Little exists on Veblen's evolutionary naturalism and secular humanism. To be sure, Stanley Daugert's *Philosophy of Thorstein Veblen* (1950) is a careful and valuable analysis of his philosophical roots and their genesis. Yet it does not focus on his broader cosmology as it relates to his view of the existing social order.[5] This is vital to understanding the intellectual and cultural milieu in which Veblen lived and worked and his response to it. Russell Bartley penetratingly comments that:

> there is an integral link between Veblen's life and his work . . . the diverse themes and problems on which he wrote and to which scholars have devoted narrowly focused attention all form part of an integrated whole reflective of an abiding intellectual need to explain human conduct. That personal need was rooted in his own particular life experience and was informed by a set of moral beliefs acquired in childhood.[6]

This approach helps put in perspective the conventional wisdom about Veblen, much of which comes directly or indirectly from Joseph Dorfman's massive intellectual biography, *Thorstein Veblen and His America* (1934), which has been subjected to severe criticism. Stephen Edgell, Russell and Sylvia Bartley, Elizabeth and Henry Jorgensen, Clare Eby, Tom Veblen, Jonathan Larson, Bill Melton, Kirill Thompsen, and others have evolved a different interpretation of some of the facts and much of the

5. Daugert, *Philosophy of Thorstein Veblen.*
6. Russell Bartley to Rick Tilman, May 10, 2004.

meaning of his life and work. As a fellow traveler in this revisionism, I attempt to solidify and synthesize this new learning on Veblen as well as offer my own views on the subject.[7]

Connections of Veblen's Evolutionary Naturalism, Social Theory, and Criticism

How was his evolutionary naturalism linked with his social science and cultural criticism? Is the latter a logical outgrowth of the former and, if so, how? These questions are not easy to answer because of the Kantian and Christian residues embedded in his early thinking and first published article in 1884. But, if what he wrote after 1892 is used as the standard, the questions become easier to answer. Although Kantianism remained in his aesthetics, the other incongruent elements largely disappeared. Christian ethics persist but they are stripped of their supernatural accoutrements. Veblen's intellectual armory never relinquished its hold on German idealism and Christianity, but only those parts of these traditions that were congruent with evolutionary naturalism survived in his thought.

A main theme in this study is that only evolutionary naturalism, or a philosophy much like it, could provide the metatheoretical basis for his social science and cultural criticism; that is, an epistemology, ontology, and set of axioms that could support and sustain the dozen or so topics around which it is organized. To illustrate and illuminate each chapter presents an opportunity to portray an aspect of Veblen's thought as an extrusion of his evolutionary naturalism. But exactly how did this extrusion occur and what are its consequences for the topics I have chosen to explore?

Given Veblen's analysis of the "make-believe" that sustains gambling,

7. Once regarded as a definitive and authoritative study of Veblen's life and thought, but now viewed by Veblen scholars as flawed, is Joseph Dorfman's *Thorstein Veblen and His America* (New York: Viking Press, 1934). More recently, correctives include Rick Tilman, *Thorstein Veblen and His Critics, 1891–1963* (Princeton, NJ: Princeton University Press, 1992), and *The Intellectual Legacy of Thorstein Veblen;* Elizabeth and Henry Jorgensen, *Thorstein Veblen: Victorian Fire Brand* (Armonk, NY: Myron E. Sharpe, 1999); Stephen Edgell, *Veblen in Perspective: His Life and Thought* (Armonk, NY, and London: Myron E. Sharpe, 2001); Russell H. and Sylvia E. Bartley, "Stigmatizing Thorstein Veblen: A Study in the Confection of Academic Reputations," *International Journal of Politics, Culture and Society* 14 (Winter 2000): 363–400. Also, see "III. An Exchange on Veblen" which features a brief dialogue between Stjephan G. Mestrovic and the Bartleys, *International Journal of Politics, Culture and Society* 16 (Fall 2002): 153–63. Clare Eby, "Boundaries Lost: Thorstein Veblen: The Higher Learning in America and the Conspicuous Spouse," in *Prospects,* ed. Jack Salzman, vol. 26 (Cambridge: Cambridge University Press, 2002), 251–93.

sport, and religion, it was inescapable that these would prove incompatible with his evolutionary naturalism. The superstitious belief in the intervention of divine providence on behalf of the individual gambler, athlete, and religionist alone discredits such beliefs and their related practices, or so Veblen would have us believe. Several aspects of them are irreconcilable with the Veblenian creed, such as the waste of time and resources by gamblers, the brutality and coercive nature of much sport, the ceremonial punctilios of religionists; and the belief in the existence of an omniscient and omnipotent deity who is personally concerned about the fortunes of individuals themselves sustained by a belief in the promise of eternal life and the salvation of the soul. Taken *en toto* as belief and practice, all reduce the possibilities that gamblers, athletes, and religionists will engage and practice what Veblen calls the "generic ends of life, impersonally considered."

Most theories of cultural lag rest directly or indirectly on the inability of cultural practices and social institutions to keep pace with an evolving techno-economic base or with the retardant effects of social structure and organization on economic growth. Veblen's theory is unusual because it incorporates both. But two-faceted or not, his view of cultural lag as an organic extension of evolutionary naturalism assumes that only natural forces could produce social disequilibrium and only naturalism could adequately explain the inability of cultural practices, social institutions, and techno-economic base to properly mesh. Disequilibrating fundamental social change is the normal state of affairs because the social order is shot through with elements that both facilitate and retard change unevenly. Disagreement and conflict exist because the uneven spread of instrumental values provides different notions of what is right and proper when impeded and inhibited by religious superstition, prescriptive status, and hierarchy. This same analysis can also be found in the theory of cultural lag provided by his contemporaries, especially Emile Durkheim and John Dewey.

Veblen's sociology of control—that is, his portrayal of the ways and means by which dominant elements in the community subjugate or manipulate the rest of it—relies heavily on an assumption drawn from evolutionary naturalism. This assumption is that individuals and groups are separated from each other by function, and not by prescriptive status. In accord with this, functionality should be the primary determinant of the way people are treated, not class, status, or heredity.

Veblen's sociology of knowledge is a logical concomitant of his philosophy in that it rests on critical assumptions about claims regarding immutable truths, transnatural verities, and value absolutes. His evolutionary perspective accepts the possibility of securing objective knowledge and identifying value constants. But he also doubts whether such can be readily obtained by those in conventional walks of life due to the ceremonial opacity of their creed. He included many intellectuals, even scientists, in his skepticism about the conventional wisdom regarding higher learning in America and the place of science in modern civilization. Scientistic cultism and institutional and cultural conformity were the possible outcome of inquiry even with the advent of the machine process and the scientific laboratory. The continuation, perhaps even the resurgence, of atavistic continuities was not as likely in scientific and humanistic circles. But in an evolutionary naturalist perspective, this possibility could not be discounted.

Veblen's anthropology when leveled against the consumption patterns, religious practices, and imperialist/colonialist claims of the powers-that-be is more revealing of the force, fraud, and superstition of the existing order than the strictures of his structural-functional critics. Indeed, he offers alternatives to existing social beliefs and practices that sociologists and anthropologists of that theoretical ilk thought essential to preservation of the social fabric of the community. Only evolutionary naturalism or a philosophy like it could provide the metatheoretical undergirding for such an assault on the status quo. Invidious ranking and behavior, religion, and ultranationalism can find no roots in such a perspective.

Veblen's conflict with the equivalents of sociobiology in his own time and ours are rooted in their differing interpretations of human nature; the sociobiologists focus on the genetic and institutional basis of human behavior while Veblen was more of an environmental determinist with a focus on the malleability and plasticity of Homo sapiens. In short, the sociobiologists and Veblen share the evolutionary naturalistic portrait. But their differing emphasis on the role of the environment in explaining the nature of human nature leads them to a disagreement as to why humans behave the way they do.

To what extent and in what ways can it be said that Veblen "progressively enriched evolutionary naturalism"? Despite his pretense at objectivity and detachment, he aimed at progressive social change, but mostly through his writing rather than through political and social activism per se. In fact, his secular humanism was often passive rather than active. It is evident, however, as I shall demonstrate throughout this study, that Veblen wanted the "recreation of non-invidious community through the instrumental use of knowledge," to use Marc Tool's felicitous phrase.[8]

8. Marc Tool, *The Discretionary Economy: A Normative Theory of Political Economy* (Santa Monica: Goodyear Publishing Company, 1979), 203.

In the last generation or so, issues that relate to the character of historical knowledge, the epistemology of history, and recovery of the meaning of past texts have become increasingly important.[9] It is the meaning of Veblen's texts and their subsequent significance that is the central focus of this study as is the relationship of these texts to those written by Veblen's contemporaries. But, ultimately, it is the reader who must decide on the present relevance of his ideas to the problems that beset America and the emerging global order.

9. See John Gunnell, *Political Theory: Tradition and Interpretation* (Cambridge, MA: Winthrop Publishers, 1979), esp. 97–98; Mark Bevir, *The Logic of the History of Ideas* (Cambridge: Cambridge University Press, 1999), offers a detailed and painstaking analysis of issues related to the recovery of meaning in historical texts.

Chapter 2

Veblen, Evolutionary Naturalism, and Secular Humanism

Introduction

Occasionally, luminaries in the humanities and the social sciences have offered brief but enlightening descriptions of their own doctrinal positions. For example, the Anglo-American poet T. S. Eliot described himself as an Anglo-Catholic in religion, a royalist in politics, and a classical in literature. The contemporary sociologist Daniel Bell declared some years ago that he was a liberal in politics, a socialist in economics, and a conservative aesthetically. Although he did not so describe himself, Veblen should be labeled as an evolutionary naturalist and secular humanist philosophically, a left-liberal in politics, an institutionalist in economics, and a Kantian-Darwinian functionalist in aesthetics.

Although more than half a century old, Stanley Daugert's *Philosophy of Thorstein Veblen* is still one of the most thorough and persuasive studies of the philosophic underpinnings of Veblen's social thought. Daugert penetratingly commented of his own book:

> The reader will discover in this book a defense of the thesis that Veblen formulated and developed a philosophy that he applied to the analysis of human problems, but he will also find that the statement of that philosophy is *partial and incomplete*. There are issues of Veblen's thought that await more complete formulation and criticism. Some of these issues are suggested, others are dealt with rather briefly. In some cases the author could not see clearly the issues involved. In other cases he found his knowledge incomplete because of the *lack of evidence* concerning the growth of Veblen's philosophy. In still other cases he found that Veblen merely suggested various issues and problems but did *not* seem to have been concerned *either to state them clearly* or to develop their implications.[1] (author's emphasis)

1. Stanley Daugert, *Philosophy of Thorstein Veblen*, vii. Although John Gambs differs

Daugert thus poses the problems and issues that any scholar faces in probing Veblen's philosophy, indeed, even that salient part of it that is evolutionary naturalist in bent. But it is not my intent to retrace Daugert's study by linking Veblen's work with the history of Western philosophy, so much as it is to connect it with the naturalism that was emerging in his time. Elsewhere, I have shown the links between his thinking and that of John Dewey, William James, and G. H. Mead; others have written extensively on his relationship to C. S. Peirce.[2] But, I will focus on the generic traits of evolutionary naturalism insofar as they can be found in his published work.

Although much has been written on his view of human nature little exists on Veblen's evolutionary naturalism and secular humanism as such, yet these three underpinnings of his thought are closely linked. To be sure, Daugert's study emphasizes Veblen's philosophical roots and their genesis, but it does not really focus on their relationship to the existing social order. It is these that remain to be explored in understanding the intellectual and cultural milieu in which Veblen lived and worked and his response to it. In the first part of this chapter, I shall define his philosophy in Darwinian and pragmatic terms a la American naturalism and then analyze aspects of Veblen's social and moral philosophy in these contexts. Then I will relate his thought to its underlying metatheoretical assumptions, that is, its epistemological, ontological, and axiomatic foundations. Finally, at the end of the chapter he will be portrayed both as a secular humanist and evolutionary naturalist with one foot in the modernist tradition and the other in the postmodernist camp. It is to the meaning of his evolutionary naturalism that I first turn.

Evolutionary Naturalism

The roots of modern naturalism lie deep in the past. As Veblen sardonically put it:

---

from Daugert in his interpretation of Veblen, he, too, recognizes the principles of evolutionary science, the emphasis upon "opaque fact," and the "uprooting of animism in economic thought." See Gambs, *Beyond Supply and Demand* (New York: Columbia University Press, 1946), 55.

2. On Dewey, James, and Mead, see Rick Tilman, *The Intellectual Legacy of Thorstein Veblen: Unresolved Issues*, ch. 3 and 4. Also, see Rick Tilman, Andrea Fontana, and Linda Roe, "Theoretical Parallels in George H. Mead and Thorstein Veblen," *Social Science Journal* 29 (July 1992): 241–58. On Peirce, see the many articles on Peirce and Veblen published in the *Journal of Economic Issues* between 1984 and 2002.

Philosophy—the avowed body of theoretical science in the late medieval time—had grown out of the schoolmen's speculations in theology, being in point of derivation a body of refinements on the divine scheme of salvation; and with a view to quiet title, and to make manifest their devotion to the greater good of eschatological expediency, those ingenious speculators were content to proclaim that their philosophy is the handmaid of theology—*Philosophia theologia ancillans*. But their philosophy has fallen in to the alembic of the idle curiosity and has given rise to a body of modern science, godless and unpractical, that has no intended or even ostensible bearing on the religious fortunes of mankind; and their sanctimonious maxim would today be better accepted as the subject of a *limerick* than of a homily.[3]

Veblen was thus well aware of the effects of cultural retardation on the human intellect as well as the persistence of atavistic continuities. In a related vein, a leading American naturalist philosopher, Peter Hare, writes:

For American naturalists, nature consists of everything there is. Nothing is "outside" nature. Whatever exists is no more or less an aspect of nature than anything else. Furthermore, nature is, to one degree or another, objective, which means that it has characteristics the content of which is not determined by our opinions of them—nature may be studied and known. Crucially, human beings are part of that knowable nature. American naturalists reject any dichotomy between nature and human beings. Human life, including its purposes, goals, meanings, values, and ideals, is wholly natural.[4]

Our search continues for an adequate definition of American naturalism to better understand Veblen's own philosophy. *The Dictionary of Philosophy* defines it as follows:

. . . naturalism, challenging the cogency of the cosmological, teleological, and moral arguments, holds that the universe requires no supernatural cause and government, but is self-existent, self-explanatory, self-operating,

3. Veblen, *The Higher Learning in America: A Memorandum on the Conduct of Universities by Business Men* (New York: Augustus M. Kelley, 1965), 38. Richard Rorty, perhaps satirizing thinkers like Veblen, wrote: "A 'mainstream' Western philosopher typically says: Now that such-and-such a line of inquiry has had such a stunning success, let us reshape all inquiry, and all of culture, on its model, thereby permitting objectivity and rationality to prevail in areas previously obscured by convention, superstition and the lack of a proper epistemological understanding of man's ability accurately to represent nature." Richard Rorty, *Philosophy and the Mirror of Nature* (Princeton, NJ: Princeton University Press, 1979), 367.
4. Peter Hare, "The American Naturalist Tradition," *Free Inquiry* 16, no. 1 (Winter 1995–96): 38.

and self-directing; that the world-process is not teleological and anthropocentric, but purposeless, deterministic . . . and only incidentally productive of men; that human life, physical, mental, moral, and spiritual, is an ordinary natural event attributable in all respects to the ordinary operations of nature; and that man's ethical values, compulsions, activities, and restraints can be justified on natural grounds, without recourse to supernatural sanctions, and his highest good pursued and attained under natural conditions, without expectation of a supernatural destiny.[5]

Accordingly, Veblen's naturalism was exclusive in that it did not include a belief in divine providence, the existence of a soul, or the likelihood of an afterlife. Thus, in his view humankind is the product of a biological template and its sociocultural environment and nothing else. Evolutionary naturalism does not resolve all problematic situations. But it does reduce them to philosophically manageable levels with possible access to empirical referents. However, naturalism as a philosophic concept or system is *supra* to scientific disciplines such as physics, chemistry, or biology. Thus Darwinian science is subsumed activity under the rubric of naturalism as it is commonly understood.

Merely to call Veblen an "evolutionist" or a "Darwinian" is to inadequately summarize or interpret his intellectual pedigree.[6] Indeed, even though many nineteenth-century thinkers were influenced greatly by evolutionary theory, their evolutionism was often more Lamarckian than Darwinian.[7] It was Lamarck's theory of evolution by adaptation to the demands of the environment and, in particular, his notion of the inheritance of acquired characteristics that most influenced Victorian intellectuals, including Herbert Spencer, not Darwin's theory of natural selection.[8] Of course, in the long run Darwin was to triumph. But according to Peter

---

5. *The Dictionary of Philosophy*, ed. Dogobert D. Runes (Totowa, NJ: Littlefield, Adams and Co., 1968), 205.

6. As to how he obtained his biologic knowledge or information—he was always observant and in a way a constant student of nature, and thus, acquired, at an early age, a good deal of knowledge of plants and animals. What he had thus learned by direct contact with things, he has since extended by reading and by intercourse with men well versed in the biologic and physical sciences. As an example, he was well acquainted with (biologist) Jacques Loeb, who was a friend of his.

Andrew Veblen to Joseph Dorfman, April 25, 1925, Joseph Dorfman Collection, Butler Library, Columbia University, New York City. On Veblen's use of Darwin, see Geoffrey M. Hodgson, "Darwin, Veblen and the Problem of Causality in Economics," *History and Philosophy of the Life Sciences* 23 (2001): 385–423.

7. See Peter Bowler, *The Non-Darwinian Revolution: Reinterpreting a Historical Myth* (Baltimore, MD: Johns Hopkins University Press, 1988).

8. J. D. Y. Peel, *Herbert Spencer: The Evolution of a Sociologist*, 134–35.

Bowler, this was not to fully occur in biology, for example, until the 1940s.[9] It is clear that Veblen regarded himself as a Darwinian, since he repeatedly invoked both Darwin's name and the doctrine of natural selection to explain institutional change and claimed that the survival of some institutions and the death of others was evidence of the exercise of the adaptive powers of the human community in the process of evolution. Nevertheless, the Lamarckian doctrine of the inheritance of acquired characteristics is often a more effective analogy to use in understanding his explanation of the internal dynamics of institutional change than the theory of natural selection. Through the institutional transmission of values and behavioral traits, acquired characteristics can both direct evolution of the institution and be "inherited." As Bowler puts it: "Darwin himself did not deny a limited role for the inheritance of acquired characters, and he was thus able to admit that the learning of new habits by the animals themselves can play a role."[10]

In Veblen's analysis, the genesis, development, and transmission of individual traits is institutionally induced, and while some institutions may prosper and flourish, others will stagnate and die. The first is best explained by the Lamarckian doctrine of acquired characteristics, while the latter is mostly aptly explained by Darwinian natural selection. Lamarck's emphasis upon the direct action of the environment rather than struggle for existence, the inheritance of acquired characteristics, the cumulative effect of use and disuse of organs, and the adaptation of organs to the environment reflected both the American emphasis on manipulative possibilities and the environmentalist tradition in the behavioral sciences. In part, Veblen was reacting against blind, nonpurposive evolution. He saw a broader social significance to human evolution—an evolution in which the inheritance of acquired characteristics affirmed their attachment to rational social behavior. His admiration of evolutionary theory did not necessarily lead to the abandonment of intellect to fortuitous circumstances but, rather, in a desire to preserve through social action humanistic ideals. As a scientist and social scientist, he looked to both evolution and environmental change for the appropriate formation of social order, reform, and progress. Yet he also recognized that humankind might be trapped by "imbecile institutions" and that what the future might hold was only blind drift. Neither the teleological intent of individuals nor the evolutionary trend of societies would necessarily consummate in a stasis of enlightenment and social harmony. Although he was strongly opposed to

9. Bowler, *The Non-Darwinian Revolution*, 201.
10. Ibid., 98.

the existing pockets of irrationality and inequity (ceremonialism), he doubted that many of these would soon disappear. Or even if they did temporarily diminish in size or perniciousness, they might reappear. In short, he was skeptical of the likelihood in the calculable future of the creation and maintenance of the best and most efficient final society.

Lamarck attributed the adaptability of species to the institutional and cultural inheritance of acquired characteristics rather than to natural selection by survival of the fittest as in Darwin, but Veblen used both approaches. Thus, although Veblen was partly Darwinian, his evolutionary theory of social change, insofar as particular institutions are concerned, may be understood more readily in Lamarckian terms. It is indeed ironic that in spite of his castigation of neoclassical economics for being pre-Darwinian, an important part of his own explanation of institutional change sounds more Lamarckian than Darwinian.[11] Veblen typically failed to acknowledge this point and a textual exegesis of his works is inconclusive on it. Still, one quotation is in order:

> The variation in race characters is very appreciable within each of these national populations; in the German case being quite pronounced between north and south. Whereas the differences which go to make the distinction between these nationalities taken as aggregates are of an institutional kind—differences in acquired traits not transmissible by inheritance, substantially differences of habituation. On this side, however, the divergences between one nationality and another may be large, and they are commonly of a systematic character; so that while no divergence of racial type may be alleged, the divergence in the cultural type may yet be serious enough.[12]

One may speculate whether in some literal sense Veblen believed evolutionary theory applied literally to the theory of evolution of the human species or whether he intended it only as an analogy.[13] I am inclined to-

---

11. On the role of Lamarckianism in late-nineteenth-century thought, see Edward J. Pfeifer, "The Genesis of American Neo-Lamarckism," *Isis* 56, no. 2 (184) (1965): 156–67, and George W. Stocking, "Lamarckianism in American Social Science, 1890–1915," *Journal of the History of Ideas* 23 (1962): 239–56.

12. Thorstein Veblen, *Imperial Germany and the Industrial Revolution*, intro. Joseph Dorfman (New York: Augustus M. Kelley, 1964), 9.

13. The quasi-Lamarckian nature of Veblen's work can be found at various points in his writing but see esp. *Imperial Germany*, 125–26. Ann Jennings and William Waller warn: "Such reifications typically launch a search for the social analogs of biological genes, or the basic units of 'trait' inheritance, and produce debates over Darwinian v. Lamarckian inheritance mechanisms . . . We find both of these concerns misguided." "Evolutionary Economics and Cultural Hermeneutics: Veblen, Cultural Relativism, and Blind Drift," *Journal of Economic Issues* 28 (December 1994): 1020–21. On these

ward the former, but it appears that Veblen, who aimed at discrediting neoclassical economists for being pre-Darwinian, also invoked Darwin for analogical purposes. Suggestions that Veblen was at times more Lamarckian than Darwinian signify that while acquired characteristics are often inherited, this is mostly a matter of cultural inheritance and institutional transmission, not genetic change.[14] Whatever the case, the evolutionary reproduction of institutions posits no fixed end in Veblen's analysis, but a continuous, cumulative process of institutional mutation that will provide no aesthetic or emotional comfort to those seeking social stasis or the arrest of cultural change.

Veblen, like Dewey, belonged to the first generation of American naturalists and this was a generation greatly influenced by Charles Darwin. Two aspects of Darwinism that greatly influenced Veblen were that in nature no finished and fixed essences exist. Thus useful knowledge of nature, which is always changing and developing, is obtainable without regard to eternal forms, which are fixed and final.[15] Veblen's work is replete with evidence of Darwin's influence and that of post-Darwinian evolutionary theory and these, along with his background in Western philosophy and American thought and culture, provide the defining characteristics of his naturalism.

Veblen, Darwin, and Naturalism

While Veblen did not actually collapse Darwinism into naturalism he believed they had a compatible metaphysical structure. In any case, his endorsement of naturalism's chief virtues was based on its opposition to philosophical apriorism whether of the revelatory supernatural kind, intuitionism, or deductivism from abstract rationalistic premises.[16] Instead,

points, also see Geoff Hodgson, "Thorstein Veblen and Post-Darwinian Economics," *Cambridge Journal of Economics* 16 (December 1992): 295–96.

14. But see Richard W. Burkhardt, "The Zoological Philosophy of J. B. Lamarck," and David L. Hull, "Lamarck among the Anglos," which are introductory essays to J. B. Lamarck, *Zoological Philosophy* (Chicago: University of Chicago Press, 1984), iv–xvi.

15. On these points, see John Ryder, ed., *American Philosophic Naturalism in the Twentieth Century* (Amherst, NY: Prometheus Books, 1994), 12.

16. Contemporary evolutionary naturalist philosopher Gerhard Vollmer describes philosophical naturalism as a conception and a program in the following manner: "(1) It calls for and charts a *cosmic view*, "worldview;" (2) It assigns to *man* a definite place in the universe (which turns out to be rather modest after all), (3) It covers *all* human capacities: language, knowledge, scientific investigation, moral action, aesthetic judgment, even religious faith, (4) Under these premises it calls for and develops in particular: (a) a naturalistic anthropology, (b) a naturalistic epistemology, (c) a naturalis-

his focus was on efficient causes that he believed had superior explanatory and predictive potency. He also held to a view of a purposeless nature that undercuts all teleological principles for there was no design in nature as a whole.[17] The historicity of the evolutionary process was assumed by Veblen but it pointed to no particular ends or consummatory processes.

To Veblen, cognitive activities, social behavior, and even moral values and aesthetic taste are evolutionary adaptations. Unless consciousness served some useful purpose it would not exist. Actually, his view of the workings of natural selection and adaptation was that they were not rigorous or severe because they were offset by cultural lag, which explained the existence of useless cultural and social debris left floating in the atmosphere sometimes to the detriment of mankind. Nevertheless, evolutionary naturalism postulated the nature of reality, truth, meaning, and value as all broadly experiential, the last three as perpetually in a state of Darwinian flux; the nature of reality, of course, is to be understood according to the constancy of natural laws, but all four are impossible to adequately understand independent of human experience. For Veblen, then, naturalism was both a presupposition and a tool for investigative inquiry, that is, a heuristic device. Evolutionary naturalism enables us to judge other metatheories, metaphysical positions, methodological attitudes, and heuristic rules. But it also has its own forms of these that his satire and mockery sometimes disguised or concealed. The citadels of orthodoxy were under siege much of the time in the corpus of his writing. But his lampooning of the conventional wisdom did not always serve to clarify either his objections to it or its actual nature and bearing on social reality.

Veblen had a naturalistic preference for continuity and flux, stability and evolution. Nor did he avoid the questions of the processes involved in the origin of new social forms and institutions, as even a casual reading of his social stages development sequence in *The Theory of the Leisure Class* will indicate. Evolutionary processes, in any case, were naturalistic

---

tic methodology of research, (d) a naturalistic ethics, (e) a naturalistic esthetics." Gerhard Vollmer, "How Is It That We Can Know This World?" in *Darwinism and Philosophy,* ed. Victorio Höslet and Christian Illis (Notre Dame: University of Notre Dame Press, 2005), 262.

17. Vollmer also comments on the evolutionary epistemology he, I believe, shared with Veblen: "Finally, evolutionary epistemology is *realistically* oriented. More precisely, it defends a *hypothetical realism,* characterized by the following: (1) *Ontological realism:* There is a real world independent (for its existence) of our consciousness, lawfully structured, and quasi-continuous, (2) *Epistemological realism:* This world is partially knowable and understandable by perception, thinking, and an intersubjective science, (3) *Fallibilism:* Our knowledge about this world is hypothetical and always preliminary." Ibid.

Table 1. Emergentist Causal Monism

Intentions are regarded as emergent properties of the workings of the materialist causes within the human nervous system. Intentions arise out of materialist causes; they are complex transformations of materialist causality. Although intentions are not completely reducible to material relations, nevertheless both the formation and implementation of intentions always involve the rearrangement or transformation of physical matter of energy. Hence any adequate explanation of the detailed processes of deliberation and reasoning must both involve and be consistent with materialist causality.	By definition, intentions involve conscious prefiguration and self-reflexive reasoning, with regard to future events or outcomes. Unintended acts lack any such conscious deliberation and prefigureation.	D. Diderot C. Darwin G. H. Lewes T. B. Veblen M. A. Bunge D. Davidson J. R. Searle

Source: From Geoffrey M. Hodgson, "Darwin, Veblen and the Problem of Causality in Economics," *History and Philosophy of Life Sciences* 23 (2001): 385–423.

and Veblen did not seem to have a strong preference for gradualism nor was he obsessed with pigeonholing tidy evolutionary sequences in explaining the cultural evolution of mankind. He did, however, focus on shared and derived, that is, environmentally induced, social characteristics without trying to either accelerate or put a brake on evolutionary scenarios. Veblen accepted the existence and maintenance of naturalistic deposits and processes as pervasive in all social reservoirs and, in logical accord with this, adopted a naturalistic perspective for use in all the social sciences. He was reluctant to accept any kind of knowledge that was not ultimately rooted in evidence or sense impressions. Although it is difficult to tell with any degree of precision, Veblen probably believed that mental states such as sensations, thoughts, desires, drives, and brain states, that is, electrical and chemical processes, were not two different sorts of things. Instead, they were just two aspects of one set of events (see table 1).

Recent scholarship emphasizes that Veblen's brand of activist naturalism is agent centered and agent empowered. This is not the reductionist atomism or passive materialism often held by either his Marxist critics or his sociobiological interpreters. But in his view much behavior is mentally controlled by habit without conscious attention. The mental capacities of Homo sapiens have changed little in the last ten thousand years according to Veblen. But overall, they are as much a product of the evolutionary process as anatomical and physiological traits.

For Veblen, the natural properties of things and purely natural knowledge were sufficient. But what did this signify about human behavior? Given the rudimentary state of knowledge regarding human genetics, cognitive science, cultural anthropology, and social psychology in the late Victorian period during which Veblen's view of learning and substantive knowledge was formed, he must have been in a quandary at times. Such issues as the strength of the genetic components of cognitive faculties and the genetic preconditions of cognitive achievements must surely have puzzled him. No doubt the language faculty had its origin in organic evolution, and human speech had played an important role in social communication and the survival of the species. But modern linguistics, fossil research, and social ecology were still in early stages of development, and Veblen's work bears the stamp of all this. An evolutionary epistemology and ontology are connected with naturalism and evolutionary theory in Veblen and with the development of science and technology and a philosophy of science. By present standards, however, this requires treating him at times as a largely historical figure whose ideas are best understood in their cultural context. Much early work on Veblen led to accusations of literalist and fundamentalist apologetics and rationalization on his behalf—some of which are, no doubt, justified.

Still Veblen differed from all but the most radical progressives in that he did not believe that the main trouble in American society came from the corrupting influence of the special interests. Or even from the swamp of outmoded belief, in short, the inertia and stultification of traditional culture with its focus on anarchic and subjectivist individualism. These were significant to him in explaining the American malaise to be sure. But the most essential shortcoming in the system was to be found in the power and income differentials produced by market capitalism and buttressed by status emulation. As he understood it, this was the difference between those who have enough wealth and property to matter and those who do not. In the 1960s and 1970s the New Left referred to these as "structural" problems and in Veblen's day they were compounded by emulatory consumption and waste and destructive nationalism/patriotism which were organically connected with the market and the system of property-relations.

Veblen recognized how the natural sciences differed from the social sciences and how both of these were separated from the humanities. But he did not suffer from the illusion that they were or could be value free, despite his occasional lapses into the use of positivist language and conceptualization. Nor did he anthropomorphize nature by assigning it the traits of the deity—omniscience, omnipotence, and benevolence.

Veblen as Evolutionary Naturalist

As a naturalist, Veblen did not support any dichotomy between nature and human beings, for human life, in his view, was merely a derivative of the biological template and sociocultural conditioning. Its ideals, values, meaning, goals, and purpose were wholly natural, having their genesis in nature, not in the supernatural. But did he believe that, while nature consists of material and nonmaterial phenomena, matter is the more fundamental in the sense that nothing nonmaterial emerges from the material? Or did he hold that there is more to be found in nature than simply matter, while refusing to acknowledge any ultimate or fundamental primary to any aspect of nature? Perhaps the former rather than the latter, but since he did not ordinarily write on philosophic subjects, as such, it is impossible to speak authoritatively.

In the collectivist, although not the libertarian wing of American naturalism, and Veblen belonged primarily to the former, knowledge should be used to further the ideas of social control. When there is the possibility of control, knowledge should be the main agency of its realization. Ideas have consequences and turning indeterminate situations into determinate ones through collective action is a main social goal and political ideal for naturalists such as John Dewey. But Veblen did not share Dewey's means such as government and unions for achieving the ends of social control. Indeed, he had little faith in the efficacy of positive government to bring about successful economic planning, income redistribution, corporate regulation, and the large-scale provision of collective goods.[18] Veblen was not an empirical collectivist and cannot be placed in the American tradition of welfare and regulatory state collectivism of the New Deal–Fair Deal–Great Society stripe. And his negative predisposition in this regard probably stems from his belief both in the frailty of the social sciences and the inefficacy of government. Thus results our inability to apply them to measures of social control as well as the unpredictability of the outcome of proposed large-scale change. Even in the naturalist paradigm a la collectivism he was odd man out.

18. See Tilman, *Thorstein Veblen, John Dewey, C. Wright Mills and the Generic Ends of Life*, ch. 3.

Value and Value Inquiry

The tradition of seeking norms as well as importing them into the texture of material life fits well with evolutionary naturalism. This is not to assume that values emerge directly from facts nor does it presuppose the givenness of all values. Veblen did not identify "is" with "ought" nor did he think of values and a theory of society as implicit in facts about the existing social order. In short, he did not confound the ought and the is because he realized that epistemologically they do not constitute a dualism; rather they lie along a continuum. The is and the ought, the world of fact and the world of value, of science and morals may not be a seamless reality, but neither are they opposites separated by a chasm from each other. By the same token, human institutions are part of a natural continuum with the noninstitutional environment and the historical past.[19] One advantage, then, of evolutionary naturalism is that it demands a sense of continuity between humans and nature, facts and values, consciousness and cosmos, science and ethical imperatives. Any reliable sense of uniqueness in the study of history and culture that we may possess is likely to be congruent with the evolutionary naturalist perspective that undergirds Veblen's inquiry.

The political scientist Norton Long, writing in the American naturalist tradition, comments in a manner roughly paralleling Veblen that evaluation,

> like explanation, generates expectations as to real-world consequences. These real-world consequences are as much the arbiter of the human usefulness of a set of values as any set of facts that test a scientific explanation. *Values*, like explanations, are human instruments and derive what validity they possess from their practical operation. However, like the chance-discovered natural tools of savages and the useful explanations embedded in ritual and myth, they need to be removed from the realm of natural evolution to that of systematic critical examination and test. The evaluatory enterprise, like that of science, can have a humanly significant, ongoing, self-corrective career.[20]

Generally speaking, Veblen's approach to our understanding of value and value inquiry is highly eclectic in terms of its intellectual lineage; combin-

19. See Daniel Boorstin, *The Genius of American Politics* (Chicago: University of Chicago Press, 1962), 179.

20. Norton Long, foreword, to Eugene J. Meehan, *Value Judgment and Social Science: Structures and Processes* (Homewood, IL: Dorsey Press, 1969), vii.

ing insights from Kant and Christianity with those of empiricism, naturalism, and instrumentalism. Yet Veblen's own values expressed throughout his writings are transcultural in the sense that he believes idle curiosity, proficiency of workmanship, and altruism in the form of the parental bent are the most redeeming proclivities in any culture. Indeed, he has little of a positive nature to say about cultures where these are lacking or submerged in a morass of force and fraud. Obviously, in view of his evolutionary naturalism these do not have transcendental origins or sanction.

To further illustrate, for Veblen the value of natural phenomena and the process of valuation are simply aspects of nature much like anything else that exists. John Ryder put it this way:

> This unique character of naturalism—its continued pursuit of knowledge while acknowledging human contextuality—has its influence when naturalists turn their attention from the consideration of reality and knowledge to the study of matters of somewhat greater concern to a wider range of people. Specifically, naturalists have written extensively on ethical theory, social philosophy, aesthetics, and religion. One of the reasons naturalists have been concerned with these areas has to do with their own theory of experience. Once experience is no longer understood as the accumulation of sense data, once it is understood as encompassing the broad range of what people do and undergo, it becomes necessary to address the most pervasive and influential aspects of experience. Naturalism, in other words, is not eliminative. Other traditions (one thinks first of logical positivism) have tended to argue away central constituents of human life such as ethics or religion. Naturalism, by contrast, looks to include them and to understand them no less than one would look to understand any other aspect of nature.[21]

In this same vein, John Herman Randall writes that experience

> is full of implicit ends or ideals, full of values, because each alike is an affair of processes, of mechanisms producing outcomes, of causes and necessary conditions of results, of means and ends. They are all alike full, that is, of things that are "better" and "worse" for other things. Nature is in truth teeming with "entelechies"; and it takes but a single flower to refute the absurd contention that there are no "values" in Nature, no achievement of ends through valuable means.[22]

21. In John Ryder, ed., *American Philosophic Naturalism in the Twentieth Century*, 21.
22. Ibid., 139.

Although this statement goes beyond most that Veblen made regarding nature and value, it appears compatible in its main thrust with his doctrinal perspective.

Veblen, of course, injected biological and social dimensions into the interpretation both of experience and of mind itself. Yet, like other naturalists such as Ryder, he tended to value science as a model for all inquiry and its related insistence that nature has objective traits, that is, traits independent of an experience or a knower. Introspection, intuition, and innatism and the corresponding claims made for them were not part of the intellectual capital that he valued and used. In any case, in the final analysis he believed that natural objects and processes should be understood in terms of their functions and purposes rather than as static and eternal realities. Like other naturalists he did not separate himself from nature or natural processes. Indeed, even his own values and valuation processes he viewed as evolutionary and cultural events that required investigation into the natural causes and consequences of human value commitments. If ignored or performed improperly, human choice could be deprived of effective status, the door opened to irresponsible intuitions and the control of nature and society that scientific understanding makes possible dehumanized.[23] As Ryder also put it:

> One of the implications most significant for understanding human life and activity of the view that there is nothing "outside" or "other than" nature is that naturalists do not endorse the traditional dichotomy between nature and human being. Human life, including its purposes, goals, meaning, value, and ideals, is wholly natural. The ramifications of this position are immense, not least because it means that the meaning and value of our lives, and the ethical ideals on which we choose to act, have their source in nature, not in the supernatural. As we will see, it becomes an important project for many naturalists to ask how meaning and ideals might arise from human life instead of descending from the heavens.[24]

In Veblen's philosophy, of course, nothing descends from the heavens. Although he believed religion might provide moral and psychological insight, it must stand the test of naturalistic sanction and practice and it mostly fails to do this.

23. Ibid., 119.
24. Ibid., 13.

Freedom Constrained by Habit

Ryder also comments:

> With respect to its conception of nature, then, naturalism distinguishes it-self in two crucial respects. The first, as we have already emphasized, is the rejection of anything other than nature. The other is that in both its "re-formed" materialist (using Sellars' terminology) and its pluralist forms, American naturalism pursues a conception of nature that avoids reductionism and rejects one of the traditional components of mechanical materialism: strict determinism. American naturalists, whether materialist or pluralist, are likely to regard nature as malleable, because while natural processes are in some respects determined, they are in other respects open; nature can be described by both law and chance.[25]

Although Veblen's naturalism usually brings him down as close to the deterministic as the volitional side of the debate, he believed that some groups and classes are able within limits to exercise free agency. It is important to note, however, that humankind, in his view, are creatures of habit and it is habit with its institutional supports and cultural reinforcements that constrains choice. Or more accurately, the economist's notion of "constrained choice" has little significance to most of humanity throughout history and prehistory because it presupposes a degree of autonomy and rationality that individuals rarely possess. Even in industrial systems with market economies and representative government, Veblen believed that the exercise of "freedom of the will" was less significant and meaningful than proponents of "individualism" imagined. Yet he held to this view without ultimately denying the efficacy of agency or volition.

The Search for Truth

"Objectivists," such as the logical positivists and their predecessors like Ernest Mach, presupposed a *truth* independent of the inquirer and not affected by the process of inquiry. But, as I shall show in the chapter on Veblen and the sociology of knowledge, this does not, on the whole, exemplify his position. For Veblen, "truth" is not only the result of applying scientific methods and "objective" analysis in the quest for knowledge. It also involves recognition of the extent to which ideological bias, doctrinal

25. Ibid., 15.

baggage, sociocultural intent, political aspirations and social status flavor and color "impersonal" objectivity. Nevertheless, "naturalism" is a faithful representation of Veblen's philosophy and cosmology and it assumes the possibility of inquiry which leads to an irreducible residue of "truth," independent of the inquirer and unaffected by the process of inquiry, however small the residue might become at the termination of inquiry. As I have argued elsewhere, he was no radical postmodernist who believes "reality" is a mere social construct.[26] As Randall put it:

> If Nature were in truth mere "flux," if she did not exhibit countless patterns of historical unification, and hosts of teleological structures of means and eventuations with a temporal spread, then human histories would indeed be wholly anomalous. Men's unification of their own history, their discovery of the significance of their own past, through knowledge or vision, would be quite impossible. So likewise would be any discovery of "the meaning" of the world, or of human life.[27]

As a student of prehistory and history, Veblen would not allow a Humeian skepticism to annihilate the meaning of the past or its significance for enriching the present and the future.

Veblen and Pragmatism

The most comfortable fit that exists between Veblen and any other influential thinker of his time is John Dewey, the Dewey of the late Progressive era and the interwar periods; roughly speaking, Dewey after 1910. The fit is not a perfect one even then as I have argued elsewhere.[28] But there are closer parallels and more convergence between Veblenian institutionalism and Deweyan pragmatism than any other leading schools of thought of the time, as Clarence Ayres claimed. Certainly Veblen is an "evolutionary naturalist" even if, strictly speaking, this does not make him a pragmatist. According to Sidney Hook:

> technically, pragmatism was developed as a theory of meaning and then as a theory of truth. In its broadest sense as a philosophy of life, it holds that the logic and ethics of scientific method can and should be applied to hu-

---

26. See Tilman, *Thorstein Veblen, John Dewey, C. Wright Mills and the Generic Ends of Life*, 259–60.
27. In Ryder, ed., *American Philosophical Naturalism in the Twentieth Century*, 136.
28. Tilman, *The Intellectual Legacy of Thorstein Veblen*, ch. 4.

man affairs. This implies that one can make warranted assertions about values as well as facts. It recognizes that the differences in the subject matter of values requires the use of different methods of inquiry, discovery, and test in ascertaining objective knowledge about them. Most daring and controversial of all, pragmatism holds that it is possible to gain objective knowledge not only about the best means available to achieve given ends—something freely granted—but also about the best ends in the problematic situations in which the ends are disputed or become objects of conflict.[29]

Hook continues, and here his comments take on a great methodological significance:

Nonetheless, reliance upon rational and scientific methods of inquiry must face certain difficulties even if the incoherent and self-refuting critiques of the validity of these methods are disregarded. First, can we choose to engage in scientific inquiry without begging the question? And if we don't beg the question, is it not as much a manifestation of nonrational faith to adopt the scientific method as it is to adopt the method of revelation or magical incantation? Briefly, our answer is that the appearance of circularity arises because the question of method is raised in the abstract, not in relation to a *problem-solving context.* Once we locate the class of problems we wish to solve, we can justify our choice of rational or scientific method by its fruitfulness in solving these problems. The problem is enormously more complicated than these lines suggest, but in the end the judgment will depend on which approach is more fruitful.[30]

What is derivable from these comments is an obvious commitment to a particular kind of methodological pluralism—namely, methodological instrumentalism wherein the social scientist, in particular, selects those means most likely to achieve the ends sought. In short, the nature of the problem will determine the means used to study it. For example, textual exegesis is the method used to ascertain the meaning of dense texts, a quantitative-statistical approach is appropriate to correlate or not certain kinds of variables; a case study method is most useful for legal issues or managerial problems in organizations. A comparative-historical approach is helpful for studying different cultural, social, and political epochs in or out of chronological sequence and, finally, the use of constructed types, which accentuates certain qualities or tendencies, facilitates studies for analytic and comparative purposes. Veblen called these latter formations

29. Sidney Hook, *Pragmatism and the Tragic Sense of Life* (New York: Basic Books, 1974), ix–x.
30. Ibid., x–xi.

"type forms" and applied them extensively in his own work, but he was familiar with the other approaches as well.[31] Probably, for his purposes the progressive enrichment of evolutionary naturalism is dependent to a very significant degree on selecting the right method and using it effectively in inquiry in both the social and natural sciences.[32] Yet he did not fully share the faith of Dewey and Hook in the efficacy of human intelligence in either understanding the social order or bringing about progressive social reform. Thus the resurgence of atavistic continuities and the possible triumph of "imbecile institutions" played a more central role in his thought than in that of Dewey, in particular.

Instrumental Valuation or the Generic Ends of Life?

One of the key issues that remains is deciding if, how, and when instrumental valuation is related to Veblen's "generic ends" of life. In short, to what extent was he an instrumentalist and a pragmatist and to what degree did he think value constants of a culturally transcendent nature provided moral imperatives for social development? Indeed, the critic may ask, are these really alternatives to each other? His relationship to the pragmatic philosophers Peirce, James, Dewey, and Mead has been developed elsewhere so that is not at issue here. Rather, the focus is on means-

31. Veblen's evolutionary methodology has been described and analyzed so often that it would be difficult to add anything new or helpful to what has been written. Readers are referred, instead, to the existing literature. Cf. Adil H. Mouhammed, *An Introduction to Thorstein Veblen's Economic Theory* (Lewiston, NY: Edwin Mellen Press, 2003), ch. 4; Kenneth Arrow, "Thorstein Veblen as an Economic Theorist," *American Economist* 19, no. 1 (1975): 5–9; A. W. Coats, "The Influence of Veblen's Methodology," *Journal of Political Economy* 62 (December 1954): 529–37; John Gambs, *Beyond Supply and Demand: A Reappraisal of Institutional Economics* (New York: Columbia University Press, 1946).

32. Also, see Veblen's own writings. "Why Is Economics Not an Evolutionary Science?" *Quarterly Journal of Economics* 12, no. 4 (July 1898): 373–97; "The Preconceptions of Economic Science I," *Quarterly Journal of Economics* 13, no. 2 (January 1899): 121–15; "The Preconceptions of Economic Science II," *Quarterly Journal of Economics* 13, no. 4 (July 1899): 396–426; "The Preconceptions of Economic Science III," *Quarterly Journal of Economics* 14, no. 2 (February 1900): 240–69; "Gustav Schmoller's Economics," *Quarterly Journal of Economics* 16, no. 1 (November 1901): 69–93; "The Socialist Economics of Karl Marx and His Followers: I, The Theories of Karl Marx," *Quarterly Journal of Economics* 20, no. 4 (August 1906): 575–95; "The Socialist Economics of Karl Marx and His Followers: II, The Later Marxism," *Quarterly Journal of Economics* 21, no. 2 (February 1907): 299–322; "Professor Clark's Economics," *Quarterly Journal of Economics* 22, no. 2 (February 1908): 147–95; "Fisher's Capital and Income," *Political Science Quarterly* 23, no. 1 (March 1908): 112–18; "On the Nature of Capital I," *Quarterly Journal of Economics* 22, no. 4 (August 1908): 517–42.

ends relationships and the nature of both means and ends in his thought. We begin with a brief analysis of these problems in the literature of American political philosophy and then turn to textual exegesis of Veblen for their elucidation.

As an approach to valuation, instrumentalism is a naturalistic theory that employs self-correction through judgment of consequences and effects. It is seriously at odds with those ethical and evaluative stances that are either fully relativistic or absolutist in nature. For example, the belief in the subjectivity of values that permeates conventional economics assumes an inaccessibility or isolation that privatizes all value criteria. The conclusion all too often drawn is that values are so subjective that when moral goods conflict, there are no standards by which they may be judged. But a method does exist by which to demonstrate that some value criteria and choices are superior to others. Of course, this is the method of self-adjustment or self-correction that makes ends congruent with means and means congruent with ends along a means-ends continuum oriented toward human growth. The relevant consequences to be judged are those upon which successful growth is contingent. As Francis Myers once put it, "The instrumentalist assumption is itself one such theory which is tested by its consequences when acted upon. It is a program of action for which the evidence is not conclusive. Yet it is different in kind from other assumptions, since, as a plan of action, it seeks a conscious control of consequences by which it may be progressively tested and modified."[33]

What is perhaps unique about this approach is its claim that there are not merely analogies to be drawn between democracy and science, but that "the parallel breaks down because, if there is an experimental social science, it is democracy itself."[34] The instrumentalist aim is thus a "politics of truth"—a self-governing, self-correcting, experimental laboratory organized for the creation of truth—a merging of the social reformer with the social scientist.[35] Ultimately, the authority of method and adherence to a method of knowing can occur only to the extent that individuals are part of a self-correcting method, and only if they take part in controlling and extending the method. But did Veblen really think political life was like this, or was it merely an unstated ideal in his thought, an unrealizable goal at that?

33. Frances Myers, *The Warfare of Democratic Ideals* (Yellow Springs, Ohio: Antioch Press, 1956), 203.

34. See Holton Odegaard, *The Politics of Truth* (Tuscaloosa: University of Alabama Press, 1971), 40.

35. Ibid., 14.

Veblen and Postmodernism

This exploration of Veblen's life and work as a progressive enricher of evolutionary naturalism involves looking at him not only in that context, but also in the environment of the broader cultural milieu in which he lived. Nowadays, this requires relating him to the movement known as "postmodernism" to which he does not quite belong. Overall Veblen does not fit perfectly within the paradigm of postmodernism, but then which thinkers really do? In any case, if David Harvey is correct in asserting that the paradigm of postmodernism itself only approached maturity in the years 1968–1972, Veblen could only represent a transition to, or partial fulfillment of, the paradigm, since he was dead.[36] That is how I intend to treat him using "postmodernity" as at best a complex, ambiguous, and slippery term. If, for example, Jacques Derrida and Michel Foucalt are fundamentally representative of postmodernism, how can it be claimed that Veblen was a postmodernist? Veblen never made the extreme claims advanced by the two French thinkers regarding value relativism and the social construction of reality. Also, he was more given to acquiescing in, if not actually advancing, the "imperialism" of science, technology, and urbanization, however critical he was of them at times. Certainly Veblen was not a prime exemplar of modernism either, but that is the point to be made. Both the more perceptive modernists of their day and the critical postmodernists of our own time should realize that, without fitting him for the bed of Procrustes, he does not exemplify either ideological camp. Rather he can be used to illustrate tendencies in both, but it is unfair to thrust him into the ideological straitjackets of either. Indeed, he is a transitional figure in modern intellectual and cultural history whose role is still being deciphered.

Thelma Lavine, although she was only summarizing what coauthors John Dewey and Arthur Bentley had in common in the early years of the twentieth century, aptly described what Veblen also shared. They all held to a

> holistic, process philosophy and are accordingly anti-dualistic, anti-foundationalist, anti-abstractionist, and anti-formalist; and in opposition to positivism and empiricism, tend to be interpretivist. [They] shared as well a broadly naturalistic, organism-environment frame; a rejection of traditional metaphysics and epistemology; an opposition to a legislative function on

36. David Harvey, *The Condition of Post-Modernity* (Oxford: Basil Blackwell, 1990), 38.

the part of mathematics and logic in relation to inquiry; and a behavioral, in opposition to a mentalistic, approach to the social sciences.[37]

Even though Lavine only intended to aggregate the commonalities of the two authors of *Knowing and the Known* (1946), Dewey's last major book, her summary, when applied to Veblen, is close to the truth. Older versions of "truth," "objectivity," and "reason" did not directly or explicitly appeal to him, but neither was he willing to completely jettison them. The alleged ethical and political "neutrality" of history and political science may have struck him as naïve, but he did not subscribe to an anarchistic relativism of knowledge, whether it be social or moral. Veblen's one foot in the postmodernist camp does not signify an identity of interest and ideas with contemporary radical relativism since his other foot, so to speak, lingered in the Enlightenment and the Industrial Revolution. At most he was a reluctant postmodernist. The question of how and whether Veblen was a postmodernist thinker is a key one, and cultural specialists have paid inadequate attention to his role in the debate over postmodernity. Some of the prime issues in this debate have to do with agency and indeterminism, anti-essentialism and anti-foundationalism, social constructionism and value relativism. Indeed, it would take another book to deal adequately with the meaning of these terms in the context of postmodernism, much less assign a well-defined role within it to Veblen.

However, it would be misleading to conclude that Veblen went the full distance to a radical postmodernism that is fully relativistic in the moral and aesthetic senses and also rooted in the conviction that political and cultural reality are merely social constructs. He knew full well that some values are more rationally defensible than others, and that brute facticity in the natural and social order must always constrain the social construction of reality. In this sense, he was still a child of the Enlightenment, albeit a critical one.

Veblen's Secular Humanism

Veblen was not a joiner, that is, a social or political activist, since he preferred to work for progressive change through his writings. Nevertheless, he did campaign for Grover Cleveland for president in 1884, on at least one occasion was involved in Populist politics, and in 1924 signed a peti-

37. Thelma Lavine, intro. to Dewey and Bentley, *Knowing and the Known* (Boston: Beacon Press, 1949), xxv.

tion endorsing Robert M. La Follette for president. John P. Diggins asked his stepdaughter Ann Sims, "Why didn't Veblen join the Socialist Party or the I.W.W.?" Ann replied that her mother frequently asked Veblen to do this, and he said, "I am not a joiner; only carpenters and furniture makers can join parts together."[38] Secular humanism of the sort to which Veblen adhered apparently did not require its adherents to perform an activist role most of the time; and, of course, Veblen did not call himself such. It is mostly by extrapolation from his writings and what is known of his intellectual biography that calling him a "secular humanist" can be justified.

That Veblen was a secular humanist of sorts is probable; but given his lack of political and social activism, what does this mean? The Council for Secular Humanism has provided "The Affirmations of Humanism: A Statement of Principles," which is a detailed articulation of its creed. The most relevant parts of it for my purposes are articulated by its author, the philosopher Paul Kurtz, in table 2.

It would be impossible here to analyze each of the twenty-one tenets of the creed, five of the least relevant of which are not listed, and relate each of them to Veblen. In any case, while some would fit his thought better than others, on the whole most of the secular humanist creed appears congruent with what is known about his social values and political ends and their philosophical underpinnings. Indeed, when the different tenets of the creed do seem to diverge from what he believed, it is perhaps because his "cosmic pessimism," if such it may be called, rests on assumptions about institutional resistance to change, that is, cultural lag, and the persistence of atavistic continuities. He anticipated the tenacity of radical structural discontinuities, massive economic waste, repressive government, international bellicosity, and the resurgence of invidious distinctions based on gender, race, religion, class, and ethnicity. Nevertheless, certain aspects of the secular humanist creed dovetail well with his broader views. It is to these that we now turn for further examination.

Veblen's evolutionary naturalism, which is the philosophical umbrella for his secular humanism, sanctions the "application of reason and science to the understanding of the universe" as well as to "the solving of human problems," with the caveats, of course, that "imbecile institutions" may delay or even make impossible achievement of such understanding, or its institutional implementation as this bears on the resolution of human problems. In particular, the persistence of animism and other atavistic continuities weakens the application of science and reason, and the poten-

38. John P. Diggins, interview with Ann Sims, December 23, 1981, Palo Alto, California.

Table 2. Part(s) of the Secular Humanist Creed

We are committed to the application of reason and science to the understanding of the universe and to the solving of human problems.

We deplore efforts to denigrate human intelligence, to seek to explain the world in supernatural terms, and to look outside nature for salvation.

We believe that scientific discovery and technology can contribute to the betterment of human life.

We believe in an open and pluralistic society and that democracy is the best guarantee of protecting human rights from authoritarian elites and repressive majorities.

We are committed to the principle of the separation of church and state.

We cultivate the arts of negotiation and compromise as a means of resolving differences and achieving mutual understanding.

We are concerned with securing justice and fairness in society and with eliminating discrimination and intolerance.

We believe in supporting the disadvantaged and the handicapped so that they will be able to help themselves.

We attempt to transcend divisive parochial loyalties based on race, religion, gender, nationality, creed, class, sexual orientation, or ethnicity, and strive to work together for the common good of humanity.

We want to protect and enhance the earth, to preserve it for the future generations, and to avoid inflicting needless suffering on other species.

We believe in enjoying life here and now and in developing our creative talents to their fullest.

We believe in the cultivation of moral excellence.

We respect the right to privacy. Mature adults should be allowed to fulfill their aspirations, to express their sexual preferences, to exercise reproductive freedom, to have access to comprehensive and informed health-care, and to die with dignity.

We believe in the common moral decencies: altruism, integrity, honesty, truthfulness, responsibility. Humanist ethics is amenable to critical, rational guidance. There are normative standards that we discover together. Moral principles are tested by their consequences.

We are deeply concerned with the moral education of our children. We want to nourish reason and compassion.

We are engaged by the arts no less than by the sciences.

Source: Paul Kurtz, "The Affirmations of Humanism: A Statement of Principles," *Free Inquiry* 24, no. 3 (April/May 2004): 2. Also, see "A Humanist Manifesto," written and signed by thirty-four American philosophers and intellectuals in 1933, which is reprinted in Corliss Lamont, *The Philosophy of Humanism,* 5th ed., foreword by Edwin H. Wilson (New York: Frederick Ungar, 1965), Appendix, 285–89.

cy of vested interests—that is, concentrated wealth, national bellicosity, and cultural inertia make unlikely the success of persistent application of "matter-of-fact" to undermine "make-believe." In an ideal setting, scientific discovery and technology can contribute to the betterment of human life, but in Veblen's view, it is best not to assume the inevitability or even the likelihood of such an environment.

He clearly preferred the more open societies of Britain, France, and the United States with all their imperfections to the authoritarianism of the European dynasties because they offered protection against the arbitrary power of ruling classes and elites. But, as a member of a small radical minority, he was sensitive to the power of repressive majorities, such as the near hysteria of large numbers of Americans during the persecutions of the Great Red Scare in 1919.

Veblen was unsympathetic toward attempts to "explain the world in supernatural terms and to look outside nature for salvation": a view that is an integral part of his evolutionary naturalism. On account of this, he undoubtedly favored separation of church and state with a high wall erected between the two.

Veblen endorsed the social nature of the economy and the collective role of society in the production of wealth. This inclined him toward a much more equal distribution of incomes in terms of "securing justice and fairness in society." Furthermore, his writing indicates his dislike of "divisive parochial loyalties based on race, religion, gender, nationality, creed, class, sexual orientation, or ethnicity." Especially in *Imperial Germany*, he made strong efforts to discredit racism and ethnocentrism, and his earlier writings, particularly *The Theory of the Leisure Class*, are replete with an indictment of class and gender. Beyond this, Veblen occasionally commented on environmental concerns such as the squandering of natural resources and "earth butchery." Although these last concerns were not as pressing as they have since become, he alluded to them in his last major book, *Absentee Ownership* (1923).

As for the "right to privacy," it is hard to imagine Veblen having any fundamental disagreement with the rights of individuals to practice contraception, abortion, divorce, sterilization, and voluntary euthanasia. Admittedly, he rarely wrote on these topics, so a textual exegesis will not demonstrate what his views really were. But it is more than likely that he believed in the exercise of reproductive freedom, access to comprehensive and informed health care, and the right to die with dignity. As for "common moral decencies, altruism, integrity, honesty, truthfulness"—his writings fairly reek with support of such. Indeed, there can be little doubt of his preference for them over their opposites, especially when he departs from social theorizing into the realm of social criticism (although there is often no decisive point of demarcation between theory and criticism in his work).

Veblen was "skeptical of untested claims to knowledge" and "open to novel ideas." As a Darwinian, he could scarcely be otherwise, for he was a process philosopher focused on the exploitation of science and technology, who denies the likelihood of social and cultural stasis. Human experience may be consummatory at times, but such consummations are historically transient, fleeting, and often ephemeral. The progressive enrichment of evolutionary naturalism socially transmuted into secular humanism will not be progressive if it sinks into eternal cultural verities and immutable moral truths.

Secular Humanism, Evolutionary Naturalism, and Ethical Inquiry

Yet, contemporary evolutionary naturalism is not the only measure of Veblen's philosophy. Another resides within the evolutionary naturalism of his own time, and to whose progressive enrichment he contributed.[39] And it is a matter of record that it changed considerably during his scholarly career. Later, I will record some of these changes and their incorporation into his thought during the period from 1884, when he published his first scholarly article in the *Journal of Speculative Philosophy,* to 1925, when his last article appeared in the *American Economic Review.*

Early philosophy in the New World was handicapped in its development by what Dewey described as "conditions of geographical isolation, social segregation, and absence of scientific method."[40] These conditions were starting to wane, their demise hastened by the spread of Darwinism, industrialism, and urbanism when Veblen was a young man. Equally important in retarding the progress of naturalism was the presence of supernaturalism in virtually all existing churches and religious institutions, including his own Lutheran heritage. Those who came to share the natu-

---

39. It would difficult to recount all the influences on naturalism that molded Veblen's version of it. But a convergent intellectual force that he was aware of was the Ethical Culture movement founded by Felix Adler (1851–1933) in 1876. Adler, who was a thoroughgoing secularist, and his followers, stressed the importance of ethical behavior independent of religious beliefs. On the birthrate and birth control, see Veblen, *The Higher Learning in America,* 200–202.

40. John Dewey, "Antinaturalism in Extremis," in *Naturalism and the Human Spirit,* ed. Yervant H. Krikorian (New York: Columbia University Press, 1944), 13.

ralistic outlook, while they may have disagreed over the details, had to contend not only with religionists, but with another brand of intellectuals as well: those who claimed to root their extra-naturalism in a higher faculty of reason or intuition. Veblen came to have a strong skepticism about this philosophical position as well as the more extreme forms of materialistic reductionism.

But in addition to the dictionary definition of "naturalism" already given, what did Veblen's evolutionary naturalism add to this intellectual tradition? As one important philosophical naturalist of their time put it, "naturalism"

is not so much a system or a body of doctrine as an attitude and temper: it is essentially a philosophic method and a program. It undertakes to bring scientific analysis and criticism to bear on all the human enterprise and values so zealously maintained by the traditional supernaturalists and by the more sophisticated idealists.[41]

In the same vein, Yervant H. Kirkorian writes:

"nature" means what empirical science find it to be and what a completed empirical science would find it to be. Empirically nature includes physical objects and living things, inclusive of human beings and their ideals. One need neither argue, to show the simplicity of nature, that differences in nature are nothing but differences in complexity of structure nor assert, to emphasize the qualitative variety of nature, that there are unbridgeable gaps. Everything that we know presents in its natural setting structural connections and qualitative differences. Moreover, the importance of the naturalist's belief that nature is the whole of reality lies not only in what it affirms but also in what it denies. It denies what a philosopher like J. Maritain maintains: "There is a spiritual, metaphysical order superior to external nature . . . above all the mechanism and laws of the material world." This order "is no part of this universe . . . it rises above the created world, the sensible, and the supra-sensible." And beyond this order there is also the order of grace and "this is entirely supernatural." *The naturalist turns away from these supernatural worlds. For him there is no supernature, no transcendental world. Beyond nature there is more nature.*[42] (Author's emphasis)

---

41. John Herman Randall, "Epilogue: The Nature of Naturalism," in Krikorian, editor, *Naturalism and the Human Spirit*, 374. Also, see Herbert W. Schneider, "The Unnatural," 123, and Randall, "Epilogue: The Nature of Naturalism," 358, both in Krikorian, ed., *Naturalism and the Human Spirit*.
42. Krikorian, "A Naturalistic View of Mind," in *Naturalism and the Human Spirit*, 243.

He concludes:

> It is true that nature as completed science is an unattainable ideal and there-
> fore in a sense transcendent; but this unattainable ideal is indefinitely ap-
> proachable and has no supernatural implications. In relation to the study of
> mind the belief that nature is the whole of reality implies that mind should
> be examined as a natural phenomenon among other natural phenomena. It
> has its origins, growth, and decay within the physical, biological, and social
> setting.[43]

This naturalistic view of mind was held by both Veblen and his influential scientist friend Jacques Loeb. Clearly, they shared with other naturalists the view that sense perception gives evidence of reality and that the
study of mind is reliant on such data; but sense data is informative, not
definitive.

Epistemologically and ontologically, did Veblen have "any rule for de-
marcating rationalization and metaphysical suppositions from hypotheses which have an initial empirical probability in their favor"?[44] Clearly,
his policy for such demarcation rested upon his distinction between "matter-of-fact" and "make-believe," with the former having epistemological
privilege over the latter. But why? Because he favored suppositions that
had empirical referents to support them. In short, he denied the existence
of extrasensory events and was skeptical, at the very least, of any claims
of extrasensory perception, particularly when the preponderance of scientific evidence was against them.

There is a strong emphasis in Veblen's thought on the unfinished state
of both the social and natural orders partly because of the interaction between the two. Indeed, even the value constants in his thought must un-
dergo adaptation to cultural and social change if they are to remain relevant to human need. The change may be glacial, as among primitive
peoples, but it is inexorable. Process, flux, and adaptation are the nature
of existence in industrial society and all adjustment to it must be viewed
as tentative or provisional. While claiming a nonnatural origin from some
other realm may rationalize moral or psychological insight for religionists, to claim origins outside the natural world has no credibility in Veblen's cosmology.

Much of the naturalist's criticism of religion is based on an opposition
to dualisms, that is, the bifurcation between ontological realms and, of

43. Ibid.
44. Edward W. Strong, "The Materials of Historical Knowledge," in Krikorian, editor, *Naturalism and the Human Spirit*, 158.

course, to subjectivisms that "treat experience as a private affair and thus make the very existence of an 'external' world or objective order a matter of doubt."[45]

Yet Veblen's version of evolutionary naturalism is not the positivist materialist reductionism[46] of dogmatic Marxism often accompanied by a solipsistic politics of isolation either.

As far as naturalism and ethics are concerned, Veblen is not always explicit regarding their linkage. Still, it is evident that the values he favors are those which he thinks will generally aid in the adaptive process: namely, critical intelligence, proficiency of workmanship, and altruism. The partly Darwinian derivation of Veblen's ethics is not simply what enables the group to survive, but to prosper and, perhaps more important, to become a properly developing community where idle curiosity, the instinct of workmanship, and the parental bent flourish. Veblen's naturalistic ethics, however derivative from the Judeo-Christian ethic of brotherly love they may be, are not rooted in the supernatural or in intuitionism. In this vein, Abraham Edel writes that

> the need for empirical interpretation of ethical terms has been strongly denied in nonnaturalistic ethics. For example, Hartmann has written: "The settlement of the matter depends upon demonstrating that there is a self-existent ideal sphere in which values are native, and that, as the contents of this sphere, values, self-subsistent and dependent upon no prior experience, are discerned a priori." In the intuitionist tradition ethical concepts become clear, certain types of action are seen self-evidently to be right or good, to have some moral or value character.[47]

As a naturalist, Edel disagrees with both the religionist and the intuitionist approaches to ethics because, in his view, the source and the test of values ultimately lie in their consequences and usefulness; and there is no reason to think Veblen felt differently. Immutable truths, eternal verities, and value absolutes have no place in the philosophy of the evolutionary naturalist that is rooted in process and flux and the adaptation of scientific discoveries and even value constants to the evolving natural and social order.

45. Sterling P. Lamprecht, "Naturalism and Religion," in Krikorian, ed., *Naturalism and the Human Spirit*, 20.

46. William R. Dennes, "The Categories of Naturalism," in Krikorian, ed., *Naturalism and the Human Spirit*, 279.

47. Abraham Edel, "Naturalism and Ethical Theory," in Krikorian, ed., *Naturalism and the Human Spirit*, 66.

Conclusion

The philosophy underlying Veblen's economics and, indeed, his social science has been repeatedly investigated and analyzed and this has resulted in several interpretations. These include Richard Teggart's claim that Veblen's philosophy and its relationship to economics is an "absolutistic, deterministic monism" and a philosophy of "pantheistic process," Derek Scott's view that it is a "materialistic monism," or Abram Harris's view that it is "institutional mutationism" embedded in a theory of progress.[48] Also, it is claimed that the core of Veblen's thought is Marxian and, therefore, underpinned philosophically by dialectical materialism, or that it is characteristically Darwinian or Lamarckian, or both, and thus a form of "evolutionary materialism." But philosophically Veblen's work is closer to pragmatism, especially as expressed in Dewey's mature thought. Like Dewey's, his thought should be labeled "evolutionary naturalism." Those residing in this ideological camp and intellectual tradition usually support its progressive enrichment. Thus "the enrichment of evolutionary naturalism" is the most accurate label for Veblen's philosophy, particularly when it undergirds his institutional economics and radical sociology. But more about this later.

How is Veblen's evolutionary naturalism linked with his secular humanism? The two are organically integrated, making it difficult even for analytical purposes to separate them. To illustrate, Veblen's conception of human nature is one that postulates humankind as a product of evolutionary forces. Humanity as it exists is the result of biological and cultural vectors. One important caveat is essential, however. Evolutionary naturalism contains a doctrine of *emergence,* that is, *novelty,* and individual and social volition may logically arise out of fundamental structural discontinuities. Veblen clearly believed in human agency, because even though he recognized the claims of the theory of universal determinism—namely, that every effect has a cause—he also knew that the human sciences were too immature to explain many of the most important causal relationships. In short, the frailties of human knowledge made the complexities of human behavior empirically ascertainable only to a limited extent. Consequently, there was no reasonable alternative to attributing some freedom of choice to both individuals and even social orders, however narrow the range of choice might be, since pure determinism postulated the possession of far greater knowledge of human behavior and analytic capacity than social scientists actually possessed.

48. See Daugert, *Philosophy of Thorstein Veblen,* 3.

Apparently, any doctrinal stance worth its salt must ultimately contain some set of premises about the nature of human nature, the composition of the social (and natural order) and its desirability, and the prospects (available means) for changing it. At this point, it should be evident as to what these are in Veblen, even though much of what he wrote is subject to ambiguous, if not multiple meanings, and his overlay of mockery and satire sometimes hinders rather than aids clarification. Nevertheless, his view of human nature, evolutionary naturalism, and secular humanism are so organically linked and integrated that they cannot be adequately understood in isolation from one another.

Veblen was not a positivist or a reductionist in the sense Jacques Loeb was despite the accusations of some of his critics.[49] Indeed, Veblen might well have responded to these critics with the same words he used to deflate the scientific pretensions of the eminent neoclassicist John Bates Clark:

> What would be the scientific rating of the work of a botanist who should spend his energy in devising ways and means to neutralize the ecological variability of plants, or of a physiologist who conceived it the end of his scientific endeavors to rehabilitate the vermiform appendix of the pineal eye, or to denounce and penalize the imitative coloring of the viceroy butterfly? What scientific interest would attach to the matter if Mr. Loeb, e.g., should devote a few score pages to canvassing the moral responsibilities incurred by him in his parental relation to his parthenogenetically developed sea-urchin eggs? Those phenomena which Mr. Clark characterizes as "positive perversions" may be distasteful and troublesome, perhaps, but "the economic necessity of doing what is legally difficult" is not of the "essentials of theory."[50]

Critics of evolutionary naturalism have mistaken it for the positivistic, reductionist, scientism that misshapes perceptions of both the natural and social world. But Lewis Mumford, a student, interpreter, and critic of Veblen's ideas writes:

> man is not born into that bare physical universe: rather, he is born into a world of human values, human purposes, human instruments, human designs; and all that he knows or believes about the physical world is the re-

49. See Charles Rasmussen and Rick Tilman, *Jacques Loeb: His Science and Social Activism and Their Philosophical Foundations* (Philadelphia: American Philosophical Society, 1998), ch. 5 and 8.

50. Thorstein Veblen, "Professor Clark's Economics," in *The Place of Science in Modern Civilization and Other Essays*, 189.

sult of his own personal and social development. The very language he uses for neutral scientific description is a social product that antedates his science. Indeed, the tendency to look upon process in the physical world as more important, more fundamental, than the processes of organisms, societies, and personalities is itself a by-product of a particular moment of human history: the outcome of a systematic self-deflation.[51]

This leads Mumford to make a comment that Veblen himself could have written:

Man's subjective and his objective world are in constant interplay: nothing that he knows about the universe can be dissociated from the facts of his own life; and no product of his culture is so detached from the larger groundwork of existence that he can impute to his individual powers what alone has been made possible by countless generations of men and by the underlying co-operation of the entire system of nature.[52]

This chapter began with the claim that Veblen is best described as an evolutionary naturalist and a secular humanist. It should be evident by now that the two are closely linked together in Veblen's life and thought. Indeed, both would have less meaning for him if they were disaggregated. For evolutionary naturalism is the formal philosophical basis for Veblen's secular humanism. It is highly unlikely that he arrived at a social and moral philosophy such as secular humanism without the underpinnings of evolutionary naturalism or some perspective much like it.

Stanley Daugert penetratingly comments about Veblen's evolutionary naturalism to this effect:

In Veblen's mind there is always this connection between the organizations of society, the kinds of environment—the complete system of "outer relations"—and the somewhat lonely, helpless individuals who are born into them, that those individuals seldom learn to resist these influences effectively. This is one of the greatest tragedies in the human drama. Veblen resigned himself to this conclusion rather early in life; an almost Schophenhaurian sense of futility can be felt throughout much of this work, as if he and his fellow men were forever battling immense, uncontrollable forces to little purpose. One may also note here the similarity of Veblen's resignation and William Graham Summer's "desperate naturalism," as one writer characterizes Summer's philosophy. Veblen's is a kind of desperate evolutionism. Perfect adjustment—equilibration, in Spencer's language—is never

51. Lewis Mumford, *The Condition of Man*, 10.
52. Ibid., 11.

achieved because of the constant cumulative change in both the "inner" and "outer" relations.[53]

Daugert concludes:

> But despite the hopeless task of achieving such adjustment, despite the many obstacles (such as the leisure class, the absentee owners, etc., who respond only tardily to the "altered general situation"), it nevertheless remains true that man continues to attempt to alter his life and the conditions of his living.[54]

Daugert thus places Veblen partly in an intellectual tradition that includes the German philosopher Arthur Schopenhauer (1788–1860) and W. G. Sumner, Veblen's sociology professor at Yale. But the metatheoretical divergences alone among the three men are considerable; as epistemologists, ontologists, and evaluators they are different. What is true, what is real, and what is valuable to them are not the same. Yet Daugert believes they share a sense of futility, resignation, even desperation. This is not the whole truth about Veblen, but Daugert has captured a feature in his intellectual physiognomy.

Veblen's quest for understanding the mind included the question of whether or not there were commonalities of thought among different ethnic groups. Did individuals from cultures alien to each other possess the same mental processes, follow the same pattern of mental development, and achieve the same intellectual level? Ultimately, he came to believe that the biological properties of the human organism as regards different ethnic and racial groups are essentially similar if not identical. Thus unification of the disparate cultural products of the mental structures of both primitive and civilized humanity was possible, however unlikely this was to occur in the near future. The constancies and the flux in social and cultural institutions reflected the basic structures of the mind as well as environmental and evolutionary forces. Veblen, on the basis of such reasoning, questioned the status of economic and religious elites and the leisure classes. He was also skeptical both of *exaggerated* claims for free will or agency and *extreme* claims for cognitive creativity.

Veblen believed that the mind developed a structuring tendency that

---

53. Daugert, *Philosophy of Thorstein Veblen*, 67.
54. Ibid. Also, see Rick Tilman, "Thorstein Veblen and Western Thought Fin de Siécle: A Recent Interpretation," *Journal of Economic Issues* 35 (March 2001): 117–38, for analysis of the influence of Schopenhauer and Veblen's European contemporaries on him.

effected perception of both the forms of natural phenomena and important forms of social relations between humans upon which power, status, and class were based. In short, the human mind imposed upon the flux of sense and experiential data those cultural and social proclivities that made for both privilege and equality. Yet he recognized the difficulties of distinguishing between the individual as a biological entity and the individual social roles he or she must play. He was skeptical of social roles assigned on the basis of either fixed genetic endowment or parental social status. His writing thus displays a generally egalitarian collectivist bias that is not without an idealistic assessment of the potential of the existing industrial order if projected far enough into the future. If Veblen's thought is to be assessed as a whole, the tension between the realization of this potential and the despair, resignation, and futility noted by Daugert must somehow be reconciled. But that is a project for another time and place.

Chapter 3

The Theory of Cultural Lag Revisited

Introduction

"Cultural lag" was once used to articulate categories, illuminate distinctions, and develop concepts in sociology while functioning as a theory itself. In Veblen's thought, it is noteworthy for its use in locating and identifying both obsolescent and forward-looking structures and in predicting the likelihood or not of the reoccurrence of certain kinds of problems and phenomena. Too, it has been relevant in situating movements and ideas along the political and doctrinal spectrum, including those of Veblen and his contemporaries. However, it is many years since the theory of cultural lag was in vogue among social scientists. By 1968 *the International Encyclopedia of the Social Sciences* listed no separate article on it, simply incorporating it into other entries such as cultural and social change. But the purpose of this analysis is not to revive interest in cultural lag, so much as to explain how it illuminates Veblen's social theory and intellectual perspective. This will be done by (1) articulating its putative meaning, (2) using it to explain Veblen's use of the concept and, to a lesser extent, his theory of social change,[1] and (3) explicating cultural lag in his explanation of cultural borrowing and international relations during the Great War.

The depth and breadth of Veblen's theory of cultural lag gives it considerable predictive and explanatory potential, not to mention providing

---

1. For a review of the contemporary literature on Veblen's theory of social and institutional change, see Rick Tilman, "Some Recent Interpretations of Veblen's Theory of Institutional Change," *Journal of Economic Issues* 21 (June 1987): 683–90. Also, see Olivier Brette, "Thorstein Veblen's Theory of Institutional Change: Beyond Technological Determinism," *European Journal of the History of Economic Thought* 10, no. 3 (Autumn 2003): 455–77; and Malcolm Rutherford, "Thorstein Veblen and the Processes of Institutional Change," *History of Political Economy* 16, no. 3 (Fall 1984): 331–48.

a basis for normative judgments about past, present, and future social orders. The words and phrases he uses to describe and illustrate cultural lag include "archaic" and "archaisms," "superstitious" and "superstitions," "rules of caste," and "ceremonial" rules of conduct. However, there is a general lack of focus on and interest in cultural lag among social scientists, which is why it is important to recognize the reasons for the declining role it plays in interpretations of his thought in the last generation or so. It is also essential to reconstruct the powerful role it played in his thinking and the often-pervasive influence it had on his interpretation of social change. Can those who share Veblen's perspective justifiably continue to anticipate the triumph of "instrumentally warranted" processes over atavistic continuities and "imbecile institutions"? For this reason among others, the drag and retardation exerted by the existing social order require further exploration.

The Meaning of Cultural Lag: Historical Overview

Veblen's theory of cultural lag is found in his ideas regarding waste, that is, conspicuous consumption, conspicuous exemption from socially useful labor, and unused industrial capacity. All of these provide him with a standard by which to distinguish between mere social change and proper social growth and development that serves the "generic ends" of human life. In short, the explicitly normative basis of Veblen's theory of social change shows that he is not merely interested in explaining why major social institutions evolve, but why they change in particular directions, good, bad, or indifferent. Ethical and moral activity and human purpose invariably find their way into institutional change and affect its outcome, and Veblen was prone to identify with certain values in applying his theory of cultural lag, despite the fact that at times his satirical prose disguised his intent.

The theory of cultural lag has an intellectual pedigree and origin much older than the genesis of formal social theory and sociology in the mid- to late nineteenth century. But, for our purposes, it will suffice to start with its formalization in the early twentieth century in the work of William F. Ogburn (1886–1959) and his contemporaries. First, however, it is essential to locate and articulate the meaning of the theory of cultural lag when it was still in common use among sociologists and played a significant role in the conceptual focus of the social sciences. The sociologist Kurt Wolff comments:

It may be defined as the period between the point in time at which one cultural element approximates a cultural goal, valued by the society or by the observer, and that point at which another element or other elements achieve such a degree of approximation; or, as the difference in the rate of change of two or more cultural elements about whose inter-relationships the observer makes no normative predication or claim either explicitly or implicitly. It is treated as a major aspect of social change, and as a causal factor in social disorganization . . . The term was coined by W. F. Ogburn, who used it to designate the time which passes between a change in "material" culture (e.g., the supply of forests) and the adjustive change in "adaptive" culture (e.g., policy of using the forests), without excluding the possibility of adaptive culture changing before material culture or non-material culture changing without the material culture changing likewise.[2]

Although Ogburn was the best-known and most-influential social scientist associated with the codification and use of Veblen's theory of cultural lag, he was not the first to recognize it. In 1914, Ulysses G. Weatherly in reviewing *The Instinct of Workmanship* wrote that:

while nowhere asserted in set terms, Veblen's leading thesis seems to be that man's intellectual faculties, as applied to technological achievement, have advanced too fast for his instinctive aptitudes. The modern civilized peoples have to deal with a complex and changing technological system with an unchanging or slowly changing endowment of instinctive capacity. Both physically and spiritually they are better suited to the conditions of advanced savagery than to those of the modern machine industry. It was not an accident that the eighteenth century craze for a "return to nature" came in the period of transition to machine production, or that in the present mechanistic age the cult of the simple life finds so large a place.[3]

Veblen was conventionally portrayed as a theorist of cultural lag. Weatherly and Ogburn were simply early exemplars of this trend. Indeed, Veblen scholars once almost universally accepted this interpretation of the central thrust of his work. To illustrate the point, William P. Glade, in an

2. Kurt H. Wolff, "Cultural Lag" in *A Dictionary of the Social Sciences,* ed. Julius Gould and William Kolb (New York: Free Press, 1964), 159. Also, see the brief discussions of cultural lag in Nicholas S. Timasheff, *Sociological Theory: Its Nature and Growth* (New York: Random House, 1964), 204–7, and Robert F. Murphy, *The Dialectics of Social Life* (New York: Basic Books, 1971), 222–23.

3. Ulysses G. Weatherly, "Review of Thorstein Veblen, 'The Instinct of Workmanship and the State of the Industrial Arts,'" in *American Economic Review* 4 (December 1914), 860–61.

important article published in 1952, summarized the significant contribution Veblen made to the theory of cultural lag and, also, the widespread interpretation of him as a cultural lag theorist by other scholars.[4] It is fair to say that this was the prevailing view of Veblen's social theory for much of the four generations during which his work has received critical acclaim and analysis.[5] This is not to suggest, however, that the term "cultural lag" has precisely the same meaning to all who write about it. C. Wright Mills pointed out long ago that it can be a "sponge word" with a highly adaptive meaning.[6]

In 1957, Ogburn wrote that he had first used the term "cultural lag" in 1914 when he was a professor of economics and sociology at Reed College and that he had fully developed it as a theory by 1915. On the source(s) from which he drew the theory, he commented as follows:

4. William P. Glade, "The Theory of Cultural Lag and the Veblenian Contribution," *American Journal of Sociology and Economics* 11 (July 1952): 427–37. However, Robert MacIver writes that technological lag

> has a clearer application than that commonly used by American sociologists—"cultural lag." "Technological lag" signifies that some part of the total apparatus of society fails to receive the technological advances or applications requisite to ensure the proper operation of other technologically advanced parts or of the whole system. A "bottleneck" in the industrial process is one type of technological lag. Another type is the failure to organize the human factors of production so that they are properly adjusted to the operative conditions introduced by the newer technology. The currency of the expression "cultural lag" is associated with the assumption that the initiation of change vests in the technological order and that the task of the cultural order is to "keep pace" with it.

Using MacIver's definition and explanation, it is evident that Veblen's theory is one of "cultural" rather than "technological" lag, but not always by a decisive margin. Nevertheless, in his analysis, cultural factors may drive technology rather than vice versa, or at least serve as powerful inhibitors of change. MacIver, *The Web of Government*, revised ed. (New York: Free Press, 1965), 350.

5. The cultural lag thesis in its various forms as important to Veblen's social theory is endorsed by most Veblen scholars. See bibliographical guides by Tilman et al. and Wasser et al.

6. Much of Mills's work on cultural lag was based on certain issues raised by Veblen and it was from this problematic that he attacked the characteristic use of the cultural-lag concept. He praised Veblen's use of the concept in these words: "Veblen's use of "lag, lead and friction" is a structural analysis of industry versus business enterprise. He focused on where "the lag" seemed to pinch; he attempted to show how the trained incapacity of legitimate businessmen acting within entrepreneurial canons would result in a commercial sabotage of production and efficiency in order to augment profits within a system of price and ownership." Mills, *Power, Politics and People*, ed. I. L. Horowitz (New York: Oxford University Press, 1963), 546. But Mills complained that in contemporary sociology the idea of cultural lag had lost its "specific and structural anchorage." It had become so generalized that it was applied to everything fragmentarily. This depreciation of meaning had been accomplished with the aid of such sponge words as "adaptive culture" and "material culture."

I am happy to discuss its origin, since I have been accused by some of taking the theory from Thorstein Veblen and by others from Karl Marx. I am quite sure there was no direct taking over of the idea from Veblen because I had never read him on this point. I had read Marx, and his materialistic interpretation of history was well known to social scientists and historians in general. This idea was a base, however, from which the theory of cultural lag was developed, but certainly neither the materialistic interpretation of history nor economic determinism is the same as cultural lag.[7]

Whatever the sources of Ogburn's theory of cultural lag, the fact remains that he is commonly credited with its formalization and its influence on other social scientists, especially sociologists. In his view:

it calls for the following steps: (1) the identification of at least two variables; (2) the demonstration that these two variables were in adjustment; (3) the determination by dates that one variable has changed while the other has not changed or that one has changed in greater degree than the other; and (4) that when one variable has changed earlier or in greater degree than the other, there is a less satisfactory adjustment than existed before.[8]

The critic may object that (2) is a flawed portrayal of Veblen's theory to begin with if such was Ogburn's intent, for Veblen rarely claimed that social or cultural variables were in "adjustment" with each other. In his view, these variables, if such they can be labeled, are usually out of "sync" with each other anyway. In short, Veblen is not an equilibrium theorist nor does he often focus on social balance, because he assumes institutional maladjustment is a chronic condition of the human order. He can be understood as a theorist of cultural lag, but his version of it differs somewhat from that held by more conventional sociologists like Ogburn and his contemporaries. Nevertheless, this treatment of Veblen as a theorist of cultural lag uses Ogburn's paradigm to establish Veblen's authenticity in this regard. But what was Veblen's theory of cultural lag and how is it related to his larger intellectual perspective? It is to these questions that we now turn.

Institutions and Obsolescence: Lag

The nature/nurture controversy arises in evolutionary naturalist circles as it also does among Veblen scholars. It is difficult to authoritatively or

7. William F. Ogburn, *On Culture and Social Change*, ed. with intro. by O. D. Duncan (Chicago: University of Chicago Press, 1964), 87.
8. Ibid., 89.

definitively place Veblen on a continuum between cultural and social determinism on the one hand and the biological/instinctual on the other. Does he lie closer to B. F. Skinner's view of human beings as tabula rasa, that is, blank slates upon which experiential reinforcement is the main determinant of behavior? Or does he view them as living things with built-in, prewired, evolutionary generated mechanisms as many sociobiologists contended? Probably neither position represents an accurate assessment of Veblen's views on human nature. Nevertheless, he emphasized the plasticity and malleability of the human species more than the biological template, although as a staunch Darwinian he could hardly ignore the latter. But how is this related to his theory of cultural lag?

Psychological aspects of Veblen's explanation of cultural lag are borrowed from Kant and James and rooted in Veblen's utilization of their term "apperceptive mass."

> All facts of observation are necessarily seen in the light of the observer's habits of thought, and the most intimate and inveterate of his habits of thought is the experience of his own initiative and endeavors. It is to this "apperception mass" that objects of apperception are finally referred, and it is in terms of this experience that their measure is finally taken.[9]

To this, Murray G. Murphey adds:

> The apperceptive mass comprises the whole of the instincts and habits, desired ends and formulated knowledge, which the mind contains and which it brings to bear upon sense data in the process of knowing. In Veblen's theory of knowledge, data is systematized in terms of this apperceptive mass with the result that sense experience is interpreted through principles, some of which are already outmoded and false. It follows then that human knowledge is never abreast of the actual situation and that there is always a time lag between environmental and institutional change. To the dictum "whatever is is right" Veblen answered "whatever is is wrong," at least to some extent.[10]

The psychological basis of Veblen's theory thus becomes more fully intelligible. In this vein, one of the most specific analyses of institutional obsolescence Veblen ever made was in *The Theory of the Leisure Class.* He begins with a definition of an institution:

9. Veblen, *The Instinct of Workmanship* (New York: B. W. Huebsch, 1918), 53.
10. Murray G. Murphey, "Thorstein Veblen: Instinctive Values and Evolutionary Science," in *Values and Value Theory in Twentieth-Century America: Essays in Honor of Elizabeth Flower,* ed. Murphey and Ivar Berg (Philadelphia: Temple University Press, 1988), 130.

Institutions are, in substance, prevalent habits of thought with respect to particular relations and particular functions of the individual and of the community; and the scheme of life, which is made up of the aggregate of institutions in force at a given time or at a given point in the development of any society, may, on the psychological side, be broadly characterized as a prevalent spiritual attitude or a prevalent theory of life. As regards its generic features, this spiritual attitude or theory of life is in the last analysis reducible to terms of a prevalent type of character.[11]

He then proceeds to a description of institutional, that is, cultural lag:

The situation of today shapes the institutions of tomorrow through a selective, coercive process, by acting upon men's habitual view of things, and so altering or fortifying a point of view or a mental attitude handed down from the past. The institutions—that is to say, the habits of thought—under the guidance of which men live are in this way received from an earlier time; more or less remotely earlier, but in any event they have been elaborated in and received from the past. Institutions are products of the past process, are adapted to past circumstances, and are therefore never in full accord with the requirements of the present . . . and each successive situation of the community in its turn tends to obsolescence as soon as it has been established. When a step in the development has been taken, this step itself constitutes a change of situation which requires a new adaptation; it becomes the point of departure for a new step in the adjustment, and so on interminably.[12]

But what are the forces and processes that actually promote social change by which cultural lag can be identified? In *The Theory of the Leisure Class*, Veblen writes:

A readjustment of men's habits of thought to conform with the exigencies of an altered situation is in any case made only tardily and reluctantly, and only under the coercion exercised by a stipulation which has made the accredited views untenable. The readjustment of institutions and habitual views to an altered environment is made in response to pressure from without; it is of the nature of a response to stimulus. Freedom and facility of readjustment, that is to say, capacity for growth in social structure, therefore depends in great measure on the degree of freedom with which the situation at any given time acts on the individual members of the community—the degree of exposure of the constraining forces of the environment. If any portion or class of society is sheltered from the action of the environment in any essential respect, that portion of the community, or that class, will adapt its

11. Veblen, *The Theory of the Leisure Class*, 190.
12. Ibid., 190–91.

views and its scheme of life more tardily to the altered general situation; it will in so far tend to retard the process of social transformation. The wealthy leisure class is in such a sheltered position with respect to the economic forces that make for change and readjustment. And it may be said that the forces which make for a readjustment of institutions, especially in the case of a modern industrial community, are, in the last analysis almost entirely of an economic nature.[13]

Veblen then writes:

An advance in technical methods, in population, or in industrial organiza-tion will require at least some of the members of the community to change their habits of life, if they are to enter with facility and effect into the altered industrial methods; and in doing so they will be unable to live up to the re-ceived notions as to what are the right and beautiful habits of life. Any one who is required to change his habits of life and his habitual relations to his fellow men will feel the discrepancy between the method of life required of him by the newly arisen exigencies, and the traditional scheme of life to which he is accustomed. It is the individuals placed in this position who have the liveliest incentive to reconstruct the received scheme of life and are most readily persuaded to accept new standards; and it is through the need of the means of livelihood that men are placed in such a position.[14]

He concludes:

The pressure exerted by the environment upon the group, and making for a readjustment of the group's scheme of life, impinges upon the members of the group in the form of pecuniary exigencies; and it is owing to this fact—that external forces are in great part translated into the form of pecu-niary or economic exigencies—it is owing to this fact that we can say that the forces which count toward a readjustment of institutions in any modern industrial community are chiefly economic forces; or more specifically, these forces take the form of pecuniary pressure.[15]

Thus, it is the amount of exposure to the economic forces more than any other factor supporting change that causes humankind to yield to the need and the demand for change.[16] Veblen, then, states:

It is to be noted then, although it may be a tedious truism, that the institu-tions of today—the present accepted scheme of life—do not entirely fit the

13. Ibid., 192–93.
14. Ibid., 195.
15. Ibid., 195–96.
16. Ibid., 196–204.

situation of today. At the same time, men's present habits of thought tend to persist indefinitely, except as circumstances enforce a change. These institutions which have thus been handed down, these habits of thought, points of view, mental attitudes and aptitudes, or what not, are therefore themselves a conservative factor. This is the factor of social inertia, psychological inertia, conservatism. . . . The evolution of society is substantially a process of mental adaptation on the part of individuals under the stress of circumstances which will no longer tolerate habits of thought formed under the conforming to a different set of circumstances in the past.[17]

This is, perhaps, the most succinct summary Veblen offers of cultural lag. But it is to William F. Ogburn's definition and analysis of it that we again turn, for "cultural lag" is an often ill-defined and slippery term that needs precise definition and conceptual clarity, if it is to be used in the revival of Veblen's now neglected analysis of archaic structures and institutions.

Ogburn, Veblen, and Cultural Lag

Almost as important as Ogburn's definition of cultural lag is, in his view, its causes which are several: (1) scarcity of invention in the adaptive culture, such as, for instance, lack of change in governmental forms; (2) mechanical obstacles to adaptive changes, that is, purely physical or mechanical obstacles to the spread of some forms of culture; (3) the relationship between the adaptive culture and the material culture; (4) the connections of the adaptive culture to other parts of culture; and (5) group pressure to enforce conformity.[18] This brief summation of Ogburn's explanation of the causes of cultural lag has its equivalents in the corpus of Veblen's work as will be shown. The question of whether or not Veblen was a cultural lag theorist can only be settled satisfactorily through a comparative textual exegesis of Veblen and Ogburn; and even this assertion rests on the premise that Ogburn was the most authoritative exemplar of the theory itself.

It does appear that Veblen uses or illustrates all of Ogburn's causes of cultural lag in his own analysis. First, there is the scarcity of invention in the adaptive culture, such as a lack of change in governmental forms. Veblen has no theory of the positive state that certainly would have fleshed out his political theory and given it a stronger correlation both with the

17. Ibid., 191–92.
18. William F. Ogburn, *Social Change* (New York: Viking Press, 1928), 257–64.

evolving techno-economic base and the interventionist state that had actually emerged during his lifetime. At a minimum, he might have developed a view of the state that anticipated its engagement in the large-scale provision of public goods, counter-cyclical macroeconomic control, and microeconomic regulation. But he did none of these things nor did he sanction state intervention into labor-management relations, occupational safety and health regulation, or any other of the major functions of the positive state. The reason was simply that he viewed government as an exploitive and predatory mechanism as well as an archaism that was seemingly incapable as yet of such equitable or efficient undertakings since it worked for the business community.

As for "mechanical obstacles to adaptive change" that is, "habit, love of the past and various utilities of the old culture," the observer will note that Veblen's work is riddled with examples of these. Cumbersome clothing, aesthetic practices founded in mere custom, and restrictive work practices on the part of unions all illustrate this.

What illustrations can be found in Veblen of Ogburn's third explanation of cultural lag, that is, circumstances where the relationship between the adaptive culture and the material culture is not very close? Numerous examples of this could be found in his time when discrimination on grounds of sex or caste abounded. The existing material culture is riddled with gender and caste bias, whereas the adaptive culture that has started to emerge and that might inhibit discriminatory practices is too weak and decentered to overcome such practices, although some changes such as property-holding, voting rights, and choice of marriage partners are significant. Nevertheless, in the longer run the gap between the adaptive culture and the material culture narrow as in the case of women's rights.

No claim is advanced here that the term "cultural lag" is an adequate description of Veblen's theory of social change. Indeed, it is simply part of his broader theory of social change, although much of the foundation of his normative judgments regarding the viability of the existing social order would be incoherent without it. The material resources, institutional structures, and organizational patterns necessary for the realization of his value constants, namely, altruism (other-regardingness), idle curiosity (critical intelligence), and workmanship (taking pains with one's work, pride in it), and deriving satisfaction from it, are all rooted in the possibilities of their realization. Lack of correlation between these value constants and their material bases of support is clearly the gist of Veblen's complaint and the source of his ruminations on cultural lag. This links his view with Ogburn's fourth explanation of cultural lag, wherein delay in adjustment is caused by lack of correlation between different parts of the interacting culture. To illustrate, Veblen's instinct of workmanship may suffer retardation or even defeat from a price system that rewards shoddy craftsmanship and inferior materials leading to the poor quality of consumer goods.

Fifth, and last, in Ogburn's list of explanations for cultural lag are circumstances where group valuations are retardant because group pressures enforce conformity and resist change. Should this discussion of cultural lag, including the fifth point, seem dated, Veblen's views can be applied in a contemporary setting to what he called "imbecile institutions" such as the Roman Catholic Church; to issues and practices like the celibacy of the clergy, divorce, contraception, voluntary euthanasia, abortion, gender discrimination (prohibition of the ordination of women), and sterilization. Veblen made no authoritative pronouncements on any of them, but it takes little imagination to ascertain what his views must have been. Certainly, for a very long time, the church hierarchy was able to muster support from millions of the faithful to suppress or at least discourage these practices. That it is no longer able to do so as effectively as it once did is evidence that cultural lag is diminishing in one of humankind's most conservative institutions.

Cultural Lag and Incremental Versus Radical Change

For two generations scholars trained by Clarence Ayres argued that Veblen believed the social order was rigidly divisible into two parts: the institutional and the technological. On one hand, the institutional related to traditional kinds of behavior that were backward looking and concerned values embedded in the past. It was static in the sense of preserving inherited beliefs, class distinctions, and status arrangements. On the other hand, the technological aspect of the life process was concerned with tools, scientific knowledge, and experimentation. To Veblen, science and technology were innovative and forward looking, and they brought about changes in the physical world that, in turn, eroded existing institutional arrangements. Activities were perceived as instrumental if they expanded the human life potential and ceremonial if they constricted that potential as measured by the techno-scientific ethos. Social value was derivative from or, at least, correlative with scientific and technological progress, while cultural lag or social disvalue was the result of an institutional failure to adapt. Thus Veblen used the concept of "cultural lag," if not the term, which to him meant a failure of ideas, habits, and institutions to keep up with the progress of science and technology, or so Ayres would have us believe.[19]

However, it appears that Ayres systematically misused and distorted Veblen's meaning of the term "institution," which is neither simply backward or forward looking but can be either/or, depending on the context in which it is used. Ayres interpreted Veblen to mean that "institutions" were always regressive, inertial obstacles to progressive change, whereas textual exegesis of Veblen makes it evident that institutions could serve a number of functions, some of which were progressive. Institutions were thus not always bad or conservative, but sometimes facilitated desirable change.

Also unlike Ayres, Veblen believed that incremental reformers would often fail because they did not adequately recognize that the state under capitalism is an instrument of the vested interests. Such changes as they supported would accomplish little, since they could not alter the basic system of power relationships within society, ultimately leaving the vested interests in command of the policy-making process. Incrementalism was also deficient because it deals with parts of a problem rather than the total problem and thus focuses on secondary issues rather than fundamentals. Since it advocates solutions that cause the least amount of disturbance to the social fabric, it ensures that no basic structural change will occur. Although Veblen sometimes proposed small-scale change in existing policies,[20] he was more likely either to endorse large-scale structural changes or else withdraw from policy prescription altogether because it was futile. His political philosophy and social theory thus lie outside the American tradition of empirical collectivism (to which Ayres belonged), whose parameters require majority designation of bona fide problems in the public realm; action by local, state, or federal governments to ameliorate these problems; and the design of remediation that will not rend the

19. "For both Dewey and Veblen the technological continuum is the locus of value; and it has this meaning because of its continuity. This continuum is identical with Dewey's 'continuum of inquiry' and Veblen's 'matter-of-fact,' and 'its significance as the locus of value—including economic value—may be understood in terms of the logical significance of the instrumental continuum.'" See Clarence Ayres, *The Theory of Economic Progress*, 219–20. Ayres, Veblen, and Dewey are usually interpreted as sharing a common theory of social value, but recent scholarship raises serious questions about Ayres's interpretation and use of both Veblen and Dewey. See, for example, Geoff Hodgson, *Reconstructing Institutional Economics Agency and Structure in American Institutionalism* (London: Routledge, 2004), ch. 16 and 17.

20. On this point, see Rick Tilman, "Thorstein Veblen: Incrementalist and Utopian," *American Journal of Economics and Sociology* 32 (April 1973): 155–69.

social fabric of the community.[21] Indeed, Veblen said little directly about these three principles of the venerable tradition.

One interpretation of Veblen is that he was prone to think that human behavior could not be changed except by force of protracted conditioning. A fundamental issue revolved around the disagreement between him and his critics about the rigidity of existing institutions and their consequent susceptibility to change. For Veblen, capitalist institutions would give way only under immense pressure from the machine process and the ethos of science, leading to iconoclastic attitudes spreading among the populace. If significant reforms were to be achieved, the human material involved must be fundamentally remolded, which would require dramatic change in existing institutions. But liberals like Dewey believed that popular governments in advanced industrial societies may achieve such change with a minimum of violence and social dislocation, provided that the method of intelligence is employed to bring it about. They thought that social institutions were flexible enough to permit the exertion of rational control, whereas Veblen believed change was often blind drift in which human rationality played a small part. Cumulative causation that exerted pressure against "imbecile" institutions was more likely, in Veblen's pessimistic view, to produce cataclysmic social upheaval or stagnation and inertia. He believed that "any community will change its habits of thought only tardily and under pressure, but in case the pressure of new conditions is extreme, uniform, and persistent a wide-reaching dislocation of the traditional habits of thought is to be looked for even in the best regulated community."[22] While Veblen anticipated an authoritarian response to industrial society in Germany and Japan, many of his critics, in their more optimistic moments, expected orderly advance toward the goals of equality and freedom.[23]

21. See Currin V. Shields, "The American Tradition of Empirical Collectivism," *American Political Science Review* 46 (March 1952): 104–20. Also, see Sid Plotkin and Rick Tilman, "Thorstein Veblen and the American Tradition of Empirical Collectivism," in *The Political Thought of Thorstein Veblen* (forthcoming), and Tilman, "Institutional Economics, Instrumentalist Political Theory, and the American Tradition of Empirical Collectivism," *Journal of Economic Issues* 35 (March 2001): 117–38.

22. Thorstein Veblen, *Essays in Our Changing Order*, ed. Leon Ardzrooni (New York: Augustus M. Kelley, 1964), 446. In Dewey's view, piecemeal experimentation undertaken with the awareness that mistakes will be made should be coupled with continuous efforts to correct these errors. Veblen agreed with this point of view as far as techno-scientific work in laboratories was concerned but never linked it with the development of a workable political technology for social reconstruction. He remained an unreconstructed radical or perhaps "radical conservative" who doubted the effectiveness of incremental reform under existing conditions.

23. As Henry Steele Commager once pointed out, "Where others [read liberals] saw

Simply to be critical of classical liberal doctrine in any of its forms, including neoclassical economics, is not necessarily to be iconoclastic. But there was no single facet of classical liberalism to which Veblen did not object. Indeed, he believed that the social and historical conditions under which it evolved had changed so drastically that it had become largely irrelevant. The residues of classical liberalism were still evident in the uncritical adherence of neoclassicists to the philosophy of natural rights and in their diluted belief in teleological purpose and an ameliorative social trend, all of which Veblen perceived as manifestations of cultural lag.

Veblen's intellectual iconoclasm is nowhere more evident than in his use of words. His rhetorical technique aims at compelling his reader to recognize the point powerfully made about him by Teresa Toulouse, that social conflict was not merely over control of land, plant, or equipment, but for control of signs as units of meaning.[24] What Veblen opposed was the language of the past that was embodied and indeed embedded in the jargon of neoclassical economics. These verbal preconceptions were barriers that prevented a more objective analysis of economic phenomena. The conventional economist's vocabulary was the product of an earlier mode of thought and stopped economists from making more accurate observations of economic activity. In brief, they were expressions of cultural lag. Neoclassical economics was a system of signs based on obsolete assumptions that mostly served the needs of the vested interests. Veblen visualized the economist's problem as a struggle to escape habitual modes of thought in order to articulate categories that the neoclassicist's vocabulary could only exclude. Cultural lag thus exemplified his indictment of the received economic wisdom.

Cultural Lag and Technological Borrowing

Perhaps the most distinctive contribution Veblen made to the study of Imperial Germany, specifically, and to economic development theory generally, was the idea of borrowing. It was the peculiar cultural traits of England that facilitated its borrowing from other social orders and that be-

---

progress, Veblen saw merely change." See Commager, *The American Mind* (New Haven: Yale University Press, 1950), 239. However, James Campbell, *Understanding John Dewey, Nature, and Cooperative Intelligence* (La Salle, IL: Open Court Publishers, 1995), argues that Dewey believed only in the possibility of ameliorism not in the inevitability of progress.

24. Teresa Toulouse, "Veblen and His Reader: Rhetoric and Intention in 'The Theory of the Leisure Class,'" *Centennial Review* 29 (Spring 1985): 249–67.

came the basis for its industrial revolution, while it was the German capacity for borrowing from England and other nearby countries that was the basis for Germany's industrial success. But what was the nature of this borrowing and the structure of the institutions that made it possible? In Veblen's analysis, many social orders cannot efficiently borrow from other cultures or else they borrow the wrong things. But Imperial Germany did not fit either category because it efficiently borrowed appropriate processes and technologies. As Veblen put it,

> being taken over ready-made and in the shortest practicable time, this new technology brought with it virtually none of its inherent drawbacks, in the way of conventional waste, obsolescent usage and equipment, or class animosities; and as it has been brought into full bearing within an unexampled short time, none of these drawbacks or handicaps have yet had time to grow to formidable dimensions.[25]

Veblen believed that the peoples in the Baltic and North Sea region had been borrowing from each other and from other cultures throughout the Neolithic, Bronze, and Iron ages. His description of this process and the institutional structures that made it possible is important in understanding his evolutionary approach:

> These people borrowed freely, both in technological and in other institutional matters, and made notable free and efficient use of all borrowed elements. The scheme of institutions, economic, civil, domestic and religious, that would fit these circumstances would be of a relatively slight fixity, flexible, loose-knit, and naïve, in the sense that they would be kept in hand under discretionary control of neighborly common sense—the continued borrowing and the facility with which borrowed elements are assimilated and turned to account goes far to enforce this conclusion. Altogether its most impressive traits are a certain industrial efficiency, particularly efficiency in the mechanic arts, and its conduciveness to the multiplication of its people; whereas its achievements in political organization or in the domain of art and religion are relatively slight. It is a civilization of workmanship and fecundity rather than of dynastic power, statecraft, priestcraft or artistic achievement.[26]

Yet Veblen points also to the failure of some cultures to advance as well as to the fallacious explanation by the eurocentric and the racist, who attribute this stagnation to the "mental incapacity of these peoples for the

---

25. Veblen, *Imperial Germany*, 249–50.
26. Ibid., 43.

acquisition of new and alien habits of thought . . . . The same view, in substance, is often formulated to the effect that these are inferior or 'backward' races, being apparently not endowed with the traits that conduce to a facile apprehension of the modern European technological system."[27] The German industrial achievement is particularly striking in Veblen's view because it consists largely of borrowing only those traits, artifacts, and processes that are conducive to rapid economic growth and leaving behind the cultural furniture that has a retarding effect.[28]

Although it is clear that Veblen thinks some cultures are more efficient at borrowing than others, it is not always apparent in his analysis why this is so. Nevertheless, he does describe the characteristics of successful borrowing.[29] While Veblen does not treat the subject of borrowing as systematically as he might have, it becomes evident that certain cultural traits are more likely than others to permit or encourage successful borrowing. He comments suggestively that the Baltic (mainly Scandinavian) pagans "were not above learning from their neighbors, or perhaps rather were temperamentally defenseless against innovation from the outside."[30] He also states that some cultures possess a larger capacity than others to acquire and use a wide range of technical information that will apparently be processed more efficiently if the population is "sagaciously addicted to industry and thrift."[31]

The introduction of a new technological scheme across a marked cultural frontier both requires and induces a "bias of skepticism" that loosens the "bonds of authenticity."[32] The intrusion of alien technology and culture into a society with fundamentally different conceptions of law, religion, and morals may cause the break-up of the importing system. Veblen's most penetrating comments explain how technological expedients borrowed from one culture can be "turned to other uses and utilized by other methods than those employed in the culture from which they were borrowed."[33]

In comparing the industrial development of Germany and Britain, Veblen's iconoclasm is again in evidence. Rather than arguing for the advantage of the British, who were the lead-takers, he claims that the disadvantages are too commonly ignored.

27. Ibid., 40.
28. Ibid., 89.
29. Ibid., 22.
30. Ibid., 21–22.
31. Ibid., 64.
32. Ibid., 39.
33. Ibid., 38.

It argues a fatal reluctance or inability to overcome this all-pervading depreciation by obsolescence. All this does not mean that the British have sinned against the canons of technology. It is only that they are paying the penalty for having been thrown into the lead and so having shown the way. At the same time it is not to be imagined that this lead has brought nothing but pains and penalties. The shortcomings of this British industrial situation are visible chiefly by contrast with what the British might be doing if it were not for the restraining dead hand of their past achievement, and by further contrast, latterly, with what the new-come German people are doing by use of the English technological lore.[34]

"The penalty of taking the lead" may strike the conventional economic theorist as idiosyncratic, but it demonstrates once again that Veblen's use of the cultural lag theory turned him in intellectually fruitful directions.

Tony Eff has identified a related facet of Veblen's theory of economic growth as the "paradox of development."[35] In *Imperial Germany and the Industrial Revolution* (1915) Veblen tried to show how Germany had taken industrial leadership from Britain at some point during the reign of Kaiser Wilhelm. According to Eff, his argument was that "adaptation precludes adoptability," which means that the existing structure of industrial use and custom slows down the introduction of more effective industrial methods. Eff refers to this as the "paradox of development," which signifies that the more mature an economic entity is, the more difficult it is for it to attain higher levels of development. High levels of cultural development slow down further cultural development. Eff's explanation of Veblen's reasoning here is that England, as the early leader in industrial development, acquired a social fabric that was attuned to the specific technology that gave it its early lead. But as technological change continued to occur, the cultural and institutional fabric became an obstacle to further development. Thus national economies, as well as more regional or localized economic entities, are best able to take advantage of new opportunities if they possess less specialized institutions.

Veblen's differences with neoclassical economists are further revealed in his critique of the effects of war on industrial economies, particularly as to how and why a war-torn economy can rapidly recover:

The material equipment in such a case of devastation will have been greatly damaged, and that is always a handicap; but the immaterial equipment

34. Ibid., 132.

35. Ellis Anthon Eff, *Veblen's Paradox of Development and Manufacturing Productivity in U.S. Metropolitan Areas, 1958–1977* (Ph.D. Diss., University of Texas, Austin), ch. 2.

of technological proficiency—the state of the industrial arts considered as a system of habits of thought—will have suffered relatively slight damage. This immaterial equipment is, far and away, the more important productive agency in the case; although, it is true, economists have not been in the habit of making much of it, since it is in the main not capable of being stated in terms of price, and so does not appear in the statistical schedules of accumulated wealth.[36]

Veblen's iconoclasm was also evident in his commentaries on the technological incompetence of American businessmen. This was in keeping with his belief that the price system is an inadequate measure of technological efficiency, in part, because the businessmen who are so adept at manipulating pecuniary mechanisms are, nevertheless, industrially inept. Veblen then reveals one source of his indictment of the technological incompetence of American businessmen.

The "efficiency engineers" have no hesitation about saying that the businessmen's management of industry has resulted in a wasteful use of the equipment and labor employed, sometimes rising to figures fairly incredible to the common layman. This waste may be more or less, but it is a systematic result of the management of industrial enterprises by the self-made businessman. At the same time the efficiency engineers' criticism takes no account of the waste involved in downright unemployment or stoppage of workmen and plant due to the pecuniary strategy of the same conservative businessmen, although this item of waste doubtless exceeds by several fold that caused by incompetent management while the plant is running.[37]

Veblen's indictment of the American business community was long standing, even predating *The Theory of the Leisure Class*. But *Imperial Germany* provides him with an opportunity to also make unflattering comparisons between the American and German business communities:

America has unrivalled material resources and a working population nowise inferior to any other for industry under the machine technology; American businessmen have had a free hand and a minimum of burdens in the way of taxes or other government exactions. The American achievement in this field within the same period has been notoriously less conspicuous and less substantial, e.g., than that of Germany since the formation of the Empire. Nowhere have the businessmen had so full and large a discretion, nowhere have they been favored by government regulations to the same ex-

36. Veblen, *Imperial Germany*, 272.
37. Ibid., 338.

tent—for the reason that nowhere have they controlled the making and administration of the laws in anything like the same degree; and nowhere have they fallen short of their opportunities by so wide a margin.[38]

Few of Veblen's countrymen, particularly those entrenched in chambers of commerce, would find this comparison flattering, but once again he has seized upon the "ineptitude" of the American business community to castigate it for its waste and impracticality. To the end of his days, Veblen remained an unrelenting critic of business culture and commercial exploitation. Even an analysis of Imperial Germany, a regime not noted for corporate dominance, was viewed by him as an opportunity to vent his spleen.

Veblen, Ayres, and Marx: Obiter Dicta

The concept of cultural lag that Veblen used to analyze social processes was used by American sociologists to account for both social change and social problems. In their view, change stemmed from science and technology, and problems resulted from the failure of institutions and organizations to keep pace. Veblen, himself, often contrasted the old institutional framework based on private property with the new machine process of industrial production, which was severely restricted, he argued, by archaic institutions. He frequently spoke of the "triumph of imbecile institutions," for the forces of technological change had taken direct effect in the industrial arts, yet touched matters of law and custom only indirectly. The principles that controlled knowledge and belief, law and morals, had lagged behind in contrast with the forward drive that occurred in industry and in the "workday conditions of living."[39] Veblen and his disciple Clarence Ayres emphasized that "invention is the mother of necessity." They believed there was an immanent logic of progress present in existing tool patterns which caused the random pursuit of idle curiosity to result in invention, provided that institutional obstacles were not too pervasive.

A primary focus in their work has been on the role of science and technology in promoting social change and on the effect of institutions in retarding it. Their view is often mistakenly portrayed as turning technology into an irresistible force with an inner developmental logic of its own

---

38. Ibid., 338–39.
39. Thorstein Veblen, *Absentee Ownership: The Case of America,* with intro. by Robert Lekachman (Boston: Beacon Press, 1967), 205–6.

that makes it the forcing bed of social change. It is thus imbued with a capacity for progress beyond human control. An opposite point of view is that human institutions can be restructured to control technological change. Men will master machines instead of being enslaved by them.

Veblen was less optimistic than Ayres that science and technology would be used in the common interest.[40] He was committed to the view that their utilization depends on who controls institutional access to them. It was not technology per se that counted, but its serviceability to the community measured without regard to invidious comparisons, status rankings, and power differentials among individuals.[41]

Veblen more than Marx places great emphasis on the change-resistant (ceremonial) qualities of social institutions and the cultural apparatus. For Marx, the time lag in the adjustment between base and superstructure is generally shorter and the adjustment more certain to occur. But for Veblen, cultural lag is an uncertain, time-consuming process wherein numerous intervening variables are likely to prolong the maladjustment between the technoeconomic base and the superstructure. There is, for example, no expectation by Veblen that the labor force will develop a socialist consciousness, much less a belief that a proletarian uprising will occur. But this absence does not imply a strictly evolutionary perspective in which social change is reduced to small increments, being advanced by pressures generated by new technology and then retarded by tradition. Veblen recognized the occurrence of sharp social discontinuities which, in his view, limited the predictive and anticipatory potency of the social sciences.

Cultural Lag in Veblen's *Nature of Peace*

Most of the scholarly focus on Veblen's theory of cultural lag features *The Theory of the Leisure Class, The Theory of Business Enterprise,* or *The Instinct of Workmanship.* But his *Nature of Peace* is emphasized here because it, too, exemplifies cultural lag in a theoretical sense. It also specifically analyzes the nature of peace in an international context that was in an extreme state of flux just prior to U.S. entry into the Great War. The possible triumph of "imbecile institutions," resurgence of atavistic continuities, preservation of rudimentary survivals and vestigial remnants and, yet, the potential for political ameliorism, are all given careful consideration by

40. Cf. Ayres, "Gospel of Technology," in *American Philosophy Today and Tomorrow,* ed. Horace Kallen and Sidney Hook (New York: Books for Libraries Press, 1968).

41. See Benjamin Smith, "The Political Theory of Institutional Economics" (Ph.D. diss., University of Texas, 1969), 82.

Veblen in this work. *The Nature of Peace* thus contains a presentation and an application of the cultural lag theory, an explanation of the sociopolitical and cultural resistance to lasting peace, and an outline of a program for peace. Even though scholars have paid less attention to this study when analyzing Veblen's theory of cultural lag, it is, perhaps, the most effective and detailed macro-political analysis that he ever engaged in.

Since Veblen did not often use the term "cultural lag," or define it as such, it is essential to provide synonymous meaning(s) for it that are, of course, dependent on the context within which he used the synonyms. What follows are some of the words and phrases he uses, taken from *The Nature of Peace*. Some are mostly terminological, the rest indicative of institutional or cultural practices; "so large and well-knit a body of archaic preconceptions to unlearn";[42] "archaic institutional scheme, of fealty and dynastic exploit and coercion";[43] "mischievous institutional arrangements, obsolete or in process of obsolescence";[44] "received preconceptions"; "order of preconceptions";[45] "medieval frame of mind";[46] "archaic affectation";[47] and "outworn institution."[48] Clearly, these are all examples of what social scientists once regarded as "cultural lag." But what is the standard as well as the catalyst for making these judgments about the promotion of obsolescence? Veblen answers that "modern technology, with its underlying material sciences, is a novel factor in the history of human culture."[49] Although it will not do to leave the analysis in this simplistic state, the basic normative thrust of his theory of cultural lag has been articulated in his own words.

In *The Nature of Peace*, Veblen has a long list of changes necessary to overcome the workings of the archaic system that he believed brought on the Great War. He also suggests changes in domestic and international policy if a more secure and stable footing that would ensure a peaceful future is to be achieved. Table 3 contains a systematic compilation of these changes and resistance to them.

Veblen's "cultural lag" is part of his larger theory of social change; in short, it is a theoretical subset of it. In his day there were several variants of social change theory in vogue and Veblen combined, or at least used,

42. Thorstein Veblen, *An Inquiry into the Nature of Peace and the Terms of Its Perpetuation,* 200.
43. Ibid., 213.
44. Ibid., 216.
45. Ibid., 197.
46. Ibid., 194.
47. Ibid., 140.
48. Ibid., 366.
49. Ibid., 314.

Table 3

Cultural Gain: Adaptive Change	Cultural Lag: Failure to Adapt
• Neutralization of Citizenship— End of National Loyalty	• National Citizenship—Drag of National Loyalty
• Free Trade—End of Industrial Isolation	• Protectionism
• Communities of Ungraded Masterless Humans	• Defense of Class Privileges and Prerogatives
• Prospective Peace Among Nations	• Unstable Equilibrium of Mutual Jealousies
• Community and Solidarity	• Diluted Rapacity and Cunning
• Advancement of Community's Cultural Heritage	• Suppression of Cultural Talent
• Disintegrating (Skeptical) Habits of Thought	• Patriotic Devotion and Superstitiously Crass Nationalism
• Insubordination to Personal Rule	• Subservience to Irresponsible Authority
• Pacific Nations—League of Neutrals—Peace by Collusive Neglect	• Warlike Dynastic Orders— Obsolete Balance of Power
• Virtual Disarmament	• National Hatred and Vain Glory
• End of Upper Class Rule for Upper Class Ends	• Preservation of Hereditary Wealth and Gentility
• Desantification and Disallowance of Rights of Ownership and Investment	• Conspicuously Wasteful Consumption of Superfluities by Captains of Sabotage
• Abolition of Feudalistic Remnants of Privilege—Establishment of Responsible Government	• Preservation of All Extra-Territorial Pretensions and Claims and Dynastic Power

Source: Thorstein Veblen, *Inquiry into the Nature of Peace and the Terms of its Perpetuation* (New York: B. W. Huebsch, 1917).

three of them. This has misled some of his interpreters into believing that these variants are so different as not to lie under the same theoretical umbrella; conversely, for others, that they are a coherent, integral part of it. Briefly, the ones Veblen utilizes are (1) the business-industry dichotomy theory; (2) the evolutionary theories, both Darwinian and Lamarckian;[50]

50. But, see Geoffrey M. Hodgson, "Darwin, Veblen and the Problem of Causality in Economics," *History and Philosophy of Life Sciences* 23 (2001): 385–423.

and (3) an equivalent of the multicausal theory later developed by Ogburn, which, as I have shown, parallels Veblen part of the time, but is also associated with American liberal sociology. It is my contention that Veblen's theory of social change is not simply one theory since he combines all three of the above. Rather, he is eclectic, his theory is sui generis and its variants are strewn throughout the corpus of his writing with no clear distinctions consistently made between them.

How does Veblen's theory of cultural lag differ from Ogburn's and, for that matter, other social scientists? Veblen believed the social system is always out of kilter with its evolving technoeconomic base and with the changing values it makes possible. In contrast to structural-functional models where equilibrium between the constituent parts is assumed to be not only possible, but likely, Veblen asserts that maladjustment is the norm and will be in the foreseeable future. Indeed, disequilibrium is so probable that continuous maladjustment is the norm while equilibrium is little more than a figment of the imagination.

Secondly, Veblen's theory of cultural lag differs from others, including Ogburn's, in that the resurgence of atavisms is always a strong possibility—the persistence of atavistic continuities is close to the norm rather than their permanent disappearance. Linked with, or at least closely related to this are, in Darwinian terms, the vestigial remnants or rudimentary survivals analogous, so to speak, to the human appendix. But the reader can learn more about these from other authors where the problems arising from Veblen's evolutionary literalism, as well as his use of biological metaphors, are discussed in the context of his theory of social evolution.

Finally, there is his skepticism regarding ameliorism as a social or political teleology, secular or religious. Ameliorists assume, of course, that maladjustments caused or exacerbated by cultural lag will be reduced in scope and intensity.[51] But Veblen, who clearly grasped the potential for

51. C. Wright Mills characterized the professional ideology of mainstream sociology as the ideology of "liberal practicality." The emphasis on practical problems of everyday life and the option of orderly change were typical of the liberal model of society. Thus the cultural-lag model was tacitly oriented in "utopian" and "progressive" directions toward changing some cultural and institutions so as to "integrate" them with the present state of technology. Mills thought this model uncritically endorsed both the natural sciences and orderly, incremental progressive change. It was rooted in the ideology of the Enlightenment and was therefore rationalistic, messianic, and politically naïve. Mills believed that the cultural-lag concept was manipulated for ideological purposes by advocates of liberal ameliorism who wrote most textbooks and dominated the major departments and professional organizations in sociology. It was Veblen's radical use of the concept, and his use alone, that won Mills's praise. See Rick Tilman, *C. Wright Mills: A Native Radical and the American Intellectual Tradition*, ch. 4.

individual and social fulfillment in technological growth and dynamism and the spread of the scientific ethos, was uncertain. Instead, regarding progress, he saw merely change, and social ameliorism was not always on the horizon.

Veblen's use of the cultural lag theory converges with Ogburn at certain points and diverges at others. It is these similarities and differences which have caused confusion as to whether Veblen was "really" a cultural lag theorist or not. But it was Veblen's actual use of "cultural lag" as a theory that is here analyzed, and there is no better method of establishing this usage than the technique of textual exegesis. It should be kept in mind, however, that Veblen never wrote a formal treatise on sociology replete with a taxonomy and the formal use of inductive or deductive logic. As Clarence Ayres argued, Veblen was a systematic thinker, but not always a systematic writer. In fact, as I have demonstrated, it is from several of his major writings that his theory of social change must be pieced together. It is not presented in its totality in any one place.

On the whole, Veblen did not anticipate social ameliorism through deliberate, collective action, especially through the agency of the positive state. Yet there are occasional exceptions to this generalization in his work that indicate that cultural lag might be offset, or at least some of its consequences mitigated. For example, in his analysis of the business cycle he wrote:

> But it is at the same time plain enough that this, in the larger sense untoward, discrepancy between productive capacity and current productive output can readily be corrected, in some appreciable degree at least, by any sufficient authority that shall undertake to control the country's industrial forces without regard to pecuniary profit and loss. Any authority competent to take over the control and regulate the conduct of the community's industry with a view to maximum output as counted by weight and tale, rather than by net aggregate price-income over price-cost, can readily effect an appreciable increase in the effectual productive capacity; but it can be done only by violating that democratic order of things within which business enterprise runs.[52]

Veblen concludes that:

> The several belligerent nations of Europe are showing that it can be done, that the sabotage of business enterprise can be put aside by sufficiently heroic measures. And they are also showing that they are all aware, and

52. Veblen, *The Nature of Peace*, 173–74.

have always been aware, that the conduct of industry on business principles is not competent to bring the largest practicable output of goods and services; incompetent to such a degree, indeed, as not to be tolerable in a season of desperate need, when the nation requires the full use of its productive forces, equipment and man-power, regardless of the pecuniary claims of individuals.[53]

Veblen was not the only economist impressed by what national economies could achieve by elimination of waste and achievement of full productivity under the press of wartime circumstances. He clearly understood the possibilities of overcoming the cultural lag brought about by business mismanagement of the economy and the systemic deficiencies of capitalism itself. Indeed, *The Engineers and the Price System* was his, perhaps tongue-in-cheek, proposal to overcome these obsolescent tendencies through a system of "democratic technocracy." At his most optimistic about the future, Veblen focused on the past as a repository of the peaceable, amiable, craftsmanlike qualities that might be revived if cultural lag were to be undermined. As he put it, "The earlier-acquired, more generic habits of the race have never ceased to have some usefulness for the purpose of the life of the collectivity and have never fallen into definitive abeyance."[54]

Interestingly, it is these archaic traits,[55] survivals from the earlier and peaceable cultural phase, that now become paramount in the proper adjustment of the existing cultural and institutional apparatus to the changing technoeconomic base of society and the spread of the ethos of science and popular government. Indeed, it is through the revival of the instincts of workmanship, idle curiosity, and the parental bent that cultural lag may be countervailed. Ironically, the remote past and the values encapsulated in it may help us "save ourselves alive" from the crises that beset Western civilization, or so Veblen suggests.

Geoff Hodgson has argued and others have hinted that Veblen did not develop a substantial or general theory of a cultural lag.[56] Perhaps they

53. Ibid.
54. Veblen, *The Theory of the Leisure Class*, 152.
55. As regards atavistic continuities, the eclectic Lewis Mumford, a former student and acquaintance of Veblen greatly influenced by him, perceptively comments: "by facing the totality of human experience, one becomes aware of elements that the fashion or habit of one's own particular epoch may arbitrarily have neglected: archaic elements, primal elements, irrational elements, neglected mutations and concealed survivals, often overlooked by the wise in their too narrow wisdom." Mumford, *The Condition of Man*, 12.
56. Geoffrey M. Hodgson, *The Evolution of Institutional Economics Agency, Structure*

are referring to Veblen's belief that there is no sharply defined point of de-marcation between science and technology and institutions anyway; plus his view that, in any case, the different parts of any social and cultural sys-tem are usually misaligned, causing waste and other kinds of slippage. Furthermore, adequate adjustment of some parts to the other parts is un-likely to ever be complete to such an extent that "lag, leak, and friction" can be entirely overcome. If this is what Hodgson and others mean, it is difficult to disagree with them. Still, there is a sense in which Veblen was a cultural lag theorist in that he was unusually sensitive to the gap that exists between the actual and the potential in society. This, for example, is indicative of his view that the existing technological base is not being ful-ly exploited for social purposes; the prime examples being unused in-dustrial capacity, wasted labor, and economic privation in his indictment of capitalist business cycles. Nor can his denunciation of animism and its doctrinal equivalents in religion, sport, and gambling as retardants in their impact on the growth of the matter-of-fact thinking which Veblen valued be ignored. In short, it would be difficult to explain Veblen's nor-mative judgments about the social order without recognizing their roots in his identification of cultural lag and its consequences.

But, using Veblen's basic assumptions about cultural lag, what are the critics of the status quo and the opponents of sociocultural retardation to do to overcome it? Occasionally, Veblen hints at the life process un-folding "in a new direction whenever and wherever the resistance to self-expression decreases." But in the modern setting this requires, briefly stat-ed, that "the resulting lag, leak and friction in the ordinary working of this mechanical industry under business management have reached such pro-portions that no ordinarily intelligent outsider can help seeing them wherever he may look into the facts of the case." In short, that the com-

---

and Darwinism in American Institutionalism (London and New York: Routledge, 204), 372. Malcolm Rutherford perceptively comments in his *Institutions in Economics: The Old and the New Institutionalism* (Cambridge: Cambridge University Press, 1994), 67, that

> Nevertheless, for Commons and, to some extent, Ayres, adaptive processes operate, although slowly and not entirely surely, on social norms. Both writers stress that so-cial norms can be resistant to instrumental prompting, but fundamentally the processes of social change are processes of adaptation that are based on an appraisal of consequences. Veblen's discussion is different. He is particularly aware of the dif-ficulties of discussing norms and changes in norms in rationalist terms. Norms often define the criteria of acceptable performance. Norms change as the goals of life them-selves change and, for Veblen, the evolution of norms is less a matter of appraisal of consequences than of shifts in the economic environment enforcing new habits of life that, in turn, bring about changes in values and goals.

mon person no longer dutifully approve the inequity and inefficiency of the situation. But Veblen warns:

> These national interests are part of the medieval system of ends, ways and means, as it stood, complete and useless, at that juncture when the democratic commonwealth took over the divine rights of the crown. It should not be extremely difficult to understand why they have stood over, or why they still command the dutiful approval of the common man. It is a case of aimless survival, on the whole, due partly to the inertia of habit and tradition, partly to the solicitous advocacy of these assumed national interests of those classes—the trading and office-holding classes—who stand to gain something by the pursuit of them at the cost of the rest. By tenacious tradition out of the barbarian past these peoples have continued to be rival nations living in a state of habitual enmity and distrust, for no better reason than that they have not taken thought and changed their mind.[57]

Conceptual inertia and theoretical obsolescence, in Veblen's view, are often the tools of vested interests and not responsive to the demands of those classes with the strongest need for social change. Cognitive adequacy is a response to such social demands and signifies that social science must itself be adapted to the increasing knowledge and resources made available by scientific and technological progress. In short, cultural lag must be overcome in the social studies.[58] The critic may inquire as to what degree John Dewey, F. J. Turner, Emile Durkheim, key figures in the

57. Thorstein Veblen, *The Vested Interests and the Common Man*, 130.
58. As to the human side of cultural lag:

[Veblen] wanted people to be aware that these "blind Forces" which he studied (cultural lag, etc.), *could* let them be destroyed as in World War I, and worse, if they left their destiny in hands of leaders—who too often may *mis*-lead. The surplus (and more) of our inordinately productive industry could become inordinately *destructive* if channeled (as now) into grasping power seekers. He saw the need for some economic "traffic cops" or other to keep benefits of machine industry from running into monopolistic channels. (An anxious student asked him what would be put in the place of the profit system—or the absentee owners—he said, "Well . . . what would you put in its place if you had a boil removed from your nose?" If the underlying citizen were not so brow-beaten by high-cost-of-living he could *use* some of this leisure made theoretically possible by large-scale industry or automation . . . What Veblen didn't like was seeing human leisure wasted on rat-race competition for unnecessary jobs—because production is planned for profit rather than for use by the people as a whole. Leisure could give so much delight and intellectual growth for all men instead of being kept for the perverted, callous persons in search of power over their fellow-men.

Becky Veblen, notes from letter to David Riesman, undated copy in author's possession.

next chapter, and for that matter Veblen, himself, were effective in their own disciplines in overcoming ideas of questionable scientific and empirical adequacy that mostly serve certain vested interests. It is to these topics that I now turn to further illustrate his evolutionary naturalism and its undergirding of cultural lag.

Chapter 4

Veblen's Cultural Lag
Applied and Compared

Introduction

Now that Veblen's theory of cultural lag has been set forth in its historical context, it will be compared with that of several of his eminent contemporaries. It played an important role earlier in the century not only in the work of Veblen, but in that of his fellow Americans John Dewey (1859–1952) and Frederick Jackson Turner (1861–1932) and Frenchman Emile Durkheim (1858–1917). The intent of this chapter is, therefore, to explain (1) the meaning(s) that the theory of cultural lag held for scholars early in the twentieth century, (2) the way that it was used by the four men to analyze the cultural and social orders they were most interested in, (3) their interpretation of the role cultural lag played in their own disciplines and their remedies for overcoming its ill-effects, and (4) Dewey's, Turner's, and Durkheim's relevance for a better understanding of Veblen.

But why compare Veblen on the subject of cultural lag with Dewey, Turner, and Durkheim? In the case of Dewey and Turner their eminence in the disciplines of philosophy and psychology, and American history, respectively, makes their views *prima facie* of interest to students of U.S. cultural studies, while Durkheim to this day remains the preeminent French sociologist. The comparison of Veblen with these thinkers not only illuminates their views on the role of cultural lag in their own disciplines and societies, but it promises to clarify the social science of the founder of institutional economics as well.

Veblen and Dewey showed little respect for the emerging disciplinary boundaries between, for example, economics, political science, and sociology in the case of the former and philosophy and psychology in the case of the latter. Turner paid only modest attention to developments in fields other than history while Durkheim sought to establish sociology as a distinct "scientific" discipline. All of which is evidence of the intellectual turmoil and developmental volatility of both the newly emerging and already established fields of academic inquiry that dotted the landscape in the late nineteenth and early twentieth centuries.

Although Veblen, Dewey, and Turner could not look back to halcyon days in America after the Civil War of their early boyhood, their country had undergone sustained industrial growth, even if this was interrupted by a severe depression that began in 1893 and lasted until the outbreak of the Spanish-American War in 1898, not to mention financial panics in 1903 and 1907. Nor were they prone to ignore the strife between capital and labor and the ensuing violence and disorder that often accompanied strikes. But it is important to note that while political crises occurred as in the aftermath of the presidential election of 1876, they did not threaten political stability the way they did in, for example, Durkheim's France during the same period. Also, although racial, ethnic, and religious conflict certainly occurred in Veblen's America, this did not destabilize the polity to the degree that conflict between clericals and anticlericals did in France. To be sure, the United States underwent a period of farmer-labor insurgency, and it successfully engaged in both labor and racial repression. But much of the labor force enjoyed a significant increase in its standard of living, and ethnic and religious conflict was mitigated by some degree of cultural and social integration. The three Americans featured here were well aware of this, however differently they may have treated it for their own purposes.

Veblen's attitudes toward progress were ambivalent, because while he viewed scientific and technological innovation as progressive, he was skeptical that moral and institutional progress were also occurring. His view can be contrasted with the intermittent claim of Dewey and Turner that social growth is occurring and that it provides the moral criterion whereby both social evolution and individual action can be called progressive. Veblen came at last to the conclusion that the atavistic continuities that he satirized might indeed overwhelm mankind, whereas Dewey and Turner usually saw a glimmer of light even in the most tragic and desperate situation—although Dewey did not believe a golden era was about to be ushered in. Durkheim believed in the potential for progress as defined by the standards of both the Enlightenment and modern science, but he was often in a state of suspended judgment about whether or not it was actually occurring on a macroscale in his own time. In any case, it is to the work of Veblen that I first turn. But, rather than focus on his economics that is so often subjected to scrutiny, it is his usually ignored political science that takes center stage.

Veblen on Science and Cultural Lag in Political Theory

The political and economic radicalism of Veblen should be balanced against his cultural and institutional pessimism; he certainly viewed science and its handmaiden technology as massive determinators and indicators of progressive change. Yet he remained cognizant of the retardant and even atavistic effects of politics, culture, and society. It is arguable that, on the one hand, he did some of the most advanced theorizing in left and progressive circles and that, on the other, he falls in certain respects into the camp of the cultural pessimists and the conservative futilitarians who believe humanity is trapped by "imbecile institutions." Nevertheless, Veblen's view of scientific progress is linked with his political theory, which, in turn, is related to his analysis of the possibilities of change of both a progressive and atavistic nature. His view of science is a modest one. Its capacity as a form of predictive inquiry is unassuming, its claims tentative and provisional at best. Yet there is no superior alternative to science as the ideational catalyst for social change. The main alternative is mostly an expression of the values and power of business enterprise, while the likelihood of political revolt against capitalist culture and institutions is indeterminant within what Veblen calls the "calculable future."

The classic text for analyzing Veblen's interpretation of the nature and function of science, is, of course, "The Place of Science in Modern Civilization," first published in 1906. In this he lays out his most detailed interpretation of science starting with a simple yet penetrating definition of it as "impersonal, dispassionate insight into the material facts with which mankind has to deal."[1] It is this which gives industrial societies an advantage over the less-developed countries. As he put it, "a civilization which is dominated by this matter-of-fact insight must prevail against any cultural scheme that lacks this element."[2]

However, Veblen does not accept the exaggerated claims made by scientistic cultists, for his own claims for science are more modest, and he does not extend them to the political theory of his day. Indeed, he differentiates between the use of science on the one hand and more "pragmatic" forms of inquiry on the other, and it is in this latter category that political theory falls.

---

1. Veblen, "The Place of Science in Modern Civilization," in *The Place of Science in Modern Civilization*, 1.
133.
2. Ibid., 2.

Its habitual terms of standardization and validity are terms of human nature, of human preference, prejudice, aspiration, endeavor, and disability,
and the habit of mind that goes with it is such as is consonant with these
terms . . . In the modern scheme of knowledge it holds true . . . that training
in divinity, in law, and in the related branches of diplomacy, business tac-
tics, military affairs, and political theory, is alien to the skeptical scientific
spirit and subversive of it.[3]

Indeed, Veblen's skepticism regarding political theory as it existed in
his day is, perhaps, more extreme than his rejection of the received economic theoretical wisdom of his time. It is, at best, merely a form of pragmatic knowledge, not a kind of scientific inquiry. As he put it:

Pragmatism creates nothing but maxims of expedient conduct. Science creates nothing but theories. It knows nothing of policy or utility, of better or
worse. None of all that is comprised in what is today accounted scientific
knowledge. Wisdom and proficiency of the pragmatic sort does not contribute to the advance of a knowledge of fact. It has only an incidental bearing on scientific research, and its bearing is chiefly that of inhibition and
misdirection. Wherever canons of expediency are intruded into or are attempted to be incorporated in the inquiry, the consequence is an unhappy
one for science, however happy it may be for some other purpose extrane-
ous to science. The mental attitude of worldly wisdom is at cross-purposes
with the disinterested scientific spirit, and the pursuit of it induces an intellectual bias that is incompatible with scientific insight. Its intellectual output is a body of shrewd rules of conduct, in great part designed to take advantage of human infirmity.[4]

Veblen continues his analysis of those disciplines which in his day still re-
sisted the blandishments of science:

The reasoning in these fields turns about questions of personal advantage
of one kind or another, and the merits of the claims canvassed in these discussions are decided on grounds of authenticity. Personal claims make up
the subject of the inquiry, and these claims are construed and decided in
terms of precedent and choice, use and wont, prescriptive authority, and the
like. The higher reaches of generalization in these pragmatic inquiries are
of the nature of deductions from authentic tradition and the training in this
class of reasoning gives discrimination in respect of authenticity and expediency. The resulting habit of mind is a bias for substituting dialectical dis-

3. Ibid., 20.
4. Ibid., 19.

tinctions and decision *de jure* in the place of explanations *de facto*. The so-called "sciences" associated with these pragmatic disciplines, such as jurisprudence, political science, and the like are a taxonomy of credenda.[5]

Veblen's objections to the political theory of his day are that it is infected with Hegelianism with its emphasis on the fulfillment of reason in history, that it is contaminated with the crudely pragmatic and utilitarian aims of vested interests, and that it contains heavy residues of natural law with its focus on final cause guiding the development of individuals and groups toward their appointed *telos* or end. These, in Veblen's view, are manifestations of the nonscientific status, that is, pre-Darwinian state, of political science as a discipline. All but the most arbitrary reading of his essay on the place of science in modern civilization indicates that until the study of politics becomes a genuine political science, it will have little predictive or explanatory power. One need only look at the failure of specialists in Soviet politics, Kremlinology, and strategic studies to anticipate changes in Eastern Europe during and after Perestroika to recognize that Veblen's earlier skepticism regarding the "scientific" status of political inquiry was still warranted. This is not to suggest, however, that he thought it impossible to predict the course of future political events including revolution itself; rather, it was that he viewed science as having made only slight inroads into the study of politics. In Veblen's view, our ability to predict and analyze the future is thus vitiated by heavy residues of cultural lag in political science.

Cultural Lag: Veblen and Dewey

Clarence Ayres urged his readers to believe that Dewey and Veblen thought that the social order was divisible into two parts, the institutional and the technological. The institutional part related to traditional kinds of behavior that were backward looking and concerned with values embedded in the past. Thus it was static in the sense of preserving inherited beliefs, class distinctions, and status arrangements. On the other hand, the technological aspect of the life process was concerned with tools, scientific knowledge, and experimentation. In the Ayresian view, both Veblen and Dewey thought that science and technology were innovative and forward looking, and they brought about changes in the physical world that in turn eroded existing institutional arrangements. Activities were per-

5. Ibid., 20–21.

ceived as instrumental if they expanded the human life potential and ceremonial if they constricted the potential. Social value was derivative from or at least correlative with scientific and technological progress, while cultural lag or social disvalue was the result of institutional failure to adapt.[6] Thus Veblen and Dewey used the concept of "cultural lag," which to them meant a failure of ideas, habits, and institutions to keep up with the progress of science and technology. The fundamental difference between them was disagreement concerning the rigidity of existing institutions and their consequent susceptibility to change. For Veblen, capitalist institutions would only give way slowly under immense pressure from the machine process as iconoclastic attitudes spread among the underlying population.

Veblen was more prone than Dewey to think that human behavior could not be changed except by force of protracted conditioning. If significant reforms were to be achieved, the human material involved must be fundamentally remolded, which would require dramatic changes in existing institutions. But Dewey believed that representative governments in advanced industrial societies may achieve such change with a minimum of violence and social dislocation, provided that the method of intelligence is employed to bring it about. He thought that social institutions were flexible enough to permit exertion of rational control, whereas Veblen believed change was often blind drift in which human rationality played a small part. Cumulative causation that exerted pressure against "imbecile" institutions was more likely in Veblen's pessimistic view to produce cataclysmic social upheaval or else result in stagnation or inertia. He believed that "any community will change its habits of thought only tardily and under pressure, but in case the pressure of new conditions is extreme, uniform, and persistent a wide-reaching dislocation of the traditional habits of thought is to be looked for even in the best regulated community."[7] While Veblen anticipated an authoritarian response to industrial society in Germany and Japan, Dewey, in his more optimistic moments, expected orderly advance toward the goals of equality and freedom.

As for Dewey, he believed that a liberal education should inculcate the

6. Clarence Ayres, *The Theory of Economic Progress* (Chapel Hill: University of North Carolina Press, 1944), 219–20. In recent years, however, Ayres's interpretations of both Veblen and Dewey have become highly controversial. See, for example, the recent exchange between James Webb and Thomas De Gregori in "Notes and Communications," *Journal of Economic Issues* 37 (December 2003), 1161–74, as well as much of the recent work of Geoff Hodgson.

7. Thorstein Veblen, *Essays in Our Changing Order,* 446.

habits of rational, critical, and reflective thought.[8] However, these were precisely the habits of thought that Veblen believed unlikely to develop given the indoctrination and propaganda characteristic of higher education.[9] For this reason, Veblen had little faith in the possibility that the good or efficient society would be the outcome of education. Cultural change cannot be achieved through the school because the school and other major social institutions are not effective agencies for reforms.[10] Instead, they are largely tools for the perpetuation of vested interest and ceremonial values—indeed, even a fulcrum for the levers of cultural lag.

For Dewey, society is a byproduct of the collective action of the rational human beings who compose the society. Veblen rejects that approach because he does not believe that social intelligence will necessarily propel humankind toward the achievement of adaptive ends. Since people are dominated by ceremonially impregnated institutions, it is naïve to suppose that significant changes can usually be achieved through the exercise of human intelligence. In Veblen's view, social organizations make the rigidly habituated individuals who exist within them unable to resist their influences effectively. Consequently, a sense of futility exists in much of his work, as if humankind were always struggling against immense uncontrollable forces to little purpose. "Proper" adjustment is rare because of the constant, cumulative change in both the individual and his or her environment. Given Veblen's position, Dewey's "method of intelligence" is unlikely to be an effective instrument of change in an environment dominated by ceremonially encrusted minds and buttressed by socially rigid institutions. This is not to suggest Veblen was opposed to the method of intelligence, but it is to state that he often doubted its efficacy under existing circumstances.

The human psyche is more environment centered and, in this sense, even more institutionalized in Veblen's analysis than in Dewey's. Knowledge or intelligence for Dewey may precede environmental changes so that the knowledge problem is solved by using ideas as instruments to effect changes: Habits and customs can be deliberately transformed. But knowledge or mental adaptation for Veblen generally follows a change in environment, so that for him the knowledge problem consists of adjust-

8. See John Dewey, *Experience and Education* (New York: Macmillan Company, 1938), and *Democracy and Education* (New York: Macmillan Company, 1916).

9. See Thorstein Veblen, *The Higher Learning in America.*

10. See Charles Chandler, *Institutionalism and Education: An Inquiry into the Implications of the Philosophy of Thorstein Veblen* (Ph.D. diss., Michigan State University, 1959), 210–12. On the relationship between Dewey's and Veblen's theories of progress, see Robert Nisbet, *History of the Idea of Progress* (New York: Basic Books, 1980), 302–3.

ment to change. This particular ramification of the cultural lag theory further illustrates the difference between the two with regard to the potency of intelligence in effecting reform. Veblen would allow that while much social change is sheer blind drift, the rate and direction of change could be affected by human thought and action. Where he and Dewey differ is over the degree to which this is true; Veblen claims a smaller role for human volition, while Dewey supports a more concerted effort to apply the method of social intelligence.

Veblen's and Dewey's belief in the potential of human intelligence to promote social change is evident in their emphasis on "matter-of-fact" procedures and the "method of intelligence." Dewey, however, was more willing "to assimilate the problems of political power and moral goods to a statement of thinking, of method, to a model of action and thought imputed to science."[11] He often treated conflict situations as problems requiring more intelligence rather than as difficulties necessitating structural change or revolution. Veblen was not as optimistic as Dewey, for he had less confidence in the ability of human intelligence to overcome its environmental constraints. His belief that the human intellect could not solve all social problems or even ameliorate many of them is evident in his emphasis on the forced adjustment of people to their scientific and technological environment. In Veblen's view, the future of humankind perhaps lay in the adjustment or not of people to the ethos of science and logic of the machine process.[12] While Veblen agreed with Dewey that people were potentially capable of rational behavior within an ideal setting, in the existing environment they remained incapable of realizing this potentiality because their behavior is controlled by institutions that promote inequity and irrationality. To illustrate, the kind of socialized emulatory behavior that suggests the impingement of community and authority on the individual rather than the workings of rational social control is more evident in Veblen than Dewey. It is futile, therefore, to appeal directly to intelligence without creating the conditions for rationality within the culture. It is only in this way that the human potential for rationality can be realized. But the dominant institutions of the culture cannot be reformed within the existing political process itself. The majority of people have been conditioned to view themselves as benefiting from their institutions and consequently regard them as morally sound and of socially redeeming value. Veblen argues that this is necessarily the case since people's

11. C. Wright Mills, *Sociology and Pragmatism* (New York: Oxford University Press, 1966), 418.

12. See Veblen, *The Theory of Business Enterprise,* esp. ch. 9.

minds are in fact controlled by the very institutions in need of change. Such is the dilemma his doctrine poses,[13] and it grows directly out of his version of cultural lag which, as we have seen, differs at key points with that of Dewey.

Dewey's Instrumentalist Philosophy as an Attack on Cultural Lag

Dewey's instrumentalist philosophy is best understood in a broader sense as "evolutionary naturalism." In this sense, it is also an assault on cultural lag whether it manifests itself in philosophy or moral and sociopolitical endeavor. Dewey's *Experience and Nature* (1925) is probably the most direct statement of his naturalism.[14] Although difficult in places to read and interpret, he forthrightly states his belief that all human experience is a natural phenomena, that is, an artifact and a process entirely interactive with the material and social environment. Humankind is thus a product of its biological make-up and its sociocultural environment and nothing else. Thinking and doing are thus embedded in a natural (not supernatural) matrix, thought processes themselves are rooted in chemical reactions and electric impulses in the body and brain and, when these cease, the organism is dead. Consciousness, itself, is thus a combination of physiological processes within the cerebral cortex plus interaction with the material and social environment. Dewey never denied that religion could be the source of important moral, cultural, and psychological insights, but he believed that all too often it was the end of inquiry not the beginning. Although he was sometimes too polite to say so, the commonality he had with religionists was a belief in the ideal, and his view was that the ideal is a purely human and social contrivance, not an emanation from on high; although as stated in his *Common Faith* (1934),[15] this aroused considerable controversy and misunderstanding among his critics.

Dewey's was a process philosophy based on Darwinian evolutionary theory that perceives humankind and its natural environment as in a continuous state of flux and change. Although certain values and behavior are more likely to aid in human adaptation than others, only change is inevitable and it is an ongoing process in which human intelligence is or can

13. See Charles C. Chandler, *Institutionalism and Education: An Inquiry into the Implications of the Philosophy of Thorstein Veblen*, 222.
14. Dewey, *Experience and Nature*.
15. Dewey, *A Common Faith*.

be the main adaptive instrument. Consequently, the most efficient way to control change Dewey called the "method of intelligence." By this he means bringing appropriate kinds of analysis to bear on particular "problematic" situations which will turn indeterminate circumstances that are problems into determinate situations that are nonproblematic, a thesis more fully developed in his *Logic: The Theory of Inquiry* (1938).[16] His methodological pragmatism also signifies that the methods used to investigate will vary depending upon the problematic situation the investigator faces. Knowing is a kind of doing and the criterion of knowledge is "warranted assertability," a point of view largely shared with Veblen, although he used different language to describe it.

Dewey was a secular humanist whose values were ultimately rooted in the realities of self and collective fulfillment of potential. If there is a locus of value in his philosophy, it clearly is to be found in "self-realization" or "self-actualization," variously labeled, and the creation of authentic, that is, genuine democratic community. *Democracy and Education* (1916),[17] his most influential treatise on learning, postulates the likelihood of achieving these traits most effectively in a society where education is an open-ended process, learning by doing is the norm, and students pursue knowledge of what interests them. But, as Dewey pointed out in *Experience and Education* (1938), this is not a classroom without structure, authority, or purpose.[18] Rather it is one where spontaneity and specified aims are pursued in an egalitarian manner without **either** anarchy ensuing or rote memory forced upon students as a main pedagogical device through rigid external discipline.

In Dewey's view, individualism was itself socially induced and determined through a process of interaction and transaction with others. The rugged individualism of laissez-faire capitalism must and is giving way to a social individualism that is more relevant to life in an industrialized, urbanized, and collectivized society, or so he argued in *Individualism, Old and New* (1930).[19] Two additional points must be emphasized. Dewey once challenged a conservative critic to give him one example of something that acts in isolation from everything else. Presumably, he is still waiting for an answer. Secondly, he once commented that humanity consists of the biological template and the socialization process and nothing else. He forcefully argued throughout a long career that political, moral, and cultural values are mostly learned because the formation of the hu-

16. Dewey, *Logic: The Theory of Inquiry.*
17. See Dewey, *Democracy and Education,* passim.
18. See Dewey, *Experience and Education,* passim.
19. Dewey, *Individualism, Old and New.*

man character is a social and cultural process in which the biological traits of the individual are submerged in a sea of institutional configurations and social conditioning. Humans are not passive recipients of whatever sensory stimulation they are subjected to, but in part they must learn not to be.

One of Dewey's most commonly voiced criticisms of both philosophy and philosophers and the mistakes made by humankind in the real world is the creation of "false dualisms." By this he means the artificial and unwarranted separation of means from ends, that is, the *is* from the *ought*, facts from values and science from morality. Dewey believes that this is not a mistake merely in the abstract. It is often a recipe for a social disaster or, at the least, an individual failing to adapt adequately to the environment.

But Dewey probably did not mean to conflate these dualisms, that is, compress them into one identity. To illustrate, means and ends should in his view be part of a continuum in which means became ends and ends means in an ongoing process which he refers to in his *Theory of Valuation* (1939) as a continuum of ends-means.[20] Ends, in particular, must not be fixed as absolute, but must remain as ends-in-view always subject to revision depending on the availability of resources for their achievement. Means-in-view must depend on the identification of ends-in-view and their own suitability for achievement by the existing means.

Theory and practice must be linked in such a manner as to actualize theoretical antecedents of practical behavior, that is, theory becomes most meaningful when it is effectuated in practice. This is not to suggest that Dewey was opposed to the pursuit of what Veblen called "idle curiosity"; rather, he thought that the ultimate test of theory is the consequences of believing and acting on it.

Not only has philosophy continuously created false dualisms such as the severance of theory from practice, it is this situation that must be overcome if philosophy is to have renewed relevance as Dewey argued in his *Reconstruction in Philosophy* (1920).[21] But, in his view, no consummatory phase of existence resides outside the mundane realm of human affairs. Thus wholesale escape from pain and suffering and uninterrupted enjoyment of eternal bliss, cultural verities, and moral stasis is unlikely. The purely contemplative state postulated as ideal in which the philosopher king oozes essence is in Dewey's view a groping in vain which inhibits

20. Dewey, *Theory of Valuation*. Also, see Abraham Edel, *Ethical Theory and Social Change: The Evolution of John Dewey's Ethics, 1908–1932* (New Brunswick, NJ: Transaction Publishers, 2001).

21. Dewey, *Reconstruction in Philosophy*.

understanding and struggle for social reconstruction. He agreed with Karl Marx that "philosophers have sought to understand the world, the point is to change it." Even more than Marx, however, he believed that the vested interests resist change because they, too, are the victims of a mindset produced by cultural lag.

Cultural Lag in Psychology

Since Dewey was a noted figure in psychology, and president of the American Psychological Association in 1900, his views on overcoming cultural lag in that discipline are also relevant.[22] However, efforts to explain his role in this regard are limited to (1) his efforts to discredit faculty psychology and replace it with functional psychology, (2) show the relevance of a reconstructed psychology for resolving practical problems by making it among other things into a moral science, (3) analyzing his attack on the ideological prejudice that interiorizes and privatizes values and morals as a way of avoiding public scrutiny of them; all this with the aim of strengthening the democratic community, and all by way of overcoming the inertia and drag produced in psychology by cultural lag.

As indicated, Dewey wanted psychology as well as philosophy and the other social sciences to become moral sciences, that is, practical sciences that admit their own existence in the world they study instead of trying to be detached from it.[23] But the very attempt at gaining knowledge of psychological reality can alter the reality itself according to Dewey:

> . . . it is . . . absurd to suppose that an adequate psychological science would flower in a control of human activities similar to the control which physical science has procured of physical energies. For increased knowledge of human nature would directly and in unpredictable ways modify the workings

22. Most recently, see Svend Brinkmann, "Psychology as a Moral Science; Aspects of John Dewey's Psychology," *History of the Human Sciences* 17, no. 1 (2004): 1–28, for an explanation of the moral significance of cultural lag in Dewey's psychology. My analysis and quotations from Dewey owe much to Brinkmann's article.

23. On morality and cultural lag in Dewey's own work, see Dewey, "The Reflex Arc Concept in Psychology," *Psychological Review* 3, no. 4 (1896): 357–70; "Psychology and Social Practice," *Psychological Review* 7, no. 2 (1900): 105–24; *Democracy and Education;* "The Need for Social Psychology," in J. A. Boydston, ed., *The Middle Works,* Vol. 10: 1916–1917 (Carbondale: Southern Illinois University Press, 1980), 53–63; "Context and Thought," *University of California Publications in Philosophy* 12, no. 3 (1921): 203–24; Dewey and J. H. Tufts, *Ethics* (New York: Henry Holt, 1910).

of human nature and lead to the need of new methods of regulation, and so without end.[24]

Thus psychology has the power to change the reality it is concerned with. In doing this it overcomes the inertia posed by the absolutistic ideas of human nature that stifle reform. Psychological knowledge can transform its own subject matter and thus society.

In Dewey's view, socially shared action is unlikely to effectively emanate from the illusory inner, mental realm, for psychology has played a huge role in perpetuating a privatization of morality:

> A false psychology of an isolated self and a subjective morality shuts out from morals the thing important to its acts and habits in their objective consequences. At the same time it misses the point characteristic of the personal subjective aspect of morality: the significance of desire and thought in breaking down old rigidities of habit and preparing the way for acts that recreate an environment.[25]

Dewey thus wants to overcome the doctrinal bias that interiorizes and privatizes values and morals. The subjectivity that supports individualist ideologies must be replaced by a view of mind and valuation that emphasizes its social and public nature. In short, we test theories in practice, and evaluate practices and institutions "by the contribution which they make to the value of human life,"[26] which can be measured by their success in overcoming cultural lag and enhancing life in the democratic community.

In "The Reflex Arc Concept in Psychology," Dewey rejects the conventional psychology of the day.[27] The main thrust of Dewey's critique of this psychology is that perception and behavior are contextually dependent. For Dewey, perception is not a form of knowledge and perception cannot be separated from cognition. Consequently, for him data are never given but are taken. Thus what are taken as data depend in part on the purpose of the inquiry. So there is no one-to-one correspondence between a given sense datum and a mental image or construct. Dewey's analysis of the reflex arc concept signifies that scientific observation is mediated by habits, intellectual constructs, perspectives, and purposes of observation. This is

24. Dewey, *The Public and Its Problems*, 197.
25. Dewey, *Human Nature and Conduct*, 57.
26. Dewey, "Psychology and Social Practice," 121.
27. Dewey, "The Reflex Arc Concept in Psychology," 252–66. For a statement of his critical views on conventional psychology as early as 1884, see "The New Psychology," 49–62.

his way of suggesting that the older psychology suffers from a lack of cognitive realism which is itself a manifestation of cultural lag.

Faculty psychology by its very nature reeked of cultural lag and perpetuated it. Dewey believed it was static, assumed natural mental harmony and thus it, too, lacked realism. The portrayal of mental processes as cognitive, conative, and affective (reason, will, and emotion) inhibited socially adaptive processes, postulated equilibrium among the faculties as normality, and explained little about either the individual or the society in which he lived. Nor was Dewey taken in by the behavioral psychology that took its place in the mainstream study of mind. We thus conclude our brief effort to trace the genesis of cultural lag in psychology plus Dewey's own efforts to undermine and replace it with a more contemporary realistic perspective.

Cultural Lag and the Frontier
according to Turner and Veblen

It is to Frederick Jackson Turner's version of cultural lag and his partial insensitivity to it by way of comparison with Veblen that we now turn. The central part of Veblen's last book, *Absentee Ownership and Business Enterprise in Recent Times*, contains his analysis of the American experience since colonial times.[28] It focuses on the self-made man, the independent farmer, and the country town. It also covers the development of the gold mining industry and the exploitation of the timberlands and oil fields, which are portrayed as they evolved under various degrees of absentee ownership. There is a rough overlap between the time-span and institutional processes covered by Veblen and the chronology and experience addressed by Turner in "The Significance of the Frontier in American History."[29] For purposes of comparison with Turner, Veblen's indictment of the frontier experience using the concept of cultural lag may be broken into five parts, including (1) economic waste and destruction of the natural environment, (2) the traits and function of self-help and cupidity, (3) the role of absentee ownership in exploiting rural America, (4) the op-

---

28. Veblen, *Absentee Ownership and Business Enterprise in Modern Times*.

29. This analysis of Turner is based on his famous essay of 1893 and *Rise of the New West* (New York: Harper and Brothers, 1906), *The Frontier in American History* (New York: Henry Holt, 1920), *Sections in American History* (New York: Henry Holt, 1932), and *The United States, 1830–1850* (New York: Henry Holt, 1935). Also, see Vernon Mattson and Rick Tilman, "Thorstein Veblen, Frederick Jackson Turner and the American Experience," *Journal of Economic Issues* 21 (March 1987): 219–35.

pression of racial minorities, especially Native Americans and blacks, and (5) the very real material progress weighed against the human costs.

Economic Waste and Destruction of the Natural Environment

Veblen's analysis in *Absentee Ownership* is replete with reference to the waste and damage done to the natural environment. The ruination of large stands of timber, the wasteful exploitation of deposits of oil and natural gas, the erosion of the valuable top-soil through rapacious farming methods, and the mining of essentially "worthless" metals such as gold all drew Veblen's fire.[30] He traced these processes of exploitation, which had their origins in the English settlements in colonial America, and which gathered force until they reached culmination in an orgy of waste and destruction in the mid- to late nineteenth century—a frenzied assault on the natural environment, aspects of which Veblen himself witnessed during his early years. The frontier experience was, in many ways, a destructive process, and future generations would have to pay its costs. Indeed, many of the wasteful practices Veblen described continued after the date given by Turner for the close of the frontier (1893) and, perhaps, assumed even greater significance then. Veblen also focused on the often dreary and ugly realities of rural existence that Turner, in his celebration of the frontier experience, was prone to downplay. A consequence of the farmer's desire to own a disproportionate amount of land, more than he could reasonably expect to till, was that the farming population was greatly dispersed. The problems of transportation and communication were greatly exacerbated because of distance, and providing education for schoolchildren and adequate medical care was exceedingly difficult in many regions. Veblen also traced a high rate of insanity among housewives to the rural isolation.

The Cupidity and Self-Help of the Independent Farmer

Veblen wrote in the same spirit as Edgar Lee Masters and Sinclair Lewis, although in more radical tones, when he asserted that:

30. See Veblen, *Absentee Ownership*, 188. "In point of material usefulness, then, it seems fair to say that the production of gold is pure waste—more especially the production of more than is currently taken up in the arts and that this waste commonly exceeds the total value of the product." Ibid., 177.

The country town of the great American farming region is the perfect flower of self-help and cupidity standardized on the American plan. Its name may be Spoon River or Gopher Prairie, or it may be Emporia or Centralia or Columbia. The pattern is substantially the same and is repeated several thousand times with a faithful perfection which argues that there is no help for it, that it is worked out by uniform circumstances over which there is no control, and that it wholly falls in with the spirit of things and answers to the enduring aspirations of the community.[31]

For Veblen, the life of the independent farmers aimed not at enjoying rural life or basking in the agrarian ethos, but through "hard work and shrewd management," to acquire a "competence" that would enable him some day to take his due place among the absentee owners of the land and so come in for an easy livelihood at the expense of the rest of the community.[32] The farmer's desire for acquisition had driven him to work hard, but he was not strongly attached to his work. As Veblen saw him, the independent frontier farmer was in greedy pursuit of a disproportionate share of land rather than in love with the land.

Veblen stated repeatedly that the incentives of self-help and cupidity of the "English-speaking colonial enterprise" resulted in a lack of "spirit of community interest in dealing with any of their material concerns."[33] In his view, it was no accident that the farmer placed little emphasis on social solidarity and harmony or that he lacked any organic sense of community. Both the system of exchange relations and the cultural ethos that underlie economic individualism worked against community solidarity. Veblen commented in the same vein that the farmers' "self-help and cupidity have left them at the mercy of any organization that is capable of mass action and a steady purpose."[34] The net result was that their possessive individualism put their own necks under the heel of both local businessmen and large absentee owners. Lurking behind the local front of small businessmen in the country town were those massive business interests whose main function in American society was to get something for nothing. As Veblen put it:

In a way the country towns have in an appreciable degree fallen into the position of toll-gate keepers for the distribution of goods and collection of customs for the large absentee owners of the business. Grocers, hardware dealers, meat-markets, druggists, shoe shops are more and more extensively

31. Ibid., 142. But, also, see Lewis E. Atherton, "The Midwestern Country Town: Myth and Reality," *Agricultural History* 26 (July 1952): 73–80.
32. Veblen, *Absentee Ownership*, 131.
33. Ibid., 132
34. Ibid., 133.

falling into the position of local distributors for jobbing houses and manu-facturers.[35]

Although Veblen often attacked absentee ownership of commerce and in-dustry, he had less to say regarding absentee land ownership than he did about the independent farmer with all of his alleged greed.

Racial Oppression on the American Frontier

Another facet of social relationships on the frontier that Turner did not emphasize was the fate of the black and the Native American. Veblen un-derscored the role of absentee ownership as a fundamental part of the plantation system in the American South and observed that: "Negro slav-ery has left its mark on the culture of that section so deeply etched into the moral tissues of its people that the bias of it will presumably not be out-grown within the calculable future."[36] In his opinion the bias implanted by "the peculiar institution" in the South was rooted as deeply there as the "equally peculiar institution of the Country Town has made the moral bias and spiritual outcome in the North."[37] The existence of slavery, of course, presupposed a slave trade, and Veblen fingered the role played by New Englanders in this nefarious traffic. As he sardonically put it:

> The slave-trade never was a "nice occupation" or an altogether unexcep-tionable investment "balanced on the edge of the permissible." But even though it may have been distasteful to one and another of its New England men of affairs, and though there always was a suspicion of moral obliquity attached to the slave-trade, yet it had the fortune to be drawn into the ser-vice of the greater good. In conjunction with its runningmate, the rum-trade, it laid the foundations of some very reputable fortunes of that focus of commercial enterprise that presently became the center of American cul-ture, and so gave rise to some of the country's Best People. At least so they say.[38]

The contrast between Veblen's portrayal of the fur trade and that of Turner is also interesting, for, unlike Turner, Veblen would have no truck with its "hideous accessories."[39] Neither would he countenance this "un-written chapter on the debauchery and manslaughter entailed on the In-

35. Ibid., 152.
36. Ibid., 170.
37. Ibid.
38. Ibid., 171.
39. Ibid., 169.

dian population of the country by the same businesslike fur trade," which he described as "one of the least engaging chapters of colonial history."[40] He concluded that, indirectly and unintentionally, but speedily and conclusively, the traffic of the fur traders converted a reasonably peaceful and "temperate native population to a state of fanatical hostility among themselves and an unmanageable complication of outlaws in their contact with the white population."[41] In summation, it is evident that Veblen saw nothing positive in the American fur trade, the plantation system, the slave trade, or white dealings with the black and Native American populations. Indeed, his interpretation of these aspects of the American experience stretching back to colonial times is congruent with his indictment of other systems based on "force and fraud," all of which represent variants of cultural lag yet parts of which Turner was prone to downplay.

Material Progress and Ideological Rationalization

Although Veblen's criticisms of the values and behavior of the independent farmer were at times scathing, he also commented that "it remains true that they have brought an unexampled large and fertile body of soil to a very passable state of service, and their work continues to yield a comfortably large food supply to an increasing population."[42] To a considerable extent, one could trace the American standard of living to the farmer's successful exploitation of science and technology. But unlike Turner, Veblen refused to see this frontier source of abundance as an unequivocal sign of progress. Combining irony with the theory of cultural lag, Veblen wrote: "Intellectually, institutionally, and religiously, the country towns of the great farming country are 'standing pat' on the ground taken somewhere about the period of the Civil War; or according to the wider chronology, somewhere about Mid-Victorian times."[43] This is consistent with Veblen's conviction that there is little about which to be complacent, for our institutions lag behind scientific and technological advance. To most Americans, to be out of date is to be hopelessly wrong, and, therefore, an admission of cultural lag was tantamount to a betrayal of American principles. However, Veblen admitted that this discussion of the American country town:

40. Ibid., 168–69.
41. Ibid.
42. Ibid., 133.
43. Ibid., 160.

unavoidably has an air of finding fault. But even slight reflection will show that this appearance is unavoidable even where there is no inclination to disparage. It lies in the nature of the case, unfortunately. No unprejudiced inquiry into the facts can content itself with anything short of plain speech, and in this connection plain speech has an air of disparagement because it has been the unbroken usage to avoid plain speech touching these things.[44]

The pungency and severity of his indictment in this, his last book, was also evidence that he was skeptical about the likelihood of structural change because the main trends in American life had become congruent with the common sense of the common man. In Veblen's view, much of what passed for common sense was a form of false consciousness that indicated a lack of awareness of objective self-interest. Veblen believed that emulatory consumption patterns and invidious distinctions had taken the place of genuine class consciousness on the part of the rural population and could be interpreted as a species of cultural lag.

Veblen was critical of classical and neoclassical economics because they were, in his view, pre-Darwinian and unscientific. Under the influence of Herbert Spencer, William Graham Sumner, and Charles Darwin, he emphasized process and claimed that economics should become an evolutionary science. Conventional economics was thus wrong in its refusal to abandon static equilibrium models and psychological hedonism and deficient in its inability to incorporate the critical-genetic method used by Veblen in his own work. In short, the important social and economic implications of Darwinian evolutionary biology were largely ignored by mainline economists, to their own and to society's detriment.

Both Veblen and Turner were Darwinians, evolutionists, and ethical relativists. Yet for Turner this provided different ideological lenses for interpreting the American experience. His America was one where the common man had already subdued the natural environment and could celebrate the triumph of American spiritual ideals at the close of the frontier. By contrast, Veblen's homeland was one where the vested interests had predictably betrayed the promise of American life and where the common man had been persuaded to join in an orgy of status emulation coagulated with patriotic gore.

Turner saw in the frontier experience a process of "perennial rebirth" and evolution to a higher order, whereas Veblen saw the frontier scene as a study in exploitation, ruination, and disorder as well as massive economic development. Veblen is best known for his iconoclasm, mordant

44. Ibid., 156.

wit, and sardonic worldview, traits that were readily evident in his writings. His opposition to capitalism, his dislike of institutional religion, and his iconoclasm have given him a reputation in some quarters for believing that whatever is, is wrong. This mindset is apparent in his view of the American past, present, and future and contrasts with the Turnerian *weltanschauung*, which acknowledges the existence of cultural lag, yet assigns it a secondary role. While few would claim Turner was a thinker of the stature of Veblen, Durkheim, and Dewey, he made or sanctioned certain innovations in the writing of American history that were significant attempts to overcome cultural lag in his own discipline. Briefly, he attempted to put history on a more scientific plane. As Howard Lamar put it:

> He has been called a pioneer social historian and the father of a "multiple hypothesis" theory of history, and, to Richard T. Ely, he was a good economist. Even Beard praised him for restoring "economic facts to historical writing in America." Howard Odum called Turner a "master" of social science in 1927, and the latest generation of social scientists and quantifiers have renewed the accolade, for they feel he used methods and approaches in studying voting behavior that resembled their own. Merle Curti, one of Turner's last students, asserts that Turner anticipated the ideas and methods James Harvey Robinson advocated in his *The New History* by some twenty years. In the last twelve years of his life Turner formulated a theory concerning the role of sections in American history, which still shapes the framework of most college history courses that treat the coming of the Civil War.[45]

Turner anticipated the approaches used by behavioral scientists and historians such as the quantitative-statistical methods used in voting behavior maps. Still, his concept of sectionalism was linked with the frontier and the flourishing of democracy and individualism on it; the ideas of free land determinism and the closing of the safety valve were more influential in the writing of American history. It is these for which he is mostly remembered and they represent attempts to both overcome cultural lag and acquiescence in it, or so his critics would have us believe.

Durkheim, Veblen, and Cultural Lag

It is to the Frenchman Emile Durkheim that we now turn, and in his conceptual focus (and Veblen's) cultural lag assumes a larger role. The role of Darwinism in the development of the thought of Veblen and Durkheim

45. Howard R. Lamar, "Frederick Jackson Turner," in *Pastmasters,* ed. Marcus Cunliffe and Robin Winks (New York: Harper and Row, 1969), 75.

is notable. For Veblen, in particular, the key to evolutionary social change lay in adequate adaptation to the environment. Much futility and waste resulted from maladaptation, while "fullness of life" was the result of successful adaptation with the human race, consisting increasingly of those who achieved effective adjustment to the environment. Indeed, both men believed that failure adequately to adapt would be followed by failure to survive and if religion, kinship, or class were impediments, they must be altered. But, in Durkheim's view, the bonds of religion and kinship were so strong that, as the population pressure of early societies increased, instead of dispersing, people specialized. Labor specialization was an adaptation that allowed more people to be supported in the same area than undifferentiated hunting-gathering. Durkheim understood that there was more than one way to adapt. Large hunter-gatherer tribes could have simply split up and dispersed and still survived. But specialization allowed them to survive without having to do this. Veblen was more certain than Durkheim that he had located the obstacles to progress; in these he included emulatory consumption, abstinence from useful labor, absentee ownership, nationalism, patriotism, and institutional religious belief and practice. Nils Gilman comments that

> Veblen's "institutions" functioned in the same way as Emile Durkheim's "cosmologies" . . . Each of these categories functions conservatively in that, as social practices, they serve to slow social change. This conservative dimension made Durkheim take a positive view of cosmologies. . . . But for Veblen, institutional conservatism was the most basic of social ills. If technological transformation was a positive good, the tenacious persistence of old ways of doing things always retarded the full flowering of new technological potentials. Society and culture, in other words, served as a brake on the full development of human life through technology. . . . Veblen maintained the more ironic, typically modernist belief that persistent institutional backwardness would inevitably dog technological change. Useless personal habits would remain because they helped service the vested interests of institutions.[46]

As an economic historian and sociologist of growth, Veblen is best known for his "institutionalism," a term coined by Walton Hamilton.[47] In

46. Nils Gilman, "Thorstein Veblen's Neglected Feminism," *Journal of Economic Issues* 33 (September 1999): 699. Also, see the comparisons of Durkheim and Veblen in Rick Tilman, "Durkheim and Veblen on the Social Nature of Individualism," *Journal of Economic Issues* 36 (December 2002): 1104–10; and "Durkheim, and Veblen on Epistemology, Religion and Social Order," *History of the Human Sciences* 15 (November 2002): 51–70.

47. See Walton Hamilton, "The Institutional Approach to Economic Theory," *American Economic Review,* Supplement 9 (March 1919): 312.

one sense, the term is misleading because it was existing institutions which Veblen criticized because of their inhibitive impact on technological change. In his theory of cultural lag, he developed the idea that institutions are sometimes inhibitory and backward looking, whereas science and technology are often dynamic and orientated toward change. The question at any point in time is whether institutions are sufficiently malleable to permit efficient exploitation of existing scientific and technological potential.[48] As the tool continuum evolves, it may become more absorptive of cultural cross-fertilization processes that bring together more and different tools, making possible new technologies. Veblen thus explains the economic history of the West by linking cultural anthropology and social history with changes in the technoeconomic base; the main variables in his explanation are the degree of institutional rigidity and cultural malleability and the dynamism and pressure exerted by technology.

In sociological jargon, "cultural lag" is a term of both use and abuse often employed with imprecision or as a term of deprecation or denigration denoting ineptitude on the part of the user. But as employed here in comparing Durkheim and Veblen, it means the period between that point in time in which one valued cultural element nears fulfillment and that point at which another element reaches the same level of development. Thus in Veblen's analysis in which cultural lag is viewed both normatively and materially, it designates technological objects and institutional cultural impediments to their growth—that is, predatory actions and institutions that lag behind "workmanlike" or "technological" ones. But, in the writings of Durkheim, focus is on stable equilibrium, not constant change. His use of a functionalist model of social change or social disorganization tends to assume "equilibrium" as a normal state of affairs so that it is disequilibrium that needs explanation;[49] whereas with Veblen the term equilibrium has little meaning or relevance since science, technology, and the cultural and institutional superstructure of society are assumed to be at odds with one another most of the time.

That Veblen believed the overthrow of the existing system of sacred values and religious belief was possible, even likely, in those social formations most under the influence of science and technology was evident. In fact, he acknowledged the persistence of the conventional dichotomy between the sacred and the secular; but he argued that this atavistic conti-

48. See Veblen, *The Theory of the Leisure Class,* 191, 195.
49. However, for a revisionist interpretation of Durkheim, see A. W. Rawls, "Durkheim's Epistemology: The Neglected Argument," *American Journal of Sociology* 102 (September 1996): 430–82.

nuity was most likely to endure in those social strata furthest removed from scientific culture and the machine process.[50]

Veblen's own break with religious belief came early, probably in adolescence, but like Durkheim he made his departure in decisive fashion. The latter came to see religious beliefs not merely as false, but rather as a bewildered and exaggerated type form of morality—that is, moral beliefs expressed in theology or mythology rather than in scientific language.[51] But for Veblen, religion, while deeply rooted in the human order, was essentially a form of false consciousness that might ultimately give away before the impact of science and technology, although because of the possibility of the resurgence of atavistic tendencies he made no predictions.

Both men rejected animism and animistic theories regarding the natural and social order. Both hoped, and Durkheim prophesied more consistently than Veblen, that religion would become ever more subject to the criticism and the control of science. In any case, they viewed society as the causal determinant as well as the cognitive and symbolic referent of religion, not as divine inspiration or revelation. Both thus possessed a sociocentric view of human behavior from which, especially in the case of Veblen, religionists could take little comfort.

Durkheim sometimes failed to distinguish between the truth of a belief and the acceptance of a belief as true, an error Veblen rarely made; to be sure, the two shared the view that beliefs had a social origin, that their efficacy comes from society, and that they have social functions that reinforce the social conscience. Where they differ is over the question of whether these beliefs reflect the realities of social life. Had he been cognizant of his views, Veblen would probably have criticized Durkheim for confusing the causes of religion, the value of the services it renders, and the truth of what it affirms.

Although both men ultimately accepted the preeminence of science, they were skeptical of human ability to distinguish between objectivity in the sense of social authority and objectivity as correspondence with the actual social and natural order. Veblen, however, believed that the impact of science and technology in industrial society was starting to increase this correspondence rapidly, but the likelihood of the resurgence of religious superstition, invidious consumption, patriotism, and nationalism always existed with their potential negative impact.

50. See Veblen, *The Theory of the Leisure Class*, ch. 8 and 9.
51. For a broad sampling of Durkheim's views on religion, see *Durkheim on Religion*, ed. and trans. W. F. Dickering (London: Routledge-Kegan Paul, 1975): ch. 1–12, and Durkheim, *The Elementary Forms of the Religious Life*, intro. K. E. Fields (New York: Free Press, 1995).

Both Veblen and Durkheim came to terms with the issue as to how religion and science make different epistemological claims and, given this, which claims are to be accepted. In the final analysis, the two men accepted the view that the claims of science are epistemologically superior and that religion will suffer a decline in terms of its accepted explanatory power. First, because the physical world is commonly regarded as best explained by the secular scientist and, secondly, because the conventional view of the religious nature of man will gradually be abandoned, although the possibility of the resurgence of atavistic continuities cannot be ignored. Also, religion will be less needed as a structural source of solidarity for society. Nevertheless, Durkheim asserts:

> The world of religious and moral life still remains forbidden. The great majority of men continue to believe that here there is an order of things which the intellect can enter only by very special routes.[52]

But the authority of science is becoming ever more firmly established and it cannot be disregarded:

> From then on, faith no longer holds the same sway as in the past over the system of representations that can continue to be called religious. There rises a power before religion that, even though religion's offspring [science], from then on applies its own critique and its own testing to religion. And everything points to the prospect that this testing will become ever more extensive and effective, without any possibility of assigning a limit to its future influence.[53]

What Veblen regarded as the persistence of cultural lag was interpreted by Durkheim to mean that the erosion and disappearance of one set of sacred observances would invariably be followed by the appearance of new entities or states to which sacred status is granted. There is, Durkheim emphasizes:

> something eternal in religion: the cult and the faith. But men can neither conduct ceremonies for which they can see no rationale, nor accept a faith that they in no way understand. To spread or simply maintain religion, one must justify it, which is to say one must devise a theory of it. A theory of this sort must assuredly rest on the various sciences, as soon as they come into existence: social sciences first, since religious faith has its origins in society; psychology next, since society is a synthesis of human consciousness; sci-

52. Durkheim, *Elementary Forms of the Religious Life*, 431.
53. Ibid., 433.

ences of nature finally, since man and society are linked to the universe and can be abstracted from it only artificially. But as important as these borrowings from the established sciences may be, they are in no way sufficient; faith is above all a spur to action, whereas science, no matter how advanced, always remains at a distance from action. Science is fragmentary and incomplete; it advances but slowly and is never finished; but life—that cannot wait.[54]

Where Veblen and Durkheim differ is over the question of whether religious belief is purely illusory; that is, a form of false consciousness or not. To the former, although religion's social function and its role as a social bonding agent are important, in the final analysis the issue is whether or not the beliefs and practices of the religionist produce desirable social consequences and whether they are epistemologically valid. In Veblen's eyes, they mostly fail to achieve these ends at least in the urban-industrial setting in which much of humanity now resides. Durkheim, too, visualizes a secular basis for moral socialization and the preservation of social order, although he decries the crisis in moral authority and social stability which the decline in religion has brought about.

Although Veblen was sensitive to those nonrational or pre-rational elements of life and culture ordinarily associated with religion, he did not accord them the respect they received from Durkheim. To the latter, religion seemed basically indestructible and an integral part of human nature and the human drama, however negative its social consequences. Thus the evolutionary naturalist Norwegian-American and the skeptical French Jew partly disagree on the social efficacy and value of formal, institutional, religious belief and practice.

Although Durkheim was clearly not a religious man, he placed great stress on the relationship between the social bond and the belief in the sacred. The indissolubility of the sacred and society in his thought can be contrasted with Veblen, who, of course, held no such views, although he was impressed by the ways in which the sacralization of culture inhibited its further rationalization. Nevertheless, Veblen used the terms "animism" and "teleological" interchangeably, whereas in common usage as well as in scholarly discourse they do not have the same meaning. "Animism" is the idea that natural objects have souls, or that inanimate objects are inhabited or infested with spirits, while "teleological" signifies evidence of design or purpose in nature and is a doctrine of final causes or purpose. Veblen knew the difference between the two, but assigned them

54. Ibid., 432.

the same general meaning hoping to convince his readers that there is no real difference between the superstition of primitives who practice animism, Catholics who believe in natural law with its teleological underpinnings, and neoclassical economists with their focus on equilibria as a norm. Durkheim, too, attempted to make Australian totemists sound like contemporary Catholics.

It was not individuality and individualism per se that Veblen and Durkheim tried to eliminate from consideration by social scientists. Instead, it was the artificial, abstracted conception of individualism that flourished in neoclassical economics and in political philosophy going back to Hobbes. Neither intended either to eradicate the individual or deify society, nor believed it possible to derive either society or what is social in the individual personality from the individual as such, or from exchange or contractual relations rooted in the aggregation of discrete individuals. In short, whatever differences existed between Veblen and Durkheim must be "set against a very large common background of premises regarding the nature of society and its priority in the development of individual mind, self, and consciousness."[55] The radical individualism that neoclassical economists claimed lay at the base of homo economicus is thus shown by Veblen and Durkheim to be social-psychologically and anthropologically inadequate. Indeed, it is a manifestation of cultural lag.

Veblen's critique of the dynastic state anticipated the appearance of fascism. To illustrate, in *Imperial Germany*, he described the Hohenzollern dynasty as a volatile fusion of modern science and technology with an antiquated class and religious structure, conservative bureaucracy, entrenched military caste, and quasi-absolutist monarchy. In the long run, Veblen believed these structures could not coexist with the new industrial system because their imperatives were not congruent with its maintenance. The opportunity of the dynasty to seize territories it coveted, and engage in other acts of unprovoked aggression, was rapidly passing away because it could neither live with industrialism nor live without it. If it lacked the forces of modernity, it lacked the military capability to prey on its neighbors. Yet with those forces its own institutions would erode under pressures exerted by science, technology, organized labor, and the demands of the underlying population for representative government and peaceable relations with other countries.

---

55. Robert Nisbet, *The Sociology of Emile Durkheim* (New York: Oxford University Press, 1974), 58.

As for fascism, Veblen did not, to be sure, predict the exact form it would take when it came, nor did he fully anticipate the personal characteristics of Hitler and Tojo and the brutal results of these policies. Nevertheless, his prophecies were uncanny for, in rough fashion, he predicted the institutional configurations and nationalist aggression of Germany and Japan long before they occurred. More to the point, however, he used his own patented brand of institutional analysis to make these predictions. Thus his success in predicting the future on these occasions was not due to random chance. Rather it was a direct consequence of using an institutional approach with its stress on uneven development, cultural lag, and the ultimate incongruity between rigidity of habit and adaptive impulse.[56] The predictable consequences of these phenomena in absolutist systems would be aggression and war unless the archaic German and Japanese political and social structures were modernized; that is, brought in line with their new industrial systems and the scientific and technological adaptive imperatives this required.

Conclusion

What have we learned from this comparison of Veblen with Durkheim, Dewey, and Turner as regards their interpretations of cultural lag in the late nineteenth and early twentieth centuries? And what import does this have for a better understanding of the social and intellectual history of their time?

In historical perspective, one conviction shared by Durkheim, Dewey, and Turner was that the positive state could serve as an effective vehicle in overcoming cultural lag. They were less sympathetic with revolutionary Marxism than Veblen was with Bolshevism because they believed that revolutions changed little; that is, deep-rooted, lasting change is usually the result of long-term social evolution. However, they were favorably inclined toward reformism and rejected the claim that the state is simply a medium of class domination. For them, the state, contrary to Veblen, could function as the vehicle for the realization of social reform. Since Veblen viewed the state as an instrument of the vested interests and cultural drag, he did not regard it in its present form as a likely medium for the

56. For a more detailed treatment of this thesis, see Colin Loader and Rick Tilman, "Thorstein Veblen's Analysis of German Intellectualism," *American Journal of Economics and Sociology* 54 (1995): 339–56.

promotion of social reform. In short, he had no theory of the positive state. But the other three, on the contrary, believed the state could engage in the social control of industry in order to achieve desired objectives; to some degree it would have to regulate industry and to a lesser degree the professions. But at no point did Veblen explicitly endorse such policies. He had little faith in the ability of government to do this equitably and efficiently since it was mostly a visible tool of the vested interests and cultural lag; in any case, he was not given to ruminating about how to make incremental changes in public policy through the auspices of the state. Veblen and his three contemporaries thus differed over the extent to which it was possible to identify and ameliorate social ills using the instrumentalities of government.

Durkheim, Dewey, and certainly Turner did not anticipate the Great Wars and totalitarian governments of the twentieth century to the extent Veblen did. They were, however, more astute in anticipating the creation of a welfare and regulatory state in industrial societies with market economies. Insofar as cultural lag provided the inertia for defending and protecting laissez-faire capitalism, they had more foresight in forecasting its downfall through political action and state intervention than Veblen. But neither Durkheim's structural functional social theory, Dewey's instrumentalist philosophy, or Turner's historiographical innovations illuminated the invidious distinctions, honorific prowess and waste as powerfully as they were satirized and indicted in Veblen's institutional economics.

The politics and ideology of Veblen, Dewey, and Durkheim can be identified in their (1) analysis and evaluation of the social role of religion and its cultural persistence, (2) identification of the loci of institutional and cultural resistance to change, and (3) persistence of atavistic continuities that impede the instrumental adaptation of the community to change. If they did not submerge themselves in the nineteenth century's faith in progress, secularism, individualism, technology, and large-scale organization, they nevertheless imbibed enough of the flavor of these to believe in the possibility of positive social change. But they also were realistic enough to anticipate the possible resurgence of atavistic continuities. Indeed, in the case of Veblen, sufficient prescience was involved to predict fascism in Japan and Germany long before it reared its ugly head.

Durkheim believed that Western society in his day was undergoing a major crisis that consisted of a pathological loosening of moral authority over the lives of the underlying population. Veblen also believed that deep-seated change was under way, but he felt that much of what passed for moral authority was itself pathological. At the very least, it was inhibitive or destructive of the "generic ends" of human existence, ends which led to "fullness of life," which he believed encompassed altruism, critical intelligence, and proficiency of workmanship, among others. Veblen's equivalent of Durkheim's explanation of the social crisis afflicting Western civilization pointed to the bonding role of status emulation and nationalization of processes which offset the centripetal effects of social disintegration. Indeed, the centrifuge of emulatory consumption patterns and patriotism were what kept society from coming apart in the face of change induced by science and technology.

The politico-ideological significance of cultural lag is evident in the work of the four men in that the stronger their focus on it, the further to the left their position on the doctrinal spectrum. To illustrate, Turner pays only modest heed to the role of cultural lag in American history until the date he assigned to the closing of the frontier—1893—because he believed the frontier acted as a safety valve until then. He was aware of the clash of labor and capital, the extermination and resettlement of Native Americans, the discriminatory practices against blacks, but he seemed reasonably satisfied with the later reforms of the Progressive Era. It would not do to overstate the degree of his acquiescence in the abuses of corporate power, the degree of political corruption and Progressive success in harnessing large-scale capitalism to the public good. But, whatever his remaining discontent with the existing social and political order, he wrote little that would indicate strong disagreement with the drift of American civilization between the end of the Progressive Era in 1917 and his death in 1932. In any case, his several failures as a historian are often as instructive as his more substantive contributions.

Turner focused primarily on the American frontier experience as both a social and geographical process, its potential impact as the frontier closed, and, for our purposes, the striving for a new balance between individualism and collectivism. Indeed, the main thrust of Progressive political and social theory was to locate a new and more equitable and efficient point of equilibrium between the individual and the larger social order.

Still, according to Turner, the American frontier experience itself had its own ways of dealing with cultural lag. As he put it:

For a moment, at the frontier, the bonds of custom are broken and unrestraint is triumphant. There is not *tabula rasa*. The stubborn American environment is there with its imperious summons to accept its conditions; the inherited ways of doing things are also there; and yet, in spite of environment, and in spite of custom, each frontier did indeed furnish a new field of opportunity, a gate of escape from the bondage of the past; and freshness, and confidence, and scorn of older society, impatience of its restraints and its ideas, and indifference to its lessons, have accompanied the frontier.[57]

Institutional novelty was preceded by the abandonment of the old ways; in short, the cultural furniture of the past gave way under the cumulative pressure of the frontier experience to practices and ideas more attuned to the present.

Cultural lag both as social practice and as an obstacle to the professional growth and development of historians began to ebb and wane. In Turner's view, progress was being made toward ascertaining both the nature of historical truth and the truth itself about the American past. As he also put it in his 1910 presidential address before the American Historical Association:

It is important to study the present and recent past, not only for themselves but also as the source of new hypotheses, new lines of inquiry, new criteria of the perspective of the remoter past.[58]

Yet, Merle Curti warns:

It would be easy to carry too far the analogy of Turner's method to that of the natural sciences. His work implies a belief that it is too soon to determine exactly, though exactness is the aim, the conditions of the historical process. He has not assumed that history can actually establish any large body of principles, or be more than one among many disciplines to aid in social control. The present task of the historian, as he has conceived it, is to make a preliminary recognition and study of the "forces that operate and interplay in the making of society," rather than to determine the ultimate laws of history. He has thus contented himself with the attempt to establish tendencies and to understand mass movements and processes in America.[59]

Like Veblen, although both men sometimes honored it more in the breach than in practice, Turner thus attempted to be historically and culturally specific. Nevertheless, the natural sciences affected the latter great-

57. F. J. Turner, "The Significance of the Frontier in American History," in *The Turner Thesis concerning the Role of the Frontier in American History*, rev. ed., ed. with intro. George Rogers Taylor (Boston: D. C. Heath and Company, 1956), 18.

58. Turner, *The Frontier in American History*, 323.

59. Merle Curti, "The Section and the Frontier in American History: The Methodological Concepts of Frederick Jackson Turner," in *Methods in Social Science: A Case Book*, ed. Stuart A. Rice (Chicago: University of Chicago Press, 1931), 357.

ly and he focused on the need for data collection and the classification of materials for the study of American history to such a degree that it probably limited his literary/historical output since he spent so much time and energy on preliminary work rather than preparing manuscripts for publication.

Turner scholars often argue that he was a pioneer in introducing to the study of American history the data and perspective of the human and physical geographer. In fact, the method of correlating political and cultural behavior with physiographic maps was taught in his seminars by 1894, influenced his students, and is often regarded as an important contribution to the techniques of historical research. The value of this should be evident to historians of methodology seeking evidence of early contributions to the heuristic understanding of voting studies, policy output, and the meaning of colonization of the American frontier. In the case of the latter, Turner interpreted our political history in sectional terms as dispersed along the frontier and leading to conflict between social and economic sections. His frontier thesis is as much an explanation of sectional political conflict and resolution as it is of largely open access to an abundance of free land and westward movement and settlement across it.[60]

Although Turner's three contemporaries were involved in related quests, they did not conceptualize or theorize about them in the same way or use the same language to describe what they were doing. For Veblen, Dewey, and Durkheim, individualism itself is understood in social terms and contexts; in fact, "individualism," as they understood it, is socially induced and manifest. The rugged individualism of the Anglo-Saxon Protestant frontiersman as an atomistic, autonomous unit advancing across and subduing the continent through his own valiance and volition clearly appealed to Turner. Although acquiescent in the close of the process, he was not always sanguine about the prospects for the future. But the greater theoretical and conceptual sophistication of his three contemporaries enabled them to interpret the human drama as both social in genesis and collectivistic in expression. Individuals and individualism may multiply and flourish, but they are first and foremost social and cultural phenomena not isolated or spontaneously generated units, facts which Turner tended to downplay in his analysis of the frontier experience.

As for Veblen and Dewey, their published work after the Great War took on an increasingly radical tone as though the existing cultural patterns and routines served as more and more of a barrier to necessary change. They clearly believed that the evolving technoeconomic base of American

60. Ibid., 353–57.

society was thrusting itself against increasingly rigid and wasteful cultural and social barriers which thwarted the development of human potentiality. The Progressive Era, in their view, accomplished only modest objectives, if that, and it left the cultural and political landscape much as it had been before the Great Crusade.

As for Durkheim, he anticipated the further development of social democracy in his native France. His methodological strictures are aimed at overcoming the impact of cultural lag in the study of sociology and, perhaps, the social sciences. His support of empiricism and induction has often been used to indict him for "positivism," but recent scholarship gives this claim short shrift.[61] He was aware of the epistemological and ontological shortcomings of Machian positivism from the perspective of the sociology of knowledge, but he was not guilty of subscribing to the grosser forms of scientism that had invaded social scientific inquiry. Nevertheless, he insisted in his own methodological discourses on precision of nomenclature, statistical verification of hypotheses when appropriate and possible, and the meticulous sifting and sorting of both hypotheses and data before deciding which to use. Furthermore, he demanded critical inquiry into the nature of any assumptions that colored or conditioned any type of inquiry that claimed to be "objective." Indeed, his objectives were to establish sociology as a rigorous scientific discipline; to provide the basis for the unity and unification of the social sciences; to synthesize the empirical, rational, and systematic basis for modern society's civil religion.[62] These were his aims in rooting out the manifestations of cultural lag in the social sciences and in French society.

Turner's way of dealing with cultural lag in the writing of history was more superficial than the efforts of the other three men to undermine it in their own disciplines because he was less knowledgeable about the metatheoretical bases of historical and social scientific inquiry. In the formal sense, however brilliant and innovative his historical writing, he was simply less educated than the other three, who possessed an encyclopedic knowledge of the human sciences. This led him to focus on social processes in a certain geographical locale, the American frontier, without adequate insight either into the nature of the processes or the genesis and degree of change needed to redirect them. The Progressive Era produced many critics who both muckraked and demanded reform, but Turner's frontier thesis with its closing safety-valve neither incited him to demand

61. See, for example, Susan Stedman Jones, *Durkheim Reconsidered* (Cambridge: Polity Press, 2001), 28–36.
62. Edward Tiryakian, "Emile Durkheim," in *A History of Sociological Analysis*, ed. Tom Bottomore and Robert Nisbet (New York: Basic Books, 1978), 188.

much beyond the conventional or explore its social and historical antecedents in any real depth. Historians in search of ways to dispel cultural lag in their own discipline found inspiration in the methodological strictures Turner actually offered, but his novel focus on the frontier and sections as the central feature in the American experience was his main contribution.

It is well recognized by historians of the behavioral sciences that Dewey contributed significantly to two disciplines, philosophy and psychology, and his contributions to the former are closely linked with his contributions to the latter. Again, however, the problems caused by cultural lag infected both disciplines in Dewey's view. The first was the persistence of "false dualisms" in both philosophy and what is recognized to have been philosophy's derivative of psychology. The divorce of mind from body, facts from values, and science from ethics led in both disciplines to disciplinary stagnation that could only be overcome by "reconstruction in philosophy," which, not surprisingly, was the title of one of Dewey's best-known books. Philosopher Peter Hare has credited him and the other pragmatic philosophers with the "progressive enrichment of evolutionary naturalism," and contemporary American philosophy, insofar as it is naturalistic, owes him a large debt.[63] Long ago, in a widely influential book, Morton White credited both Dewey and Veblen with providing intellectual leadership in the "revolt against formalism," which was correct.[64] But White failed to recognize the radicalism particularly of Veblen during the Progressive Era and lumped such different thinkers as Oliver Wendell Holmes, James Harvey Robinson, and Charles A. Beard with Dewey and Veblen as though all were approximately equal in their repudiation of formalism in their own disciplines. White misgauged the degree of their alienation from the effects of cultural lag as manifest in their own disciplines as well as what was necessary to overcome it. In philosophy, Dewey, in particular, believed that theory and practice were sharply separated from each other and this greatly hindered efforts to overcome the drag exerted by archaic or obsolete institutions, a drag manifested by professional philosophers who could or would not reconcile or mesh the evolving technoeconomic base and scientific ethos with their own practice, or for that matter with existing social reality. Much the same could be

63. Peter Hare, "The American Philosophical Tradition as Progressively Enriched Naturalism: Comments on Arnold Berleants, 'Metapragmatism and the Future of American Philosophy'" (presented at the meetings of the Society for the Advancement of American Philosophy, Bentley College, Waltham, Massachusetts, March 1995).

64. See Morton White, *Social Thought in America: The Revolt against Formalism* (Boston: Beacon Press, 1949).

said for viewing Dewey's work in psychology as an attempt to overcome cultural lag in that discipline; as, for example, his famous article on the reflex arc concept, which aimed among other things at discrediting the dominant faculty psychology that he regarded as outdated.[65]

Dewey, as a social theorist, often relied on both the theory and the phenomena of cultural lag to illustrate points he wanted to make about existing social, political, and moral problems. Of course, he hoped through his writing to illuminate the causes as well as point to remedies that would close the gap between advancing science and technology and a retarding culture and its social institutions. But he was well aware, in spite of his exaggerated reputation for optimism, that much of what cultural lag encapsulated consisted of atavistic continuities that were not likely to be overcome without persistent social effort over long periods of time. Like the more obviously pessimistic Veblen, he doubted that some "imbecile institutions" such as the papacy and monarchy were eradicable at least in the calculable future. It is not too surprising to read his essay "Anthropology and Ethics" and discover that atavistic continuities began with primitive cultures and persist in their detrimental impact on the way modern humans view themselves and their social and natural environment in industrial culture. Once again, Veblen is cited in the bibliography of Dewey's essay, although this time it is *The Instinct of Workmanship* (1914). Dewey writes that:

> Many writers tend to exaggerate the differences which mark off the more primitive cultures from those with which we are familiar today. Accordingly, when similarities are found they are disposed of as "survivals" of early ideas and customs. As a matter of fact, there is hardly a phase of primitive culture which does not recur in some field or aspect of life today. For the most part, tradition does not operate and "survivals" do not occur except where the older beliefs and attitudes correspond to some need and condition which still exists. To put it briefly, the reign of animistic ideas, of the magic ceremonialism which are sometimes considered to be exclusively or at least peculiarly primitive, is due to modes of feeling, thought, and action which mark permanent traits of human nature psychologically viewed. The important phenomenon is not survival, but the rise of scientific, technological, and other interest and methods which have gradually and steadily narrowed the extent and reduced the power of what is primitive in a psychological sense.[66]

65. Dewey, "The Reflex Arc Concept," 357–70.
66. Dewey, *The Later Works*, vol. 3, ed. Jo Ann Boydston (Carbondale: Southern Illinois University Press, 1984), 11.

Cultural lag and the persistence of atavistic continuities are threatened by the advance of science and technology, but as Veblen is fond of pointing out their extinction is not inevitable. The judgments Dewey made regarding the difference between the "good" and the "bad" in institutions can be linked with Veblen's theory of cultural lag in which institutions as opposed to science and technology are perpetually in a state of maladjustment with each other.

Veblen's view was that selective adaptation can never keep pace with the constantly changing social situation; hence "each successive situation of the community in its turn tends to obsolescence as soon as it has been established. When a step in the development has been taken, this step itself constitutes a change of situation which requires a new adaptation; it becomes the point of departure for a new step in the adjustment, and so on interminably."[67] While Veblen in his more pessimistic moods felt that they were destined to remain so, Dewey claimed that through use of the method of intelligence the lag could be significantly reduced. He remained more optimistic about the possibilities of progressive change than did Veblen, particularly toward the end of the latter's life.

Then, too, in addition to cultural lag as an explanation, Veblen gave other reasons for the generic inadequacy of classical liberalism and the specific deficiencies of neoclassical economics. Veblen viewed neoclassicism as a subset of classical liberalism sharing most of its assumptions. It should be kept in mind, though, that the neoclassicism referred to is that which existed between about 1890 and 1925, that is, mainstream economics after the marginalist revolution, but prior to Keynesian macroeconomics and the Robinson-Chamberlain–inspired "monopolistic" or "imperfect" competition revolution in microeconomics.

Veblen's criticisms took two forms: one was scientific and had to do with the requirements of a properly evolutionary science of economics; the other was political and moral and had to do with the direction in which he thought society should evolve. First, economics in his view was still largely pre-Darwinian in that it used utilitarian theory as its criterion of choice and clung to an outmoded hedonistic psychology. It was teleological in that it unrealistically postulated certain processes such as equilibrium as normal and taxonomic insofar as it is substituted classification for casual explanation. It erred, methodologically, in using deduction to draw conclusions from unrealistic axioms. The static bias exhibited by neoclassicism made it seem unlikely to Veblen that it could develop into

67. Veblen, *The Theory of the Leisure Class,* 191.

an evolutionary economics capable of explaining cumulative change in terms of causal sequence.

Veblen's criticism of neoclassicism also focused on its neglect of the origins of consumer tastes and preferences, its use of marginalist explanations of income distribution and, implicitly, with the static role it assigned to the state. In his view, neoclassicists simply took consumption patterns for granted without inquiry into the origins of consumer values. They ignored the emulatory nature of much consumption and, also, the role of advertising and salesmanship in whetting consumer appetites. Marginal productivity theory claimed to explain the distribution of income without explaining how the factors of production, such as land, labor, and capital, came to be owned as they were; acquisition was thus identified with production. Although he was not explicit about the role the state should play in the economy and doubted that much deliberative change could be achieved through collective action, he implied that the adaptive processes of the community to a changing environment might require the state to perform other than the static, limited role assigned to it by the neoclassicists. Thus Veblen had negative attitudes toward classical liberalism and the powerful strains of it deposited in neoclassical economics. He found them wanting on both scientific and moral grounds, and since they were the conventional wisdom among economists, it is not surprising he remained a marginal figure in his own profession. Veblen's contemporaries Durkheim and Dewey, however, reached the apex of their professions attempting to overcome the negative effects of cultural lag in their own disciplines. Although Turner's fame was less deeply rooted either in broad learning or novel intellectual achievement, it has not been transient.[68] Once again, Veblen is odd man out. Political scientists pay him little heed and he has remained a marginal figure in mainline economics in large measure due to his penetrating but iconoclastic criticisms of it and the novel alternatives he offered to it by way of overcoming cultural lag.

68. The exploration of the biological idea of adaptation and the social-scientific use of cultural lag remains an important inquiry, and the fact that they play a less significant role in the corpus of Turner's writing is, perhaps, more evidence of how the most influential American historian of his time lacked the theoretical acumen and intellectual depth of three of his contemporaries.

Chapter 5

On Sports, Religion, and Gambling

Introduction

Veblen treats religious observances, gambling, and sports as activities that not only influence the rhythms of daily life but that also reveal the predilections of American thought. In particular, he claims like other left social critics that they inhibit secularization, rationalization, and humanization of American life. Also, they promote superstitious, predatory, and invidious beliefs as well as authoritarian social relations. These claims are a logical outgrowth of his evolutionary naturalistic assumptions regarding the cosmos and cultural existence within it and his preference for a society where beliefs and social practices encourage its progressive enrichment.

Veblen believed sports, religion, and gambling had much in common not only in the psyche of the common person as manifest in their behavior, but conceptually in his own theoretical approach. Their commonalities are a belief in divine providence and personal intervention by the Deity, the metaphysics of make-believe, the transcendence of the supernatural with its accoutrements of soul and afterlife and, of course, "luck" in the case of sports and gambling. *The Theory of the Leisure Class* is studded with his analysis of all three, and they are important in his delineation of the "predatory" and "peaceful" types of human beings.

Veblen, also, saw important similarities in the social origins, functions, and consequences of sports, religion, and gambling. Furthermore, he saw them as having an interlocking, interweaving, and interpenetrating relationship particularly in his own society and societies like it. As superstructural phenomena, they had an important inhibitive effect on the secularization, rationalization, and humanization of the industrial order. But why and how?

First, they downplay or ignore "matter-of-fact" knowledge in that divine providence is called upon to intervene on behalf of the participant. God's aid is intermittently, if not continuously, invoked on behalf of the individual to favor him in his trials and tribulations, in the heat of battle, so to speak. Veblen believed that this was an atavistically continuous process that could be traced far back in human existence. Sportsmen, religionists, and gamblers had at least this in common, but in his view they and the organizations and institutions to which they belonged were also engaged in the production of massive waste.

Waste of time, waste of resources, and multiplication of both these through social emulatory processes of an invidious nature that entices others to join compounds the wastefulness of sporting, gaming, and churchgoing. Veblen implies the possibility of using these resources in other more socially productive ways, although he was not very specific about the alternative ways in which the resources might be used. Nevertheless, it appears that in his "industrial republic" if sports, gambling, and religions did not disappear, they would at the very least be far less obtrusive.[1]

Animism, natural law, and providential teleology were all mocked by Veblen because among other things they defy matter-of-fact knowledge. The secular, scientific, factual, empirical and, presumably, more socially useful knowledge prized by Veblen is defied by the mindset of sportsmen, gamblers, and religionists at least when engaged in or preoccupied with the pastimes of the arena, casino, and church. In short, a belief in the supernatural with its focus on divine intervention on behalf of the favored characterizes their relationship with providence. From Veblen's purely secular perspective all three are an exercise in make-believe, and it is to the analysis of sport that we now turn.

Veblen and Cady on Sport

Veblen was not the only scholar with a humanistic bent to write on the social role of sport; so, too, did Edwin H. Cady, Andrew W. Mellon Professor in the Humanities at Duke University and a noted authority on American literature. His *Big Game: College Sports and American Life* was an example of a growing trend in the late 1970s toward the analysis of sport

---

1. The social surplus supporting the three is mostly derived from rent, interest, profit, and government expenditures were one to attempt to calculate a la Baran and Sweezy in *Monopoly Capital*, the entire surplus of the existing order. See Paul Sweezy and Paul Baran, *Monopoly Capital* (New York: Monthly Review Press, 1966), appendix.

by academic social scientists and humanists.[2] A brief recapitulation of this book is in order because it serves as an illuminating contrast with Veblen's own analysis of sport.

Cady's primary focus is on intercollegiate athletics. His main concern is maintenance of academic control over the Big Game so that it may perform its own positive functions, which in his view are many. For example, he says sports are a substitute for war or at least for predatory and aggressive behavior; they provide us with a sense of identity and community; they permit us to engage in fulfillment of the deep-seated human need for ceremony, ritual, and play; and, the athlete learns self-discipline and other-regarding values that can be transferred to other walks of life when his playing days are over. Cady also alleges that the individual's sense of isolation in an atomized society can be overcome through participation, an otherwise boring existence enlivened, and disorientation caused by constant social change reduced. Cady claims all this and more, and there is some truth in his assertions, but not as much as he imagines. Indeed, one could point to numerous examples of intercollegiate athletics producing just the opposite of what he claims.

There is a powerful streak of moralism in Cady which causes him to overemphasize the voluntaristic aspects of human behavior. He thinks we can literally "will" to clean up the "mess" that exists in intercollegiate athletics. Unfortunately, this causes him to give insufficient emphasis to the structural and institutional constraints that make reform an unpromising task. Although aware of existing ties between booster clubs and the business community and alumni groups and coaching staffs, he thinks that men of strong moral fiber will be able to set all this right, if only they will exert themselves. He fails to explain how men of such character will ever be able to work themselves into positions of power in the first place. One can share some of Cady's values without accepting his optimism about the possibilities of reform.

He calls for reform of the Big Game and suggests that where corruption or other deficiencies exist, we need "continuous elastic, tireless war" against them, but he fails to outline strategy for conducting such a campaign. However, true to his calling as an incremental reformer, Cady establishes academic criteria by which to measure the viability of the Big Game. He asks that intercollegiate athletic programs genuinely represent the character and qualities of the undergraduate student body, as well as

2. Edwin Cady, *The Big Game: College Sports and American Life* (Bloomington: University of Indiana Press, 1976). The most useful brief analysis of Veblen on sport is John M. Hoberman, *Sport and Political Ideology* (Austin: University of Texas Press, 1984).

match the traditions and character of the larger academic community. He also argues for structural integration of athletics into the academic institution, rather than letting them remain a separate, if related, enterprise. But the size and complexity of the modern multiversity which often contains numerous incongruent professional and vocational orientations, conflicting social values, diverse student body and faculty, and greatly varying educational goals make it difficult to apply his criteria in a systematic way. However, until better criteria are developed and articulated his suggestions are worth remembering.

Cady stresses the fact that athletes from underprivileged backgrounds are able to benefit from a college education that would not otherwise be available to them. This is, no doubt, true but given insufficient emphasis is the fact that huge numbers remain in the ghetto, the barrios, and Appalachia without much assistance from collegiate athletic programs. Cady is only mildly critical of the inequalities that exist in American life and does not adequately recognize that these inequalities are rooted in capitalist institutions and property relations which the Big Game and its supporting agencies do little to change. Indeed, in a significant way, they may have a reinforcing effect on the status quo as the proponents of a "cultural critique" of capitalism have tried to show, including Veblen himself.

Authoritarianism in America has many resources in industrial work relations, religious traditions, military activities, and so on. Cady fails to recognize that another important source of authoritarianism is intercollegiate athletics. The use of systematic intimidation, verbal degradation, and even physical coercion are by no means unusual. Cady's analysis ultimately fails because it pays inadequate attention to the structural underpinnings of the system which continue to produce social pathologies. It is to these underpinnings and the pathologies they produce, according to Veblen, that we now turn, and it will soon be evident that the contrast between him and Cady is dramatic.

Sports and the Sporting Proclivity

Veblen's attitude toward sport, insofar as it aimed at invidious distinction and honorific prowess and served the purposes of the predatory elements in society, was negative. It is important, however, to keep in mind that Veblen did not object to physical culture itself as a means of maintaining physical and mental health and well-being through various forms of exercise. What he objected to were the ceremonial and predatory purposes to which athletic contests, organized games, and sporting events are put; namely, destructive emulation, invidious comparison, and waste of valuable resources and time which could be used to better advantage.

Although Veblen sets forth an anthropology of sport in his writing on primitive peoples, his comments on sport in the modern world, especially Britain and the United States, are more to the point:

> Addiction to sport of one kind and another and preoccupation with sports-manlike interests and values has spread from the levels of gentility down through the body of the population, until this category of dissipations has become almost the sole ground of common interest on which workingmen meet or hold opinions. It is safe to say that one-half the volume of printed matter daily put out for popular consumption is devoted to sports; a classification aiming to include all ramifications of the sporting interest would probably rate the proportion somewhat higher.[3]

Veblen continues:

> The mere direct waste of time and substance involved in this ubiquitous addiction to sports and their adoration need perhaps stir no one's apprehension. That much of dissipation may nowise exceed the salutary minimum; though persons with a predilection for artistic and intellectual dissipations may be moved to deprecate addiction to dissipations of this crude and brutalizing nature. What is more to the point here, however, is the fact that this preoccupation with the emulative and invidious interests of sportsmanship unavoidably has an industrially untoward effect on the temper of the population, bends them with an habitual bias in the direction of trivial emulative exploits and away from that ready discrimination in matters of fact that constitutes the spiritual ground of modern technological efficiency.[4]

But Veblen has saved his most deep-rooted indictment for last:

> It is not so much that this pervasion of the British population by sports-manlike preoccupations wastes the products and the energies of the industrial system, as that it perverts the sources from which the efficiency of the industrial system is to come. Its high consequence as a means of destruction lies in its burning the candle at both ends. Again it is to be noted that the generation and establishment of such a pervasive and stubborn habitual bent takes time, and that to get rid of it would also require time, stress and experience.[5]

---

3. Veblen, *Imperial Germany*, 148.
4. Ibid.
5. Ibid., 149.

In short, the lasting damage done by sport is its diversionary impact and retardation of the psychic and physical energies needed for the long-run flourishing of the industrial economy.

Most interesting is Veblen's interpretation of "the predatory temperament," for males who reach maturity, he says, "ordinarily pass through a temporary archaic phase corresponding to the permanent spiritual level of the fighting and sporting men" who—"punctilious gentleman of leisure" and "swaggering delinquent" alike—show "marks of an arrested spiritual development." Of course, Veblen himself was from the dolicho-blond stock and took pride in his Norwegian heritage, if not in his descent from Viking pirates and marauders about whom he occasionally wrote. Sporting activities present an opportunity for "histrionic" displays which Veblen obviously feels are inane and unredeemed by any compensatory function.[6] Veblen views "the sporting character" as "a rehabilitation of the early barbarian temperament." At the same time, however, this archaic mentality has been stripped of "the redeeming features of the savage character" which "assist in human self-preservation," "the culture bestowed in football gives a product of exotic ferocity and cunning," qualities which "are of no use to the community except in its hostile dealings with other communities."[7]

"The addiction to sports . . . in a peculiar degree marks an arrested development of [a] man's moral nature."[8] Veblen interprets immoderate devotion to sport as an emotional barbarism: "The ground of an addiction to sports is an archaic spiritual constitution—the possession of the predatory emulative propensity in a relatively high potency. A strong proclivity to adventuresome exploit and to the infliction of damage is especially pronounced in those employments which are in colloquial usage specifically called sportsmanship."[9] Few observers have concluded, as Veblen did, that competitive sport is little more than an exercise in honorific prowess and invidious emulation.

The critics and contemporary sociologists of sport who are familiar with Veblen's work perceive him as a radical critic of sports, both competitive and noncompetitive. For him, sport was merely another unjustifiable form of competitive ranking of human beings based on invidious distinctions that are wasteful of time and economic resources. In short, it is a realm of human endeavor rooted in force and fraud. As he put it:

6. Veblen, *The Theory of the Leisure Class*, ch. 10. The material cited in footnotes 7–9 is the same as that in Hoberman above.

7. Ibid.

8. Ibid.

9. Ibid.

Sports—hunting, angling, athletic games, and the like—afford an exercise for dexterity and for the emulative ferocity and astuteness characteristic of predatory life. So long as the individual is but slightly gifted with reflection or with a sense of the ulterior trend of actions—so long as his life is substantially a life of naïve impulse action—so long as the immediate and unreflected purposefulness of sports, in the way of an expression of dominance, will measurably satisfy his instinct of workmanship. This is especially true if his dominant impulses are the unreflecting emulative propensities of the predaceous temperament.[10]

There is little evidence in Veblen's work, however, that he was opposed to physical culture as such, but this claim needs qualification. Indeed, one of his most penetrating comments was that "football bears much the same relationship to the physical culture that bullfighting bears to agriculture."[11] Implicit in this remark is his view that football was so invidiously organized, ostentatiously displayed, and brutally conducted that it defeated the very ends of physical culture for which it was ostensibly created.

But his Frankfurt critic Theodor Adorno's own deep revulsion for sport was rooted in his inability to see it as anything other than a manic syndrome of pathological attitudes and instincts.[12] Indeed, Adorno's only complaint about Veblen's portrayal of sport as a savage and even vicious primitivism was that it failed to grasp sport's sadistic and masochistic elements. Athletes enjoy having pain inflicted on them by others, particularly in a fascist state! To quote Adorno in more detail:

> According to Veblen, the passion for sports is of a retrogressive nature: "The ground of an addiction to sports is an archaic spiritual constitution." But nothing is more modern than this archaism; athletic events were the models of totalitarian mass rallies. As tolerated excesses, they combine cruelty and aggression with an authoritarian moment, the disciplined observance of the rules—legality, as in the pogroms of Nazi Germany and the people's republics. Veblen's analysis, of course, should be expanded. For sport includes not merely the drive to do violence to others, but also the wish to be attacked oneself and suffer. Only Veblen's rationalist psychology prevents him from seeing the masochistic moments in sports. It is this which makes sports not so much a relic of a previous form of society as perhaps an initial adjustment to its menacing new form.[13]

10. Ibid., 260.
11. Ibid., 261.
12. See Theodor Adorno, "Veblen's Attack on Culture," *Studies in Philosophy and Social Science* 9, no. 3 (1941): 394–95.
13. Ibid., 395.

Adorno interpreted sports in any form as leading to a cult of the body, and as fulfilling a masochistic desire to inflict suffering on oneself or a sadistic impulse to inflict it on others. He also wrote in his essay on Veblen that "Modern sports, one will perhaps say, seek to restore to the body some of the functions of which the machine has deprived it. But they do so only in order to train men only the more inexorably to serve the machine. Hence, sports belongs to the realm of unfreedom, no matter where they are organized."[14]

Veblen's attitude toward competitive sports and athletics was, of course, to focus on their waste and predation. But sociologist Bernard Rosenberg complained, on the contrary, that Veblen did not understand that sports and athletics provide physically wholesome escapes from boredom. However, he admitted that:

> Veblen would probably reply that this was all well and good for the boys with their surplus energy, but quite unbecoming for an adult who should work not to better his own physique but to perfect the body social. Veblen can see no critical difference between primary and secondary conflict; he does not speculate about a moral or an amoral equivalent of war. It is enough the exhibitions of prowess are irrational. Ergo, they may be written off as undesirable. The issue of athletics gives us a clue to *The Leisure Class* which, from this angle, could be viewed as a brief in favor of total rationalization.[15]

Rosenberg's criticism was that Veblen favored "total rationalization" of society and this meant total elimination of all economic waste regardless of sources or consequences.

But what if Veblen believed that only aesthetic and athletic endeavors that aimed at invidious comparisons be eliminated and that those activities that are noninvidiously meaningful to the common man be retained? Unfortunately, both Rosenberg and Adorno overlook the latter possibility in their eagerness to convict Veblen of wanting complete rationalization of all social processes and institutions. For some forms of physical culture and aesthetic endeavor may be noninvidiously meaningful to the underlying population and can be spared in what may not, after all, have been Veblen's drive toward total rationalization of social life.

14. Ibid.
15. Bernard Rosenberg, *The Values of Veblen: A Critical Appraisal* (Washington, DC: Public Affairs Press, 1956), 76.

Sports and War

Much of Veblen's analysis in *The Theory of the Leisure Class* links both the historical and prehistorical exercise of force and fraud with sports and war. The sociologist Stjephan Mestrovic uses Veblen as a vehicle to indict a society with which both are disenchanted:

> Government becomes a predatory occupation, marked by scandals, conspicuous waste, and *militarism*. Addiction to sports becomes a substitute for and an extension of this *warlike* tendency, even as the leisure class engages in ostentatious moralism to compensate for its own immorality [author's emphasis]. Religious devotion takes on the character of fetish worship, no different from the reverence for economic activity. But genuine morality, the habits of self-abnegation and love of neighbor, diminish in intensity. As illustration, one has only to refer to the "Me" generation and its almost complete lack of guilt or moral conscience regarding the damage caused to the environment and the so-called Third World by its voracious and conspicuous consumption. The leisure class retards cultural development and conserves what is obsolescent.[16]

Like Veblen, Mestrovic spares no major institution or cultural practice from his criticism. There is no pandering to vicious sporting practices merely because they are profitable to predatory interests in the community, for example, motor sports, professional football, and boxing. There is no sparing of incompetent government officials and no neglect of religious practices which are rooted in social pathology. Finally, there is a focus on the "Me" generation whose intent is the accumulation of wealth and income and whose moral values are rooted in possessive individualism, avaricious transactionalism, and voracious consumerism. But they are the product of the system, and symptomatic of what is wrong with it, as well as catalysts in its continuation.

Most prophetic in view of what has since happened is Mestrovic's account of what is now widely regarded as an avoidable tragedy and its setting was, of course, a sporting event:

> Let me review some observations from a football game between the Texas Aggies and the University of Texas Longhorns that Veblen would probably have noticed as well. The night before the game, a pile of logs five stories high was lit and students danced around it consuming alcoholic beverages

16. Stjephan G. Mestrovic, *The Barbarian Temperament toward a Postmodern Critical Theory* (London and New York: Routledge, 1993), 5.

and intoxicants. This primitive ritual is referred to as "Bonfire," and is re-fracted in countless "tailgate" parties across America in localities that are not as willing or are not as able as Aggies to destroy trees.[17]

Mestrovic continues:

> Prior to the opening of the game—which is signaled by a cannon blast—the team mascot is paraded. The Aggie mascot is a collie named Reveille. Reveille is treated as a sacred object, a focal point of veneration, like some Hindu sacred cow: if she decides to sleep on a student's bed, the student must move to make room for her. She will be buried in a sacred burial ground reserved for her predecessors and successors, surrounded by sacred grass on which one must not walk. She watches over the game, as if she could lend magical power to the Aggies. As the fans anticipate a touchdown or goal, the male cheerleaders stand beneath the goalpost clutching their genitals in some masochistic ritual that is performed in full view of the middle-class spectators. Every score is marked by cannon fire, followed by the cheerleaders getting on their knees and bowing to Reveille. The half-time entertainment usually consists of a military marching band or civilians who engage in military maneuvers. Now I ask any unbiased reader: how are all these rituals really different from the barbaric rites documented by Malinowski [and Franz Boas] and other anthropologists [and used by Veblen for purposes of illustration]?[18]

The numerous deaths, injuries, and mutilations resulting from the 1999 "Bonfire" apparently only made sports fans in and around the academic community at College Station, Texas, more determined to keep their pregame ritual, adding authenticity to Mestrovic's indictment of the whole episode.[19] Indeed, Mestrovic, a penetrating social critic, is at his most volatile in using Veblen to analyze the sporting life and the culture that surrounds it in American society:

> The results of sporting events are routinely presented by the news media by using the *vocabulary of war,* that such and such a team was mauled, demol-ished, terminated, and so on by the opposing team [author's emphasis]. Even the most cursory examination of how sports fans behave at a match— the profanity, aggression in the stands, the trash that is left behind, the liquor that is consumed—suggests that Veblen's barbaric habits operate despite the civilized constraints and rules that are supposedly imposed on the con-test. Even the pre-game cheerleading and exercises in civil religion and oth-

17. Ibid., 25.
18. Ibid.
19. Ibid., 37.

er devotional observances testify to the barbaric temperament, if one examines these events from Veblen's perspective.[20]

Given the massive amounts of space in the sports pages devoted to reporting the results of dog and horse racing, Mestrovic would be remiss in not commenting:

> Veblen adds that "the case of the fast horse is much like that of the dog" because "he is on the whole expensive, or wasteful or useless—for industrial purpose." And, "beyond this, the race horse proper has also a similarly nonindustrial but honorific use as a gambling instrument." One has only to think of the famous Kentucky Derby, a host of smaller derbies, and the multi-billion dollar horse-gambling industry to find contemporary illustrations.[21]

The waste of material resources and leisure time on the sporting activities of animals can also be attributed to a belief in the intervention of divine providence on behalf of bettors, and not merely to a need for social diversion and "recreation," and Veblen links this with religion to which we now turn.

Religion

Veblen scholars occasionally comment on his hostility to religion, which they correctly see as linked with his antisupernatural bias against religious metaphysics and animism. However, little serious or systematic textual exegesis has ever been done of his two most detailed analyses of institutional religious belief and practice found in *Absentee Ownership* (1923) and his introduction to *The Laxdaela Saga* (1925). Perhaps this is because of the inflammatory and polemical nature of his comments; more likely, however, it is due to the focus his critics and analysts have on certain conceptual and theoretical aspects of his writing on the sociology of religion. Whatever the case, it is to the exegesis of the thirty or so pages he actually wrote on institutional religious belief and practice that I now turn.

First, I will exegete Veblen's analysis of the Roman Catholic Church in medieval Iceland, then critique his attack on Christianity in the United States after the Great War. Readers familiar with Veblen's views on the epistemology and sociology of religion will see a considerable consisten-

20. Ibid., 24–25.
21. Ibid., 280–81.

cy between the two historical epochs, widely separated in time, in the way he treats the role of religion and the institutional church. Let us begin with the Laxdaela Saga, which is believed to be rooted in Iceland in the tenth and early eleventh centuries. Veblen, in describing its sociohistorical context, writes that:

> Here it is necessary to note that while the Viking Age prepared the ground for the Christian Faith and the Feudal State, there were at the same time also certain institutional hold-overs carried over out of remoter pagan antiquity into the Christian Era; hold-overs which also had their part in the new dispensation. Chief among these was the blood-feud; which appears to have suffered no impairment under the conditions of life in the Viking Age. At the same time it appears that in principle, and indeed in the concrete details of its working-out, the habits of thought which underlie the blood-feud were not obnoxious to the interests of Holy Church or to the Propaganda of the Faith. Familiarity with its underlying principles and its logic would rather appear to have facilitated conversion to the fundamentals of the new Faith. The logic of the blood-feud, with its standardized routine of outlawry and its compounding of felonies, lends itself without substantial change of terms to the preachment of Sin and Redemption; perhaps in an especially happy degree to the preachment of Vicarious Atonement. So that this ancient and ingrained familiarity with the logic of the blood-feud may even be said to have served as an instrument of Grace. And as might fairly have been expected, the institution continued in good vigor for some centuries after the conversion to Christianity. In a certain sense, at least permissively, it even enjoyed the benefit of clergy; and it eventually fell into decay under the impact of secular rather than religious exigencies.[22]

Veblen has succinctly argued that religion is so intertwined with affairs of state and cultural practices that its larger impact cannot be adequately understood apart from these. He also comments on the relatively late date at which Iceland and Scandinavia was converted to Christianity and brought into the political system that characterized feudalism.[23]

Veblen shrewdly notes the moral, cultural, and political changes the conversion to Christianity both required and helped bring about:

> The conversion of these peoples to the ritual and superstitions of the new Faith was swift, facile, thorough and comprehensive, both in the temporal and in the spiritual phase of it, but more notably so in the latter respect. In-

22. Veblen, *The Laxdaela Saga,* translated from the Icelandic with intro. (New York: B. W. Huebsch, 1925), viii–ix.
    23. Ibid., vii–viii.

deed the gospel of Sin and Redemption was accepted by them with such alacrity and abandon as would argue that they had already been bent into a suitable frame of mind by protracted and exacting experience of a suitable kind. And on the side of the temporal reorganization, as concerned the revolutionary change in their civil institutions, they made the transition in only less headlong fashion. And in both respects the submission of these peoples to this new order of allegiance was notably abject.[24]

Throughout his introduction to *The Laxdaela Saga*, Veblen is scathing about the behavior of the powers-that-be and their relationship to the Church:

Increasingly as time passed, the ethics of the strong arm came to prevail among these peoples and to dominate men's ideals and convictions of right and wrong. Insecurity of life and livelihood grew gradually more pronounced and more habitual, until in the course of centuries of rapine, homicide and desolation it became a settled matter of course and of common sense that the underlying population had no rights which the captains of the strong arm were bound to respect. And like any other business enterprise that is of a competitive nature this traffic in piracy was forever driven by its quest of profits to "trade on a thinner equity," to draw more unsparingly on its resources of man-power and appliances, and so cut into the margin of its reserves, to charge increasingly more than the traffic would bear. Until, between increasing squalor and privation on the material side and an ever increasing habituation to insecurity, fear and servility on the spiritual side, this population was in a frame of mind to believe that this world is a vale of tears and that they all were miserable sinners prostrate and naked in the presence of an unreasoning and unsparing God and his bailiffs. So this standardized routine of larceny and homicide ran through its available resources and fell insensibly into decay as the State and Holy Church came in and took over the usufruct of the human residue that was left.[25]

24. Ibid., vii–viii. Earlier in his introduction to the *Laxdaela Saga*, he anticipated these changes when he wrote:

This new gospel of abnegation, spiritual and temporal, was substantially alien to the more ancient principles of that pagan dispensation out of which the North-European peoples had come; but the event goes to show that in principle the new gospel of abnegation was consonant with their later acquired habits of thought; that their more recent experience of life had induced in these peoples such a frame of mind as would incline them to a conviction of sin and an unquestioning subjection to mastery. The discipline of life in the Viking Age appears to have been greatly conducive to such an outcome. And the Laxdaela reflects that state of society and that prevalent frame of mind which led the Scandinavian peoples over from the Viking Age to the Medieval Church and State.

Ibid., viii.
25. Ibid., x.

Ultimately, his indictment becomes still harsher, although he is careful to utter caveats about the realities of the existing order which qualify his condemnation of it:

> The subsequent share of Holy Church and its clerics in the ulterior degradation of the Scandinavian peoples, including Iceland, was something incredibly shameful and shabby; and the share which the State had in that unholy job was scarcely less so. But these things come into the case of the Icelandic community only at a later date, and cannot be pursued here. The medieval Church in Iceland stands out on the current of events as a corporation of bigoted adventurers for the capitalizing of graft and blackmail and the profitable compounding of felonious crimes and vices. It is of course not intended to question that this medieval Church all this while remained a faithful daughter of Rome and doubtless holy as usual; nor is it to be questioned that more genial traits and more humane persons and motives entered into the case in a sporadic way. It is only that the visible net gain was substantially as set forth. In abatement it should also be noted, of course, that there is no telling what else and possibly shabbier things might have come to pass under the given circumstances in the conceivable absence of Holy Church and its clerics.[26]

Veblen continues:

> For a nearer view of that tangle of corrosive infelicities there are an abundance of documents available; [which] will show how the fortunes of that people [Norwegian] from the advent of Christianity onward, swiftly tapered off into a twilight zone of squalor, malice and servility, with benefit of Clergy.[27]

Should the reader find this process of exegesis and quotation tedious, the main justification for it lies in the provision of the genuine flavor and style Veblen used late in his career (1925) to express his contempt for the medieval church. It should not be assumed, however, that his attitude toward Roman Catholicism in the twentieth century was more favorable. He still regarded it as an "imbecile institution" and its clergy as superstitious and stupid.

Veblen also illustrates the religious bias of the medieval editor-author of the Laxdaela Saga, upon which much of his analysis is based. He thus uses the Saga to illustrate the environment in which the Icelandic people came to live after the introduction of Christianity and feudalism. It is ap-

26. Ibid., x–xi.
27. Ibid., xi–xii.

parent that the stories related in it are most meaningful in the particular culture in which they allegedly took place, and this requires understanding of both the ideological and religious bias of the clerk-editor and Veblen's, at times, vitriolic evolutionary naturalism.[28] But Veblen's attack on medieval Catholicism should not be disregarded as a glib dismissal of the Dark Ages. His analysis of religious dogma, hierarchy, sacramentalism, asceticism, and manipulative superstition is also a warning about the possible resurgence of such atavistic continuities.

Religion: The Case of America

Most interesting in his analysis of institutional religion in the United States, however, is his clever analysis of commercial advertising and business exploitation as analogue to ecclesiastical employment of the like or perhaps the reverse. As Veblen put it in *Absentee Ownership* (1923):

> Writers who discuss these matters have not directed attention to the Propaganda of the Faith as an object-lesson in sales-publicity, its theory and practice, its ways and means, its benefits and its possibilities of gain. Yet it is altogether the most notable enterprise of the kind. The Propaganda of the Faith is quite the largest, oldest, most magnificent, most unabashed, and most lucrative enterprise in sales-publicity in all Christendom. Much is to be learned from it as regards media and suitable methods of approach, as well as due perseverance, tact, and effrontery. By contrast, the many secular adventures in salesmanship are no better than upstarts, raw recruits, late and slender capitalizations out of the ample fund of human credulity. It is only quite recently, and even yet only with a dawning realization of what may be achieved by consummate effrontery in the long run, that these others are beginning to take on anything like the same air of stately benevolence and menacing solemnity.[29]

Veblen continues:

> No pronouncement on rubber-heels, soap-powders, lip-sticks, or yeast-cakes, not even Sapphira Buncombe's Vegetative Compound, are yet able to ignore material facts with the same magisterial detachment, and none has yet commanded the same unreasoning assent or acclamation. None other have achieved that pitch of unabated assurance which has enabled the publicity-agents of the Faith to debar human reason from scrutinizing their pro-

28. Ibid., xiii–xiv.
29. Veblen, *Absentee Ownership,* 319.

nouncements. These others are doing well enough, no doubt; perhaps as well as might reasonably be expected under the circumstances, but they are a feeble thing in comparison. Saul has slain his thousands, perhaps, but David has slain his tens of thousands.[30]

Veblen has often been criticized for disguising his real views on various controversial and divisive subjects with his tongue-in-cheek style and massive barrages of irony and satire. Suffice it to say that few, if any, of the critics ever surpassed his candor and wit in their denunciation of institutional religion belief and practice, especially what he had to say about Roman Catholicism.[31]

Shortly, thereafter, Veblen returns to his business-religion analogue commenting that:

> That the same principles of sales-publicity are found good and profitable for the traffic in spiritual amenities and in these material comforts should serve to show how deep and pervasively the scheme of deliverance and rehabilitation is rooted in the merciful gift of credulous infatuation. It should redound to the credit of the secular arm of sales-publicity rather than cast an aspersion on those who traffic in man's spiritual needs; and should go to show how truly business-as-usual articulates with the business of the Kingdom of Heaven.[32]

Interestingly, since Veblen did not often use a quantitative-statistical approach, he then provides data to buttress his case regarding the similarity of commerce and religion:

30. Ibid., 319.
31. Ibid., 319–20.
32. Ibid., 320. Veblen then stretches toward the apex of his vitriolic mockery when he writes:

> But as viewed objectively and as seen in any other than their own dim religious light, these admirable feats of manifestation have been after all essentially ephemeral and nugatory hitherto; very much of a class with those lunch-counter sample-packages that are designed to demonstrate the expansive powers of some noted baking-powder, in miniature and with precautions. They are after all in the nature of publicity-gestures, eloquent, no doubt, and graceful, but they are not the goods listed in the doctrinal pronouncements; no more than the wriggly gestures with which certain spear-headed manikins stab the nightly firmament over Times Square are an effectual delivery of chewing-gum. *Bona-fide* delivery of the listed goods would have to be a tangible performance of quite another complexion, inasmuch as the specifications call for Hell-fire and the Kingdom of Heaven; to which the most heavily capitalized of these publicity concerns of the supernatural adds a broad margin of Purgatory.

Ibid., 322.

Of such sacred sales-publicity concerns operating as certified agents for this marketing of supernatural intangibles, the Census of 1916 enumerates 203,432 retail establishments occupied exclusively with the sale of such publicity to the ultimate consumers; of whom there is one born every minute, and who are said to be carried on the books of these retailers to the number of 41,926,854. It has been confidently estimated, on the ground of these data, that the effectual number of paying customers will be approximately 90,000,000; regard being had to the very appreciable floating clientele and the great number of effectual consumers attached to and associated with the customers of record. The stated value of "church property" is $1,676,600,582. These tangible assets are exempt from taxation.[33]

Finally, Veblen feels that some concluding remarks are in order regarding the wastefulness of religion and this time he adds Protestants to his list of abusers of humanity:

The man-power employed in this work of the Propaganda is also more considerable than those engaged in any other calling, except Arms, and possibly Husbandry. Prelates and parsons abound all over the place, in the high, the middle, and the low degree; too many and too diversified, in person, station, nomenclature, and vestments, to be rightly enumerated or describe,—bishops, deans, canons, abbots and abbesses, rectors, vicars, curates, monks and nuns, elders, deacons and deaconesses, secretaries, clerks and employees of Y.M.C.A., Epworth Leagues, Christian Endeavors, etc., beadles, janitors, sextons, Sunday-school teachers, missionaries, writers, editors, printers and vendors of sacred literature, in books, periodicals and ephemera. All told—if it were possible—it will be evident that the aggregate of human talent currently consumed in this fabrication of vendible imponderables in the nth dimension, will foot up to a truly massive total, even after making a reasonable allowance, of, say, some thirty-three and one-third per cent, for average mental deficiency in the personnel which devotes itself to this manner of livelihood.[34]

On these block quotations, with apologies to the reader, hang the law and the prophecy about American religion in the early twentieth century insofar as it can be linked with commercial enterprise and advertising exploitation—or so Veblen claims. Yet altruism was one of his favorite human virtues. But in his view it was not by necessity derivative from religion, and only sustainable by such, but a generic human trait fully compatible with a secular order.

33. Ibid., 323–24.
34. Ibid., 324–25.

Religion: Critics and Conclusion

Veblen's often-caustic attacks on institutional religious belief and practice did not go unnoticed by his critics, particularly those with religious convictions of their own. To illustrate this point, B. W. Wells, in reviewing *The Theory of the Leisure Class*, argued that it contained "a vicious attack on Christian ideals."[35] Veblen's own generic denunciation of "imbecile institutions" surely included the Roman Catholic Church, evangelical Protestant sects, and Eastern Orthodoxy at the least; it may also mean the Reformation Protestant denominations, including the Lutheran, Veblen's own religious heritage. His critics focused their rebuttals on several aspects of his analysis by proclaiming the validity of formal religious belief, that is, the truth of the Christian creed, the stabilizing role that religion plays in the community, and the moral anarchy that must ultimately result from a relativism such as Veblen's.

To illustrate, the critic perhaps most interested in Veblen's views on religion was the one critic who wrote a large study of him from an explicitly religious perspective, namely, Lev Dobriansky.[36] His Thomism provided the metaphysical, epistemological, ontological, and ethical fulcrum for his analysis of Veblen's critique of religion. It is evident that the latter's disbelief in God, soul, and afterlife was offensive to Dobriansky as were his satirical comments regarding the institutional church and the intelligence of the clergy. Even more germane to his analysis, however, was his conviction that Veblen's value relativism could only lead to moral anarchy and, ultimately, to a political despotism unrestrained by the ethical dictates of the natural law.

Interestingly, most radicals had little to say about Veblen's attitudes toward religion, undoubtedly because they largely shared these attitudes and believed, in any case, that they had other more important theoretical issues to debate with him. Social democrat Bernard Rosenberg, however, stressed the system-maintaining qualities of institutional religious belief and practice that he believed Veblen did not adequately understand. The structural-functional mode of analysis popularized by Talcott Parsons and his students influenced Rosenberg to the point that he did not think that a viable social order could exist without the system-maintaining properties of the church or ersatz religion. He believed Veblen failed to take religion into consideration adequately, since he did not propose any

35. B. W. Wells, Review of "The Theory of the Leisure Class," *Sewanee Review* 7 (July 1899): 373.
36. Dobriansky, *Veblenism: A New Critique*.

institutions or social processes to take its place. It was thus ironic to find secularized liberals and radicals like Rosenberg criticizing him for his failure to understand the social equilibrating role of the church in society.[37]

Veblen viewed the Christian Church as an "imbecile institution" and its clergy as suffering from a "fifty-percent deficiency" in "average intelligence." If he is harsher on Roman Catholicism than on Protestantism, it may be suggested that he had little of a positive nature to say about the latter either. His biographer Joseph Dorfman tells us two anecdotes that are revealing of his personal feelings about religion. Once, when queried as to which church he belonged, he said "the Lutheran"; when asked why, he commented tongue-in-cheek "because the nearest one is fifty miles away." On another occasion, he remarked that he couldn't see much difference between magic and religion! In his view, religious denominations are chain-stores, the local church being a retail outlet.[38] He, nevertheless, recognized its importance in human affairs, insisting on its intimate relationship with the economic and social organization of society and accompanying social psychology.

Still other comments he made regarding evangelical Protestantism and fundamentalism can be quoted from his major writings. Yet the Roman Catholic Church bears the brunt of the satirical mockery and ideological venom, not too surprisingly since it was so publicly visible during his years in New York City when he taught at the New School for Social Research and served as an associate editor and contributor to *The Dial.*

Although the primary focus in this chapter is the role of religion in medieval Iceland and the United States in the 1920s, some mention must be made of Veblen's anthropology of religion. *The Theory of the Leisure Class* is replete with references to the relationship between religious belief and practice and class structure. Religion is portrayed primarily as a leisure-class function confined to a priestly caste, as a form of superstition based on make-believe and as a waste of resources and time. Since *The Theory of the Leisure Class* has so often been subjected to textual exegesis and interpretation, *Imperial Germany and the Industrial Revolution* is briefly used here to show that Veblen was consistent in his analysis of religion.[39]

In primitive cultures, Veblen finds that religious belief and practice is interwoven with other facets of human existence:

> If magical and superstitious practices, or such of them as are at all of material consequence, are with virtual universality to be traced back through the

37. Rosenberg, *The Values of Veblen: A Critical Appraisal,* 49, 76.
38. Dorfman, *Thorstein Veblen and His America,* 479–80.
39. Veblen, *Imperial Germany.*

channels of habituation to some putative ground of serviceability for human use, it follows that the rule should work, passably at least, the other way; that the state of the industrial arts which serve human use in such a culture will be shot through with magical and superstitious conceits and observances having an indispensable but wholly putative efficacy.[40]

Veblen continues:

In many of the lower cultures, or perhaps rather in such of the lower cultures as are at all well known, the workday routine of getting a living is encumbered with a ubiquitous and pervasive scheme of such magical or superstitious conceits and observances, which are felt to constitute an indispensable part of the industrial processes in which they mingle. They embody the putatively efficacious immaterial constituent of all technological procedure; or, seen in detail, they are the spiritual half that completes and animates any process or device throughout its participation in the industrial routine. Like the technological elements with which they are associated, and concomitantly with them, these magically efficacious devices have grown into the prevalent habits of thought of the population and have become an integral part of the common-sense notion of how these technological elements are and are to be turned to account. And at a slightly farther shift in the current of sophistication, out of the same penchant for anthropomorphic interpretation and analogy, a wide range of religious observances, properly so called, will also presently come to bear on the industrial process and the routine of economic life; with a proliferous growth of ceremonial, of propitiation and avoidance, designed to further the propitious course of things to be done.[41]

Veblen concludes that:

These matters of the magical and religious ritual of industry and economic arrangements among the peoples of the lower cultures are sufficiently familiar to all ethnological students, and probably they also are so far a matter of common notoriety that there is no need of insistence on their place and value in these lower cultures. They are spoken of here only to recall the fact that the large and consequential technological elements involved in any primitive system of industry have commonly carried such a fringe of putatively efficacious, though mechanically futile, waste motion. These naïve forms of mandatory futility are believed to belong only on the lower levels of culture, although it should not be overlooked that magical and religious

40. Ibid., 26.
41. Ibid., 26–27.

conceits still exercise something of an inhibitory influence in the affairs of industry even among the very enlightened peoples of Christendom.[42]

Those viewing my emphasis on Veblen's animus toward the various branches of the Christian Church as diverting attention from his negative attitudes toward non-Christian religions will be relieved to be told of his penetrating essay on Jewish intellectuals and his critique of Zionism, which features Veblen's fears of what a Jewish state might be like, a prediction of its character on a par with his anticipations of fascist Germany and Japan. Also, in *The Instinct of Workmanship,* and elsewhere, Veblen identifies the beginnings of exploit with the emergence of a priestly class, culminating in his indictment of the ancient biblical tribes for their violence, destruction, and general mayhem.[43]

By now the reader has tasted the flavor of Veblen's anthropology of religion and recognized his claims that religion often impedes economic rationality, that it consists mostly of make-believe and wasteful ceremonial practices, and that it is a tool of the vested interests. They may also conclude that Veblen views it as an atavistic continuity that surfaces and resurfaces in culture after culture and impedes the efficiency of modern industry, albeit in altered form.

Gambling

Introduction

In the third part of this chapter, the reader should not lose sight of the claim advanced earlier that Veblen linked sport, religion, and gambling together because he saw marked similarities in their social and cultural genesis and consequences. It is to gambling that we now turn and for our understanding of it, two traditions in the philosophy of leisure are relevant. The classical perspective dating from the Greeks, particularly Aristotle, views leisure as time free from the obligation to work. It is a condition of existence in which activity is engaged in as an end-in-itself, that is, for its own sake. Pure contemplation is thus leisure's most sublime form. In order to engage in contemplative activity, individuals must possess the ability to both reason logically, that is, generalize from the particular, and

---

42. Ibid., 27.
43. See "The Intellectual Pre-eminence of Jews in Modern Europe," in *Essays in Our Changing Order,* ed. Leon Ardzrooni (New York: Augustus M. Kelley, 1964), 219–31; and *The Instinct of Workmanship,* ch. 4.

intuitively understand the good. But knowledge is to be used not for personal or material gain but for self-illumination. Its purpose is self-actualization, not social aggrandizement. Thoroughly class and gender biased and supported in terms of material provisioning by slavery, this tradition of leisure has little in common with the one analyzed by Veblen in *The Theory of the Leisure Class*. Veblen wrote that leisure connoted "nonproductive consumption of time"[44] and believed that leisure-related resources were ordinarily employed not for contemplative purposes nor intellectual activity, but for purposes of conspicuous consumption and display, conspicuous waste, and exemption from useful labor. He felt that the question to be asked regarding all expenditure is "whether it serves directly to enhance human life on the whole—whether it furthers the life process taken impersonally."[45] He also writes that "in order to be at peace with himself the common man must be able to see in any and all human effort and human enjoyment an enhancement of life and well-being on the whole. In order to meet with unqualified approval, any economic fact must approve itself under the test of impersonal usefulness—usefulness as seen from the point of view of the generically human."[46]

Legalization of gambling in Nevada by an act of the state legislature in 1931 was an important catalyst for the massive growth of the "gaming" industry after World War II, although it existed on a substantial subterranean scale before the Depression. The city of Las Vegas was incorporated six years after the publication of *The Theory of the Leisure Class*, and its first large hotel-casino was built twelve years after Veblen's death in 1929. As powerful as Veblen's predictive ability was, within his frames of reference, he did not foresee Las Vegas. However, gambling in its various forms dovetails into his analysis of leisure as sheer waste and futility and as often possessing emulatory significance.

Veblen suggests in his distinctive satirical prose that gambling is evidence of superstition and animism in the human community. In his time, the main forms of gambling, most of which he discusses, were church raffles, often illicit card games, horse-racing and other animal sports, and, of course, athletic events.[47] He noted that the higher leisure class found prizefighting distasteful,[48] something that may have changed, in view of the exorbitant ticket prices charged for championship boxing matches in Las Vegas to people who buy the best seats.

Apologists for the casino industry argue that gambling has its own "en-

44. Veblen, *The Theory of the Leisure Class*, 43.
45. Ibid., 99.
46. Ibid., 98.
47. Ibid., ch. 11.
48. Ibid., 271.

tertainment value," that it provides an emotionally exhilarating experience especially if one wins and, in any case, provides relief from boredom. And, from the perspective of neoclassical economics, what people do with their resources should be a private matter and not subject to critical public scrutiny or collective censure. However, a century after publication of *The Theory of the Leisure Class* it is important to note that most states permit several forms of gambling; and lotteries, which have worse odds than roulette, are run by most states to raise public revenue.

Historically eliminated by most states from the late nineteenth century through the first half of this century, state revenues from lotteries now often replace, but do not supplement, other revenues. In London, lottery revenues helped to finance renovation of the British Museum and Library. The very presence and issue of gambling on the internet means that gambling is virtually global. Legalized gambling (and gambling addiction) spread internationally through the device of the corporation makes important the analysis of Veblen's claims about it.

C. Wright Mills, in his introduction to the 1953 edition of *The Theory of the Leisure Class,* uses the term "crackpot realism" to describe various illusory beliefs and practices exposed by Veblen. We argue that the coexisting status of corporate "gaming" and the epidemic of gambling addiction by governments and individuals (Las Vegasization), amounts to what Mills termed "crackpot realism." More to the point, we call gambling legalization "jackpot realism," which ignores the ugly side of the trend to the advantage of the people in power.

First, we will explicate Veblen's theory of gambling through textual exegesis of his chapter, "The Belief in Luck," which is sandwiched between chapters on "Modern Survivals of Prowess" and "Devout Observances." Veblen's discussions in these three chapters overlap as easily and obviously as sports books with football betting and dozens of wedding chapels that pack the Las Vegas Strip. We will show the relevance of Veblen's theories to legalized gambling today, including its emulatory features. We argue that Veblen was essentially correct in his diagnosis of its socially pathological justifications and consequences, the latter meaning that gambling "industries" do not serve Veblen's prescribed ends of life.

Prowess, Belief in Luck, and Devout Observances

"As it finds expression in the life of the barbarian, prowess manifests itself in two main directions—force and fraud."[49] Veblen's discussion of prowess ties into analysis of gambling in at least two ways. First, sports

49. Ibid., 273.

books and athletic events, for which heavy betting occur, are an inextricable aspect of our casinos and culture. Second, "force and fraud" is not off the mark in describing the power of "gambling" lobbyists or casino developers and the advertising used by government lottery directors and corporations to entice gamblers.

Veblen points to the egoistic role of self-aggrandizement present in sports betting when he comments that not only does the stronger side score a more signal victory, and the losing side suffer a more painful and humiliating defeat, in proportion as the pecuniary gain and loss in the wager is large.[50] The wager is seen as "enhancing the chances of success for the contestant on which it is laid."[51]

Veblen did not use the term "gambling addiction," but he talks about "the spiritual basis of the sporting man's gambling habit."[52] His thoughts on athletics link his theories to current trends. In the following segment from Veblen's "modern survivals of prowess" chapter, we substitute the word "gambling" for the term "athletic sports": Addiction to [gambling], not only in the way of direct participation, but also in the way of sentiment and moral support, is, in a more or less pronounced degree, a characteristic of the leisure class; and it is a trait which that class shares with the lower-class delinquents and with such atavistic elements throughout the body of the community as are endowed with a dominant predaceous tend.[53]

In a subsequent paragraph, Veblen writes: "The prevalence and the growth of the type of human nature of which this propensity is a characteristic feature is a matter of some consequence. It affects the economic life of the collectivity both as regards the rate of economic development and as regards the character of the results attained by the development."[54]

Veblen thus links gambling with the history and prehistory of the human race and claims that it possesses transcultural significance; "the gambling propensity is another subsidiary trait of the barbarian temperament. It is a concomitant variation of character of almost universal prevalence among sporting men and among men given to warlike and emulative activities generally."[55]

Veblen explained that at the root of the gambling urge was a belief in luck, that is, personal intervention on one's behalf by Divine Provi-

50. Ibid., 277.
51. Ibid.
52. Ibid., 294.
53. Ibid., 271–72.
54. Ibid., 272–73.
55. Ibid., 276.

dence.[56] According to Veblen, this provided the main psychological impetus for taking part in games of chance; yet the proclivity for gaming did not depend entirely on a belief in intervention by the Almighty. The excitement of the game itself and the cultural stimulation of the environment in which gambling takes place also lent support to the gambler's flagging spirits should he or she succumb to a losing streak.

Veblen's main point was that the gambler believes in luck and that this quasi-religious belief is what motivates the persistent bettor who may actually find it necessary to invoke the name of the Deity, or a mascot, to further his or her fortunes with the gambling apparatus. Not surprisingly, Veblen argues that these beliefs and their social consequences are not culturally isolated phenomena nor socially encapsulated practices. Rather, they are connected with other values and processes that collectively damage the social fabric of the community. He advances an important thesis regarding gambling's social impact, "this trait also has a direct economic value. It is recognized to be a hindrance to the highest industrial efficiency of the aggregate in any community where it prevails in an appreciable degree."[57]

Veblen points to the animistic and superstitious aspects of the gambler's psyche which are linked with a belief in the potential intervention of Divine Providence in the game. As he put it:

> In its simple form the belief in luck is this instinctive sense of an inscrutable teleological propensity in objects or situations. Objects or events have a propensity to eventuate in a given end, whether this end or object point of the sequence is conceived to be fortuitously given or deliberately sought. From this simple animism the belief shades off by insensible gradations into the second, derivative form or [sporting man's] phase, which is a more or less articulate belief in an inscrutable preternatural agency.[58]

Veblen's theory of gambling emanated from his theory of the leisure-class phenomenon, yet it was also practiced by working-class delinquents who represented an arrested, that is, socially stunted stage of individual and social development. Gambling, in his view, lowered the collective industrial efficiency of the community through dissipation of its mental acuity and its physical and emotional energies. It was also objectionable because it wasted other material resources that conceivably could be put to better use. Although the animistic and superstitious proclivities of hu-

56. Ibid., ch. 11.
57. Ibid., 276.
58. Ibid., 280.

mankind have been present throughout the history and possibly much of the prehistory of the race, the pursuit of gaming reinforces and intensifies them in industrial society, or so he claimed.

Status Emulation in Large-Scale Gambling

In what follows, the gambling industry is analyzed from a Veblenian perspective, modified, of course, by an updating of the social environment to take account of structural tendencies and economic changes that have occurred since his lifetime. Specifically, this means using Veblen's theory of status emulation to analyze the use of economic resources in the economy and the patterns of consumption prominent among gamblers insofar as they can be shown to be linked directly with the structural imperatives of the gaming industry. Our hypothesis is the following: where legalized gambling spreads on a large scale in market economies and cultural environments roughly similar to our own, it will reproduce those same patterns. Thus, the expansion of legalized gambling in its corporate form to the Pacific rim countries, the North Atlantic community, the Commonwealth, and the more advanced parts of the Third World can be expected to produce parallel effects to those gambling has induced in Nevada.

We now inject Veblen's ideas regarding invidious emulation as it pertains to the provision and consumption of forms of gambling. Gamblers, be they professionals, regulars at local casinos, or simply tourists casually trying their luck, are part of either status-enhancing or status-detracting emulatory processes that may expand or weaken the sense of self-worth or general esteem. The common thread for the range of gamblers is their use of leisure time that appears to an observer to be both entertainment and wasteful of resources. Assuming that losing money is less wasteful than winning, the entertainment factor and the waste factor would seem to be inversely related to one another. Although traditional one-armed bandit slot machines have been largely supplanted by games of chance in video slot machine form, the former, especially if played continuously as gambling addicts do, is an apparatus that requires waste of physical and mental resources, as well as time.

What may appear on the surface to be spontaneous self-indulgence in a frivolous game of chance by the gambler, may in reality be a serious form of emulatory rivalry. What matters is not simply winning but playing those games which are most status enhancing—sitting at the baccarat table to provide visible evidence of ability to pay is more invidious than motives for playing the nickel and quarter slot machines. It is ostentatious display of pecuniary prowess that is most likely to enhance social status.

Appearances of status enhancement for the "high rollers" are, ironically, increased especially when they lose large sums of money. The hotels in which the high rollers play will often "comp" or compliment these gamblers with deluxe accommodations, service, and entertainment.

As important as amounts of money wagered, styles of dress that accompany betting, and sophistication of the game are, it is also the esteem of peers that matters and this may require getting on the "bandwagon." Playing those games most often indulged in by the group upon whose esteem one's sense of self-worth most depends affects gamblers' choices. For most players, this means joining the game(s) most appealing to companions and other players who really "count" in the hopes of making oneself socially acceptable to them. Yet, a self-anointed elite may ultimately come to deride those mediocre and vulgar gamblers, on whose bandwagon almost anyone can climb; or sneer at those whose play produces "Veblen effects" based on price or the amount waged. Repulsed by these philistines, the wagering "snob" emerges as evidence of the cultural superiority of the aesthete. Bored by convention and contemptuous of both *nouveau riche* and unwashed herd, the snob distances himself from both by engaging in games of chance that only the culturally competent can appreciate and enjoy.[59]

In time, the snob may become bored with gambling snobbery or come to feel that the practice of it is not sufficiently status enhancing. He will thus move on to what has been described as "counter-snobbery."[60] This is a more simplistic and austere lifestyle in which gambling may be considered irrelevant or, at best, as a casual, inexpensive activity that aims only at impressing the counter-snobs with its innocuousness The status aspirations and pretensions of other groups no longer matter—only the needs of the counter-snob for more social deference and then only from other practitioners of counter-snobbery.

Regarding most casino employment, the relative status of jobs generally correlates with the invidious hierarchy of the range of games available. Employees who rove slot machine areas providing coins are at the bottom of the invidious job ranking. Slot players are among the masses of casino gamblers, but like all gamblers, they are served "free" drinks, and this points to one of the gender-specific casino roles. Beverage servers are traditionally female and usually must have certain physical attributes to be employed (for example, being shapely with good legs) and look attractive

59. Harvey Leibenstein, "Bandwagon, Snob and Veblen Effects in the Theory of Consumers' Demand," *Quarterly Journal of Economics* 64 (May 1950): 183–207.

60. Robert L. Steiner and Joseph Weiss, "Veblen Revised in the Light of Counter-Snobbery," *Journal of Aesthetics and Art Criticism* 9 (March 1951): 263–68.

in their often skimpy uniforms or costumes. In the patriarchal system of mega-resorts, cocktail waitresses, not unlike "showgirls," as casino trophies, are offered to gambling customers for vicarious consumption. The place of cocktail waitresses in the invidious employee ranking may depend on the gamblers they serve. For example, serving drinks to bingo players cannot enhance status as much as serving to poker players. The chance for making good tips, called "tokes" in Nevada, is one factor because "classier" gamblers will generally tip more. Cocktail waitresses have higher income and, perhaps, more invidious status among casino personnel who are traditionally women, such as "keno girls" and "change girls."

Blackjack dealers make up a disproportionate number of workers on the casino floor. On the Las Vegas Strip, they are required to audition as part of the hiring process. Physical appearance in the dealer's uniform is a criterion for employment. Their status among the casino-floor hierarchy may be roughly the same as cocktail waitresses, and their income is also enhanced by tips. However, dealers have at least two levels of superior employees constantly watching their moves on the job. Supervising blackjack dealers are the "pit bosses" (usually male), probably another supervisor, and experts on spotting cheaters monitoring partly hidden security cameras. Pit bosses have invidious distinction on the casino floor. Opportunities are narrow for casino workers to accrue experience in games with more limited numbers of gamblers, such as baccarat. Specially trained attendants to the most glamorous game hold a prestigious position in casino-floor hierarchies.

This suggests that one could develop a taxonomy of status enhancement as it pertains to games of chance; an empirically verifiable inventory of social deference or honor ranging from cheap slots with lower payout rates to intermediate status production through blackjack ("21"), poker, craps, or roulette; and finally, to more sophisticated games of both skill and chance where the well-attired indulge themselves at baccarat.

Generic Ends of Life Defeated?

Gambling takes place in a localized cultural setting characterized by institutionalized and legalized fraud and manufactured consensus. To illustrate, note the use of the euphemism "gaming" as opposed to "gambling," which serves as a symbol to divert attention from the bad odds and consequences for the gambler. Veblen says that "collective interest is best served by honesty, diligence, peacefulness, goodwill, an absence of self-seeking, and an habitual recognition and apprehension of casual sequence, without admixture of animistic belief and without a sense of dependence on any preternatural intervention in the course of events."[61] Gambling, sport, and religion all defy his proscriptions and prescriptions in much the same ways.

Mega-resort casinos and government-run gambling were rare in Veblen's day. Games of chance, "wagering," as he has called it, or betting, included the church bazaar or raffle, betting on sports and games, horse and dog racing. A question for the student of legalized gambling who wishes to apply Veblen's standards to the contemporary situation is how the industry as a whole, or discrete segments of it in isolation, contribute to the growth of noninvidious community. And, perhaps more important, what is the larger social and moral impact of gambling likely to be as it continues to spread? The same query could be made regarding sport and religion.

Gambling and the tourism based on it quickly came to dominate the Nevada economy. The correlation between the spread of legalized gambling and the existence of certain socially pathological traits is remarkable and strikingly evident in the empirical studies on the subject. Although no definitive claims are made here for the existence of causal relationships, it is difficult to otherwise explain the incidence of homicide, suicide, other violence, juvenile delinquency, substance abuse, gambling addiction, obesity, broken families, mental illness, respiratory disease, and work alienation. Of course, these are present without legalized gambling, but, at the very least, the work patterns and social values induced or attracted by the gambling industry exacerbate existing social problems. It is no accident that Nevada has a disproportionately high number of its residents clinically depressed, incarcerated, or on probation or parole.

To many public-choice theorists and neoclassical economists of the libertarian stripe for whom consumer tastes are a given, the genesis of the gambler's revealed preferences (for sport and religion also), as expressed in the fleshpots of casino resorts (or the arena or church), is irrelevant. For these economists, value is merely subjective preference and measurable only by price in the exchange mechanism of the market. They ignore the fact that capitalist property relations, a price system, and a market mechanism are part of a larger power system which benefits ownership and control groups disproportionately from the want creation that it calculatingly fosters. Large sums are expended by corporate gambling interests on political campaigns, advertising, and salesmanship. If Veblen's "generic ends of life" are used as a formal standard, the wants and desires fos-

---

61. Veblen, *The Theory of the Leisure Class*, 227.

tered by these corporate leaders are not in the long-term "best interest" of individuals or society.

Veblen wrote of the "generic ends" and of fullness of life impersonally considered.[62] These are closely tied to the "instincts," or more accurately, "proclivities" for altruism, proficiency of workmanship, and critical intelligence for he believed that any viable scheme of life must be rooted in values that will sustain the community. Those values that are not community sustaining, he referred to as "pecuniary" and "sporting." It is clear that these are based primarily on the use and validation of force and fraud, especially the latter. Although Veblen's readers cannot be certain when he is engaged in speech acts with serious valuative import, he does use language and concepts that are undeniably value laden. Indeed *The Theory of the Leisure Class* is littered with phrases such as conformity "to the generically human canon of efficiency for some serviceable objective end,"[63] and "actions and conduct as conduce to the fullness of human life."[64]

Though defenders claim gambling provides needed public revenue and diversion from monotony as use of leisure time, gambling precludes enjoyment of other potentially consummatory behaviors. Using Veblen's standard on leisure time experience as a means to measure "fullness of life, impersonally considered," it follows that gambling is development primarily in a pathological sense. For, given his criteria, it is neither culturally elevating, conducive to the growth of critical intelligence, likely to increase empathy or altruism, or likely to enhance proficiency at socially useful work.

Much the same can be said regarding commercial sport, both intercollegiate and professional, and much of institutional religion, although in the realms of "self-abnegation" (humility) and altruism an exception might be made as Veblen himself acknowledged as late as 1910.[65] Nevertheless, if realization of his generic ends of life is the standard by which social institutions and practices are to be judged then gambling, compet-

---

62. Ibid., ch. 11.

63. Ibid., 259

64. Ibid., 311.

65. Nevertheless, "He always spoke of churches as 'bunk houses.'" Motier Fisher to Joseph Dorfman, February 26, 1932, Dorfman Collection. "One chance remark I recall his having dropped with one of his genial smiles, when the eagerness of the Christian notions to civilize the heathens had been alluded to, namely, that "the heathens had some rights too"—apparently to retain, if they preferred, their own modes of life. Whether this expressed a real opinion or a cynical mood at a given time I shall leave undetermined." Andrew Estrem to Joseph Dorfman, February 12, 1930, Dorfman Collection.

itive sport, and religion[66] are all to be found wanting and for much the same reasons. Rather than lead to the progressive enrichment of life, over-investment in them as practiced and uncritical indulgence in their meta-theoretical claims, will lead to the progressive arrest of evolutionary naturalism in societies where they achieve hegemony—or so Veblen would have us believe.

66. Although Veblen never denied the potential of Christianity for the provision of moral and psychological insights, he did not perceive it as a fully independent and dynamic variable. Instead, it was more often a strategy of superstitionists, institutional rationalization, clerical domination, and often class subjugation. In his view, it was an error to sever it from its sociocultural roots or political consequences. Probably, for Veblen, personal identity insofar as it is relevant to his social theory consists of prag-matically, socially, and culturally definable particulars. Veblen, of course, denied the existence of the soul "save as the name for verifiable cohesions in our inner life," which he preferred to think of as naturalistic. See William James, *Pragmatism* and four essays from *The Meaning of Truth* (Cleveland: Meridian Books, 1969), 70.

Chapter 6

Aesthetician of the Commoners

Introduction: Veblen's Naturalism as Ethics and Aesthetics

Much of what is said here about Veblen's philosophy is based on extrapolation from what he actually wrote. Since he authored no formal philosophical treatises on either naturalism or Darwinian evolution, no closure as to its meaning can be definitively or authoritatively invoked for his thought. Assessment of it as partial and incomplete is thus essentially correct and this includes his ethics and aesthetics:

> Man's ethical values, compulsion, activities, and restraints can be justified on natural grounds without recourse to supernatural sanctions, and his highest good pursed and attained under natural conditions, without expectation of a supernatural destiny.[1]

In this sense, Veblen's ethics were naturalistic: I am not referring to his personal deportment but to the values he used in assessing the behavior of social aggregates and those he prescribed and proscribed regarding the behavior of the community and ways to improve it. A naturalistic ethics may also be defined as:

> any view according to which ethics is an empirical science, natural or social, ethical notions being reduced to those of the natural sciences and ethical questions being answered wholly on the basis of the findings of those sciences.[2]

---

1. Dagobert D. Runes, ed., *Dictionary of Philosophy* (Totowa, NJ: Littlefield, Adams and Company, 1968), 205.
2. Ibid., 206.

Using this narrow, scientific definition of ethics Veblen would be on the fringes of naturalism. Clearly, much of his thinking on prevailing ethical standards and particularly the veracity of nations, classes, social orders, and the prescriptions of intellectuals is rooted in what he called "the generic ends of life, impersonally considered," which encompasses altruism, critical intelligence, and a proficiency of craftsmanship. Although the origins of these values and value-practices are partly Christian and Kantian, Veblen adopted a historical and cultural perspective that judged them by their social and political as well as individual consequences. He viewed them and their opposites as transcultural value constants which could be studied like any other extrusion or intrusion of nature:

Are Veblen's aesthetics naturalistic? The commonly assigned meaning of aesthetic naturalism is that the proper study of art is nature.

> In this broad sense, artistic naturalism is simply the thesis that the artist's sole concern and function should be to observe closely and report clearly the character and behavior of his physical environment. Similarly to philosophical naturalism, aesthetic naturalism derives much of its importance from its denials and from the manner in which it consequently restricts and directs art. The artist should not seek any "hidden" reality or essence; he should not attempt to correct or complete nature by either idealizing or generalizing; he should not impose value judgments upon nature; and he should not concern himself with the selection of "beautiful" subjects that will yield "aesthetic pleasure." He is simply to dissect and describe what he finds around him.[3]

Given this dictionary definition of meaning assigned to aesthetic naturalism, it is evident that Veblen falls within its ambit in the sense that he focuses on the empirico-natural aspects of human aesthetic endeavor. That is, it is the verifiable waste, display, and status-enhancing and status-seeking aspects of art that he analyzed and satirized in *The Theory of the Leisure Class*. Individious display, honorific prowess, and exaltation of the archaic and inefficient are the points of his attack and all point to the defeat of the generic ends of life in the real world.

In any case, the moral and the aesthetic are interpreted by Veblen in a naturalistic way by reducing them to states of affairs within the empirical world. Science, ethics, and aesthetics are undergirded and reinforced by evolutionary naturalist epistemology; in short, ultimately grounded in it and its ontology after the Kantian and formal Christian elements in his early thought were relegated to the background. This enabled the bioso-

3. Ibid., 205–6.

cial and historico-anthropological genesis of the remainder to play a vital role in asserting their supremacy.

To Veblen the meaning of "value" is not personal or individual in a private or exclusive sense. For he possessed a naturalistic conception of the origin and status of values—valuation whether moral or aesthetic is a natural phenomenon and simply part of a naturalistic cosmos. Value does not belong to a subjective realm of judgment, but is the consequence of social interaction; in short, it is the result of cultural discourse and socioeconomic relations. The solipsistic and egoistic articulation of value narcissistically construed leads primarily to personal soliloquy and claims of "inner experience."[4] But in Veblen's view the function of "value" even in market exchange is not an isolable or solitary event, but is subject to public scrutiny and has social consequences.

Veblen did not advocate directly instrumentalizing art as a class weapon, but the significance of his aesthetics for the achievement of a more egalitarian society is evident. He was skeptical of Marxian views of class consciousness and also critical of claims regarding the uniqueness of American social structure. But Veblen never succumbed to reductionist demands for stripping art of its moral and cultural autonomy and simply reducing it to a tool of class warfare. Indeed, he probably believed that the aesthetic sense is partly hereditary; that is, a feeling for beauty is quasi-instinctive. In short, harmony of color, symmetry of form, and visual rhythms, although they are expanded and embellished by the life experience of the individual, partly precede it through their embedding in the biological template and perceptual apparatus.

The question is whether or not life, if lived according to Veblen's aesthetics, would be austere, somber, drab, and joyless? After all, evolutionary naturalism does not prescribe or sanction any particular set of aesthetics forms or values despite its apparent compatibility with, say, Winslow Homer's art. Could Veblen's aesthetic world include "a joyful celebration of workmanship where beauty comes simultaneously from designed-in ease and simplicity of production and execution of design?"[5]

4. This is a paraphrase of John Dewey, *Experience and Nature*, 171–73, but it is also characteristic of Veblen's thinking on the subject of value. Like Dewey's, Veblen's brand of evolutionary naturalism does not maintain the sole legitimacy of science as a mode of knowledge and as a frame of reference by denying cognitive meaning to metaphysics and ethics and arguing against any connection between science and common sense. For he believed both that the truth about nature was independent and yet accessible to us and, in any case, that inquiry was by its very nature cooperative. Veblen did not believe that knowledge can be adequately understood in abstraction from the processes that produce it and the consequences that flow from acting on it.

5. I am indebted to Bill Dugger for these comments.

To what extent could the underlying population share his practical aesthetics and achieve some reasonable degree of realizing them while celebrating an authentic craftsmanship that avoided both expensive and unnecessary ostentation and futility? No return to the archaic and nostalgic, nor honorific display of ability to pay would be necessary. Instead, the genuinely useful and enduring, the waste and want avoiding workmanship of life-cherishing humanity would not only prevail but dominate. Still the meaning and relevance of the life-enhancing qualities of Veblen's aesthetics need clarification, which they will now receive.

Enter Oscar Wilde

Since the claim is made that Veblen provided an aesthetics for the commoner, for purposes of illumination and illustration it is juxtaposed to claims that aesthetics are primarily for the cultural elite. Britisher Oscar Wilde (1854–1900) well illustrates the contrast between an aesthetics for the masses and an aesthetics for the elite. Wilde believed in the aesthetic incompetence of the masses to be sure, but attributed this in some measure to their cultural illiteracy resulting from lack of training and indoctrination; also, a consequence of lack of exposure to objects of beauty; in short, to the educational deficiencies and social environment of their upbringing.[6]

In his essay, "The Soul of Man under Socialism,"[7] written in 1891, Wilde delineated both mass aesthetic and cultural incompetence along the following lines: Public opinion is malinformed about art and literature, some of this is due to differing aptitudes and interests, part of it because journalists and the press pander to popular tastes which are rooted in cultural vulgarity. This situation is made worse by artistic recognition of popular authority and a corrupt descent into mass favoritism by artists themselves who abandon or lower their own standards to please public opinion and thus achieve little more than uniformity of type.

6. In the same vein, Pierre Bourdieu, French sociologist and aesthetician influenced by Veblen, writes that: "The denial of lower, coarse, vulgar, venal, servile—in a word, natural—enjoyment, which constitutes the sacred sphere of culture, implies an affirmation of the superiority of those who can be satisfied with the sublimated, refined, disinterested, gratuitous, distinguished pleasures forever closed to the profane. That is why art and cultural consumption are predisposed, consciously and deliberately or not, to fulfill a social function of legitimating social differences." Pierre Bourdieu, *Distinction: A Social Critique of the Judgment of Taste,* trans. Richard Nice (Cambridge: Harvard University Press, 1984), 7.

7. Oscar Wilde, "The Soul of Man under Socialism" (1891), in *"De Profundis and Other Writings,* with intro. Hesketh Pearson (New York: Penguin Books, 1986), 17–53.

But to Wilde's credit, he does not blame the masses and the press alone for debasement of cultural standards. He also attributes the deterioration of aesthetics to censorship by political despots of earlier times such as Louis the Fourteenth, king of France from 1660 to 1720, and religious leaders such as the renaissance popes. According to Wilde, such autocracies often lead to cultural despotism and this means the destruction of aesthetic individualism, the monotony of repetition, and the rigid conformity to rules.

Wilde castigates the renaissance papacy for the arbitrary and capricious nature of its treatment of art and artists. In short, although sometimes accomplishing positive achievements of a cultural or aesthetic nature, the popes often imposed their own values in irresponsible and dictatorial ways through their personal imposition of debased aesthetics. In any case, Wilde doubted the aesthetic value of much of the religious art and sculpture which dominated European cultural life in the late medieval and renaissance period; and the role of the Roman Catholic Church and its hierarchy and patronship were to blame. In summation of Wilde's message in "The Soul of Man under Socialism," both mass and elite are guilty of debasing human aesthetic existence. As he puts it, the Prince tyrannizes over the body, the pope tyrannizes over the soul, and the people tyrannize over body and soul alike![8] It is to Veblen's aesthetics of countervailence that we turn for the antidote to these pathologies.

Enter Veblen

His aesthetics have been subjected to scrutiny by aestheticians and social scientists from various doctrinal perspectives.[9] Yet even the most sys-

8. Ibid.

9. See esp. Rick Tilman and Robert Griffin, "The Aesthetics of Thorstein Veblen Revised," *Cultural Dynamics* 10 (November 1998): 325–40, and Tilman and Cyrill Pasterk, *The Intellectual Legacy of Thorstein Veblen*, ch. 5; cf. Robert L. Steiner and Joseph Weiss, "Veblen Revised in the Light of Counter-Snobbery," *Journal of Aesthetics and Art Criticism* 9 (March 1951): 263–68; J. E. Chamberlain, "Oscar Wilde and the Importance of Doing Nothing," *Hudson Review* 25 (Summer 1972): 194–218; Chandra Mukerji, "Artwork: Collection and Contemporary Culture," *American Journal of Sociology* 84 (September 1978): 348–65; Paul Dimaggio and Michael Useem, "Social Class and Arts Consumption: The Origins and Consequences of Class Differences in Exposure to the Arts in America," *Theory and Society* 5 (March 1978): 141–62; and Edward O. Laumann and James S. House, "Living Room Styles and Social Attributes: The Patterning of Material Artifacts in a Modern Urban Community," *Sociology and Social Research* 54 (April 1970): 321–42. Also, see Martha Banta, *Taylorized Lives* (Chicago: University of Chicago Press, 1993); John M. Jordan, *Machine Age Ideology: Social Engineering and American*

tematic efforts in this regard emphasize the ambiguity of his aesthetic ruminations. Given Veblen's stature as an American critical theorist and the debate over the nature of his aesthetics,[10] there is no need to justify further efforts to both clarify and extrapolate from them.

What follows is an outline of an aesthetics of countervailence that might be noninvidiously meaningful to the common person—that is, what Veblen usually referred to as "the underlying population," who lived in what he hoped might someday become "an ungraded commonwealth of masterless men," or an "industrial republic." Our focus will be on Veblen's view of human experience, especially that kind of experience that is commonly labeled "aesthetic," and the nature of the noninvidiously beautiful in his aesthetics. His distinction between pure and applied art forms, and the importance of this distinction to the common person, and the social basis and artistic consequences of the noninvidious as exemplified in the painting of the American artist Winslow Homer (1836–1910) is also emphasized. I will address these problematic areas in his social theory and aesthetics and articulate a noninvidious aesthetic of countervailence which can be logically derived from Veblen's social theory (and Homer's painting), or which is at least congruent with them.

Veblen does not object to beauty per se, only to the tendency for the leisure class to transform beauty, ostentatiously displayed, into prestige claims. As a practical matter, would there not be far less beauty in the world—in the form of art or music, say—if there were no leisure classes to cultivate and consume it as a way of drawing invidious distinctions? Would art not be deprived of its historic patrons and consumers? More would-be artists would inevitably regard art as a sure path to destitution—and this path would be even surer than it is now—and so they

---

Liberalism, 1911–1939 (Chapel Hill: University of North Carolina Press, 1994); and Eileen Boris, *Art and Labor: Ruskin, Morris and the Craftsman Ideal in America* (Philadelphia: Temple University Press, 1986).

10. See Joseph Dorfman, *Thorstein Veblen and His America,* 239, 497–98. "Veblen's work is drab, colorless, a world of measured matter-of-fact, the dullness of the slums, a perception resembling that of the Ash-can school of painters. D. W. Prall, who knew Veblen in California, told me once that Veblen had seemed to him unusually insensitive to the colors about him, indifferent, for instance, to the colors and paintings on the wall. There was a bleakness in Veblen's world, but it was a part of American experience that was unavowed in the official myth." Lewis Feuer, "Thorstein Veblen: The Metaphysics of the Interned Immigrant," *American Quarterly* 5 (Summer 1953): 99–112. Veblen's friend Max Handman once sent him a postcard with a picture of an austere classroom at the University of Salamanca. On it Handman wrote, "This fifteenth-century classroom would please you in its simplicity." Max Handman to Veblen, June 29, 1924, Veblen Collection, Carleton College Archives, Northfield, MN. Also, see Theodor Adorno, "Veblen's Attack on Culture," *Studies in Philosophy and Social Science* 9, no. 3 (1941): 389–413.

would pursue other lines of endeavor, such as commercial design. While theoretically Veblen might hold to an aesthetic ideal that includes a standard of pure beauty, in the absence of a leisure class and its patronage, his standard would leave less pure beauty to admire, except, perhaps, for that furnished by nature. Or would it?

Veblen understands that leisure-class perceptions of beauty are affected by considerations of status, invidious distinction, and pecuniary value; moreover, since other classes emulate the leisure classes in these matters (presumably accepting the leisure class's corrupted conception of beauty), how is Veblen's standard of "pure" or "generic" beauty to be socially and culturally identified and distinguished from its aesthetic competitors? Apparently, his answer is that we retain a general conception of pure beauty as a result of centuries of "habituation" and cultural conditioning. But, in societies resting more recently on "canons of pecuniary decency," habituation to leisure-class values and conceptions occurs, as the notion of pecuniary emulation implies. If this is so, then the standard of "pure" beauty may be philosophically sound but socially and culturally irrelevant, since it has no locus in a social order unless a leisure class exists to function as its cultural guardian.

Veblen says that some commodities and artifacts possess intrinsic beauty despite also serving as items of status enhancement. Such objects are held to be beautiful because—and we use Veblen's words here—they have "an antecedent utility as objects of beauty." But in what way does an object of beauty have utility as an object of beauty? What is its usefulness? If its utility lies in its quality of being admirable for its beauty, does this not make it an object of predatory desire, a trophy of someone's prowess and, eventually, an item whose *real* utility is in conspicuous display for purposes of status enhancement? In other words, if an object is intrinsically beautiful, does not this fact alone defeat or undermine this quality by making the item desirable mostly as an object of ceremonial ostentation?

In the end, if we want to assess Veblen's aesthetic values, we must ask: what would society look like if his values were to become the prevailing standards? Even if Veblen has aesthetic standards apart from sheer functionality, is it not likely that his society would be austere and somber? His relentless commitment to use values or serviceability provided him with a utilitarian standard by which to assess critically the operations of capitalism. But does it not also lock him into a drab and joyless society which he embraces as an alternative? All of the diversions that characterize our society are missing—and so is much of the zest and fun. It is a dull and spiritless society, resulting in large measure from Veblen's functionalist aesthetics which decolor human existence, *or so it has been argued.*

Since, in the final analysis, most of the foregoing objections to Veblen's aesthetics can be traced to his claim that the leisure class is the hegemonic producer, filter, and recipient of aesthetic values, it behooves his apologists to suggest a *means* as well as a *theory* of aesthetic countervailence. Of course, these must be logically extrapolated from the main thrust of his social theory in order to be authentically Veblenian. Perhaps a systematic extrapolation from *The Theory of the Leisure Class* will provide a more coherent and specific restatement of what he might have accomplished in his classic work, if he had not diverted his energies to lampooning the aesthetic nausea and ostentatious waste of turn-of-the-century predatory culture. In any case, a summation and evaluation of a Veblenian aesthetics that would be noninvidiously meaningful to the common person is in order, *so as to counter the criticisms just made.*

Veblen on the Nature of Aesthetic Experience

Veblen as a Darwinian naturalist had to regard experience, including aesthetic experience, as a process occurring in a natural environment, mediated by a collectively shared symbolic system. Of course, there is a great difference between attributing mediation to a universal, creative spiritual power and attributing it to the habits acquired by interacting with a natural/social environment. For Veblen, humanity is best described by the latter, creatures of habit, or "habit of mind" as he liked to put it. Thus, when he suggested that aesthetics goes beyond the physical art product and becomes a shared experience, that meant incorporating and transforming a biophysical and a cultural environment. Furthermore, Veblen believed, as did his mentor in aesthetics Immanuel Kant, that the experience we call "perception" deals with constructed objects and meanings, not mere brute sensations. Indeed Veblen's psychology of aesthetics was too sophisticated to rely on the unmediated, brute sense data of crude British empiricism or to retreat to the mysticism of classical rationalism, that is, unmediated, self-evident, intuited concepts, although he was not fully consistent in this respect as we shall later show.

British empiricism presents a retrospective associational psychology by arguing that art is our custom of "holding up the mirror to nature" and the immediate objects we sense in our past experiences. Yet, Kant and C. S. Peirce refute this appeal to past experience only and argue that perception is constructed by anticipation of possible future experience. Unfortunately, for Kant and Peirce possible experience acquires meaning only if there is a specific object to be experienced. This leaves unexplained the manner in which we develop just those meanings and values we are to assign to the object. The advance that William James, Noah Porter (under whom Veblen studied Kant at Yale), and Veblen make is to show that the feelings of immediate experience communicate series of relations and meanings which we select for the further transformation of our physical and cultural experience. Therefore, Veblen does find meaning in immediate sensation, yet not retrospectively as the British empiricists or in a loosely associational way as J. S. Mill, but as a process and prospectively as he learned from Porter and James.

As did his one-time Chicago colleague John Dewey, he severely criticized the sort of utilitarianism/behaviorism which sees the human organism as a passive or reactive stimulus-response mechanism. Rather, it proactively organizes experience into an integrated order. Veblen thus regarded the organism as the locus of life activity, an enhancer of experience and an explorer of its world.[11] Veblen knew that before the generically beautiful can be apprehended it has to be mauled and tampered with by the meddlesome activity of the mind, which is itself an evolutionary cultural product. His Darwinism thus compels him to recognize the human situation as inherently problematic: something ever to be inquired into and critically analyzed for purposes of reconstruction. Aesthetic inquiry, then, is an aspect of evolutionary natural selection and adaptation, although in Veblen's application of Darwin lies the recognition that there is much cultural furniture and debris floating around the social landscape that is neutral or indifferent as regards the evolution of the human species. The very existence of leisure-class aesthetics alone is massive evidence of this, since not all aesthetic endeavor has significance for adaptation or survival. Veblen thus agreed with Dewey that we select from a field of experience and meaning, rather than consciously pursuing an unrelenting course of progressive evolution.

Without specifically saying so, Veblen believed that a genuine aesthetics, that is, a noninvidiously meaningful aesthetics, would involve the artistic appropriation of the ideal possibilities for human life and a creative endeavor to live with those meanings and values that aid in the adaptation and survival of the species. At a minimum, the aesthetic ex-

11. On these points, see Thomas M. Alexander, *John Dewey's Theory of Art, Experience and Nature: The Horizons of Feeling* (Albany: SUNY Press, 1987), xviii. Although Alexander is discussing Dewey's theory of aesthetics, there are many points of convergence between Dewey and Veblen, including this one.

perience for him and for the common person should not require abstract training or cultural sophistication, but merely involve an appreciative awareness of immediate meanings. Probably, the underlying population needs little besides a rudimentary sense of color and form and the tridimensionality of space to appreciate artifacts of aesthetic significance. At least this is congruent with his estimate of the aesthetic capabilities of the industrial and farming classes who lacked formal aesthetic training, but were nevertheless potential consumers of the noninvidiously beautiful. However, the nearest he came to estimating the aesthetic potential of the common person was in discussing the arts and crafts societies. Here he touches only obliquely on the periphery of mass aptitudes for the understanding and appreciation of artistic beauty.

Veblen and the Applied Arts

Veblen did not attempt to rank the arts in any sort of hierarchy; instead, he merely distinguished between pure and applied art. This, in turn, is clearly reflected in his attitudes toward the arts and crafts societies, attitudes shared by other critics. Indeed, much criticism as well as praise has been leveled at William Morris, John Ruskin, and their disciples and followers in the arts and crafts societies in America and England. Eileen Boris has written:

> Thorstein Veblen ridiculed the unreadable books and uncomfortable chairs that his colleagues and their families made or bought during the city's handicraft renaissance. However commendable their protest against industrial ugliness, it nevertheless resulted in an archaic Romanticism, a sophisticated aestheticism which—in Veblen's mind—actually reinforced leisure-class values. Though Veblen understood that "the absolute dearth of beauty in the Philistine present forces [Morris' followers] to hark back to the past," he argued that no former mode of production could fulfill their "insistence on sensuous beauty of line and color and on visible serviceability" better than modern, machine technology. Handicraft merely encouraged honorific display.[12]

Veblen was not surprised that the arts and crafts industry showed itself to be more effective an institution in transmitting existing cultural values

---

12. Boris, *Art and Labor,* 51. Veblen was highly eclectic in his literary tastes. To illustrate the point, Becky Veblen commented that: "Thorstein quoted Ben Franklin a lot—But I can't think of an appropriate quote. 'The best was always good enuf?' 'What's worth doing at all is worth doing well?'" Becky Veblen to John P. Diggins, February 26, 1981.

and work habits than in challenging the status quo. In any case, he was undoubtedly skeptical of using preindustrial forms to build an alternative culture that had the power to break the hold of the dominant one. Nevertheless, Veblen recognized that the direction of modern design was toward functionalism and away from the inorganic ornamentation and clutter of the Victorian era. His social aesthetics probably also led him to believe that an elegant, efficient design could reduce self-centeredness by opening the mind to disinterestedness and altruism. He thus preferred a secularized aesthetic that addressed social as well as artistic problems, as opposed to the fetishism of art for art's sake which titillated the expensive appetites of the leisure class.

Veblen also knew that industrial design could be a weapon for or against class conflict and social ills, that is, it could be used to promote, conceal, or ameliorate them. While many other aestheticians either denied the need for the adoption of a class perspective, or avoided it altogether, he took delight in explaining the hegemonic role of the aesthetics of the American leisure class in our social order. Veblen, of course, believed also that the sexual and class divisions of social life must be challenged; in short, the sex role expectations, the sexual division of labor, and the class structure must somehow be altered, although he did not say how. In these respects, Morris, Ruskin, and the arts and crafts societies in both England and America thus ultimately failed in their quest for a new social order of nonalienated "craftspeople," as Veblen was painfully aware.

Pecuniary beauty was in conflict with what Veblen referred to as the "canon of serviceability." The exemplification of this conflict was found in the work of Ruskin and Morris, in their efforts at reviving craftsmanship in its hand-wrought, aesthetic forms. As Veblen put it:

> The visible imperfections of the hand-wrought goods, being honorific are accounted marks of superiority in point of beauty, or serviceability, or both. Hence has arisen that exaltation of the defective, of which John Ruskin and William Morris were such eager spokesmen in their time; and on this [they] ground their propaganda for a return to handicraft and household industry. So much of the work and speculations of this group of men as fairly as comes under the characterization here given would have been impossible at a time when the visibly more perfect goods were not cheaper.[13]

Veblen then focuses on the work of Morris's Kelmscott Press as regards binding materials, paper, type, illustrations, and quality of workmanship. From the perspective of brute serviceability, these books are found want-

13. Thorstein Veblen, *The Theory of the Leisure Class*, 161–62.

ing for they have been sacrificed on the altar of the canons of pecuniary decency. They have "excessive margins and uncut leaves, with bindings of a painstaking crudeness and elaborate ineptitude," and "argue ability on the part of the purchaser to consume freely, as well as ability to waste time and effort."[14] The reader senses Veblen's disapproval of the whole project, but he then writes tongue-in-cheek that he is only commenting as to the economic value of this tradition of aesthetic practice, not passing judgment on it as to its artistic worth. In short, he is not depreciating it, but simply trying to characterize its effect on the production and consumption of goods.[15] Veblen's own stress on the value of community serviceability, and his disapproval of conspicuous consumption, waste, and exemption from useful labor, however, make it evident that there can be no ultimate reconciliation in his thinking between the canons of pecuniary decency and those of serviceability, even in the realm of aesthetics. Nevertheless, he perceives the painstaking "instinct of workmanship" even in the sometimes farcical productions of Morris.

Veblen also presented his aesthetics in the *Journal of Political Economy,* which he edited from 1896 to 1904, through a review of *Chapters in the History of the Arts and Crafts Movement,* written by his friend and colleague Oscar Lovell Triggs, secretary of the Industrial Art League. Triggs's attempt to rehabilitate the antiquated handicraft ideal was described by Veblen as a product of romanticism, "archaism," and "lackadaisical aestheticism."[16] Both on business grounds and for reasons of economy the scope of the movement is limited "to those higher levels of consumption," where economy is not important and "the goods must be sufficiently expensive to preclude their use by the vulgar."[17] The industrial art method is too costly for modern business purposes and its products too expensive for general consumption. Veblen also writes that business exigencies "demand spurious goods"—that is, goods must cost less than they appear to; and "modern, that is to say, democratic, culture . . . requires low cost and a large, thoroughly standardized output of goods."[18] He reminds us that the machine process is here to stay, its progress is likely irreversible, and as a cultural phenomenon it will root out whatever institutional remnants stand in its way. Any part of the social fabric alien to its inexorable working is, at the most, "an exotic [without any] change of life beyond the hot-

14. Ibid., 163.
15. Ibid., 162.
16. Thorstein Veblen, *Essays in Our Changing Order,* 198.
17. Ibid., 196.
18. Ibid.

house shelter of decadent aestheticism."[19] He thus concludes that any vital reform movement in aesthetic ideals must fall in line with the exigencies of the machine technology, which is indispensable to modern culture. In Veblen's view it was only through reaping maximum economies of scale, which lower per-unit costs of art, especially applied art, to affordable levels, that the underlying population could have access to aesthetics at all. This in turn would accentuate the role of the machine process as the essential means for achieving this, although its impact on other aesthetic facets of society might be negative.

Veblen's Aesthetics: Tension between Kantian and Darwinian Influences

Veblen probably did not mean to suggest that there exists a secret, fixed, or complete essence of beauty which every aesthetic experience aims for, but only duplicates in an incomplete manner. Rather, given his pervasive Darwinian outlook, the timeliness of a species highly adaptable to a variety of environments is more congruent with Veblen's aesthetics than the *timelessness* of a platonic essence or a Kantian thing-in-itself. Unlike Kant, he did not believe that beauty or any other part of the natural order is inaccessible—or devoid of meaning for lack of a clear, experienceable object.

But this still does not adequately explain in what sense the "generically" beautiful is beautiful. Veblen does not believe in fixed essences or in unique binding categories standing objectively behind every aesthetic experience. Yet he fails to explain why gold or certain types of cloth are regarded by most cultures as intrinsically beautiful, although he asserts that this is so. Like other aestheticians of his time, Veblen had difficulty distinguishing between those values embedded by nature in the human psyche and those which were merely cultural effects. At times he seemed torn between explaining universal or generic categories such as beauty in cultural or environmentally determined ways or simply acquiescing in objects as intrinsically beautiful without further explanation. The tension in his thought between the Darwinian and Kantian influences meant that, while he could repudiate a metaphysics which sought to externalize a timeless present or beauty, he never abandoned the notion of the generically beautiful which is a consequence of evolutionary adaptive processes.

For Veblen, expressively organized works of art are developments of

19. Ibid., 197.

habits of perception connected with the world of cultural meanings. For this reason they are easily usurped by culturally hegemonic elements who would employ them for invidious ends. Art may thus serve primarily as a vehicle for status enhancement. Aesthetic endeavor can certainly become the shared experience of a community at a particular moment in time. But it can also remain little more than a reflection of differences of power, wealth, status, and leisure-class dissipation of scarce resources.

Veblen knew that art could help illumine experience with consummatory ideals that genuinely bind human beings together in shared pursuit of the good life and the good society. Yet he also knew the waste, futility, and frustration of the generic ends of life that could result from leisure-class aesthetics. Art could be expressive, creative, and emancipatory, but it could also be stultifying and inhibitory and dissipate into emulative rivalry. For Dewey, value, including aesthetic value, may be defined "as the situational functioning of intelligence relating ends and means toward consummatory, continuous experience,"[20] while Veblen complained that much aesthetic behavior is rigidly constrained and channeled into habitually consummatory experiences contaminated by emulatory intent and invidious achievement.

The Career and Art of Winslow Homer

Often regarded as the greatest U.S. artist of the nineteenth century, the New Englander Winslow Homer (1836–1910) first received critical acclaim for his paintings of the Civil War. Recognized as a painter and illustrator at an early age, Homer began to work in watercolor as well as oil after the war, creating often dramatic but naturalistic images of American life. Later in his life he painted a series of seascapes, and he depicted social life through the effects of outdoor light in nature on color and form.

Early in the Civil War, Homer went to the front as a war artist for *Harper's Weekly*. His first significant painting of the war, *Sharpshooter*, reveals his grasp of the war's essential modernity. Unlike conventional paintings of war, Homer shows a lone figure enabled by the new rifle technology (telescopic sight) to shoot at a distance and in isolation from his target. Homer's paintings of the war were also strongly egalitarian in character. Scenes of camp life emphatically illuminate the physical and psychological misfortune of the common soldier.

After the war, Homer surveyed the mixture of social classes, while also

20. As quoted in Alexander, *John Dewey's Theory of Art, Experience and Nature*, 28.

revealing his skill as an astute observer of social practices, including gender bias. Many of his paintings show women riding, swimming, or playing the popular game of croquet (in which women competed directly with men). Homer also portrayed employed women in occupations such as teaching and factory work. He spent about two years in a small fishing village near Newcastle-upon-Tyne, England, and the oil paintings and watercolors from this period mostly focused on the working women of the village, those whose lives were dominated by the tiresome and tedious gathering of bait, fixing of nets, hauling and cleaning of fish, in addition to their household chores. Much of this artwork is ideologically congruent with Veblen's attacks on the "barbarian status of women."

Winslow Homer as Noninvidious Aesthetician

At this point in the development of our thesis, it is essential to offer examples of noninvidious aesthetics to support our hypothesis that Veblen's aesthetics, if they are to be noninvidiously significant, can neither be reduced to mere criticism of ostentatious display nor simple functionalism where form follows function. Rather, painting must also become the medium through which the "generic ends of life" can be enriched. In the case of Homer, his work encompassed pastoral subjects such as two farm boys challenged by a bull in the field, the nostalgia invoked by schoolboys playing "snap the whip" in a spring field by a red school house, or local loafers on a cold day sitting around a stove in the country store enjoying the warmth and swapping yarns.[21] It also took in social realism of the sort that focuses on the sadness and tragedy of impoverished black women in the South picking cotton, the dark side of Homer's visions of women struggling for their new day, or a lone black sailor perilously adrift in a mastless, rudderless small boat surrounded by sharks in bloody waters. There is much in the dry, ironic pessimism which critics have viewed as Darwinian, which suggests that Homer's artistry, both his subjects and his portrayal of them, is noninvidious in the Veblenian sense—except, of course, that Veblen rarely bothered to extrapolate from his basic assumptions about leisure-class aesthetics to articulate adequately the meaning of noninvidious art forms.

21. See Michael Kimmelman, "Spinning American Myths without Sentimentality," *New York Times* (June 21, 1996): B1, B10. Also, Robert Jensen, *Marketing Modernism in Fin-de-Siécle Europe* (Princeton: Princeton University Press, 1994), for an analysis of the European art scene during Homer's career, most of which was spent in the United States, but which, nevertheless, shows European and even Japanese influence.

The noninvidious features of Homer's work also parallel Veblen's thought in that, unlike many French impressionists and neoclassicists, Homer did not paint for upper-class glorification or eminence of their personages, that is, for reasons of invidious comparisons. Of course, from a Veblenian perspective, there are two stages of aesthetic development for Homer: pre- and post-Darwinian. His earlier local-color paintings appear to stress humanist values important to Veblen. The later Darwinian dynamic scenes have, indeed, many striking parallels with the ideas and values so important to the socio-historic aesthetic views of Veblen. Homer's work, therefore, provides a Zeitgeist aid, lending a visual dimension to aesthetic interpretations of the founder of American institutional economics. It also might be construed as noninvidiously meaningful to the common person.

What is intriguing about the thematic values of Homer is his stress on evolution from the turbulent, sometimes shark-filled ocean waters; for example, his painting of the Jamaican on the dinghy correlates well with Veblen's emphasis on disciplined, objective adaptation for survival of the fittest. Homer's use of cliffs in other sea and landscapes appears to indicate cultural levels to which humanity can ascend in its evolution from the ocean depths. Alternatively, the cliffs may suggest future death and transition. As with Veblen, his artistic themes often center around the struggle for adaptation to variable changing and threatening conditions of nature. To these, of course, Veblen adds social conflict, that is, the conflict of classes and groups.

In the negative sense, there is little in Homer that would sanction ostentatious display or waste or emulatory processes aimed at status enhancement; nothing that would glorify honorific prowess or predatory behavior. In a positive way, there is much that sanctions peaceable and harmonious social relations, physical comfort, and domestic tranquility. However, Homer is referred to here not because he is a perfect exemplar of Veblenian values such as critical intelligence, altruism, and proficiency of workmanship. Rather, he illustrates social realism and idealism that are congruent with life in an "ungraded commonwealth of masterless men." Homer is also efficient in his use of subject and detail. Surely, much the same could be said about dozens of other artists, artistic mediums, and schools of aesthetic endeavor. Our point here is not to identify a "Veblenian" school of aestheticians as such, for one does not exist. Instead, we aim at locating the expression of aesthetic values logically derivable from or at least congruent with the main thrust of *The Theory of the Leisure Class*.

Aesthetics and Veblen's Evolutionary Naturalism

In view of the claim made throughout this study that Veblen's world-view was that of evolutionary naturalism, it is essential to explain the naturalism of his aesthetics as well. Very briefly, he viewed the common man as a product of his culture with the potential to engage in aesthetic experience. How did humankind get this way? Eliseo Vivas writes:

> A naturalistic theory of aesthetics should recognize at the start that the distinct modes of human experience which analysis can discover today have emerged in the process of human development out of relatively less differentiated activities of men in nature. The gradual differentiation that has taken place is only a matter of degree, coming about slowly as psychological aptitudes develop under the pressure of biological and cultural exigencies. We seek, however, to analyze the structure of each mode of experience considered as distinct from the others. But this means isolating conceptually what is not completely distinct in reality, and this in turn imputes to the conceptual isolate a uniqueness and self-sufficiency which in reality no one mode of experience can claim.[22]

Vivas's comments are in keeping with the evolutionary naturalistic tenor of Veblen's thought and with its holistic nature, which simply treats aesthetics as a human activity not too different from any other. Vivas then makes this inquiry:

> What . . . distinguishes the aesthetic from other modes of human experience? There are three other modes from which we must distinguish it—the moral or practical, the religious, and the scientific. Human activity is never purely one thing or the other, but it does tend to become one thing or the other, and to the extent to which it is human it is never anything else besides.[23]

Vivas then asks for clarification of the meaning of aesthetics and offers this lucid statement in the naturalist tradition as an answer:

> The denial of the legitimate presence of moral meanings in the aesthetic experience current in some contemporary theories, the anaemic character of art for art's sake, the assertion of the purity of art, the enactment of the intrinsic value of art into an absolute value, these notions and other doctrines

---

22. Eliseo Vivas in Yervant H. Krikorian, ed., *Naturalism and the Human Spirit*, 96.
23. Ibid.

and attitudes draw what specious justification they have from the implicit assumption of an "aesthetic man," devoid of any other interest whatever.[24]

Veblen's naturalistic perspective, his distrust of false dualisms, and his view of the continuity between experience and nature are well summed up here, although Vivas speaks more overtly in a Deweyan vein.

The Problematics of Veblen's Aesthetics Restated

In the final analysis, art, in order to be consistent with the rest of Veblen's social theory, must arise from the natural interaction of an organism with the world and from the cultural interaction of members in a society according to their different occupational roles and interests. Yet do residues of intrinsic and generic beauty, fixed and final essences, remain along with those of nineteenth-century racialism and instinctivism to occasionally mar his thought with their incongruence?

Veblen's early belief that races may have different innate psychological proclivities, and his use of the term "instinct" in an ambiguous and archaic manner, have often elicited negative comments from critics. His employment of the notion of transcultural or "generic" beauty seems equally incompatible with the radical theory of cultural and institutional determinism which he actually held. But these seeming incongruities in his thought do not require abandoning the larger theoretical matrix in which they are embedded.

Ultimately, however, it is the extent to which Veblen's aesthetics can be made noninvidiously meaningful to the common person upon which his reputation as a aesthetician must rest. The four questions raised earlier can only be rebutted if his aesthetics can be demonstrated to be noninvidiously meaningful or at least have this potential. To reiterate briefly, these questions were as follows: Since the leisure class largely controls the production and dissemination of aesthetics, if there were no leisure class to cultivate and consume them, would there be less beauty in the world for the plebeians to enjoy? If the leisure class were largely to disappear or lose its hegemony over aesthetics, where would Veblen's standards of pure or generic beauty be socially located, and would the lack of such a locus lead to a more general aesthetic impoverishment? If an object such as gold or fabric is intrinsically beautiful, as Veblen suggests, does this fact not defeat or undermine this very quality by making the item desirable

24. Ibid., 116.

primarily as an object of ostentation? In the final analysis, if we want to assess Veblen's aesthetic values as conventionally interpreted, what would society be like if they were given full rein? Would this not lead to a decoloring of human existence, that is, a culture of austerity and dreariness largely devoid of aesthetic stimulation?

If the answer to these questions as they are posed is largely in the affirmative, of what aesthetic value is Veblen's own analysis except as a largely negative criticism of leisure-class aesthetics? Even conceding the brilliance and subtlety of much of what he says regarding the pecuniary canons of beauty and waste, what is his own positive contribution to aesthetics, aside from ruminations essentially like those emanating from the functionalist school(s) to the effect that aesthetic value results when "form follows function"? Since Kantian notions of intrinsic or generic beauty are largely incompatible with most of the rest of his larger social theory, what is salvageable from his theory of aesthetics as it is articulated in *The Theory of the Leisure Class*?

All four of the objections to Veblen's aesthetics are less compelling if it can be shown that he was a proponent of noninvidiously significant aesthetics for the plebeian. First, such an aesthetics would shift the locus for producing, disseminating, and preserving art from the leisure class to the underlying population. The existing cultural elite, at a minimum, would have to share its privileged access to aesthetics with others. Second, instead of concentrating access to and control over pure or generic beauty, access would become more dispersed and fragmented and thus more egalitarian. Third, if intrinsic beauty were more widely accessed, it would less often be ostentatiously displayed for invidious purposes. At the very least, it would be significantly less effective for purposes of status enhancement. Finally, if the noninvidiously beautiful in cultural effects and social objects were more widely displayed, the barracks-like existence and assembly-line organization of employees, devoid of any serious aesthetic comforts, Veblen is accused of favoring, might be avoided. All of this is, of course, premised on the assumption that a noninvidiously significant aesthetic can be distilled from Veblen's writings, and that an aesthetic countervailence could find effective social and political contours and force.

Specifically, how do invidious and noninvidious aesthetics differ from each other? What values and forms distinguish the two? What topics, themes, issues, and problems set them off from each other? Hopefully, some suggestions have been made here which will help provide answers to these questions. But it should be remembered that Veblen wrote no formal essays on aesthetics of a noninvidious nature, so that much of what has been set forth here is based on extrapolation from what he actually said in analyzing the leisure class.

Conclusion

No systematic explication or analysis of Veblen's aesthetics appeared in English until 1996, probably because it was assumed he was merely lampooning or else making a Tayloresque assault on the aesthetic values and practices of the English and American leisure classes.[25] Are such two-dimensional literary efforts naïve? The truth is more complex, for Veblen is neither a mere satirist nor a functionalist, although there is much of both in his aesthetic ruminations. Instead, it is his view that transcultural values of a generic nature exist as a consequence of the human evolutionary process. Beauty is a distillation of these values that even the aesthetically illiterate can understand and enjoy.

Veblen's theory of aesthetics is *sui generis*. Had he been a more conventional thinker, Veblen might have subscribed outright to Kantian aesthetics or to pragmatic functionalism. But he found neither of these approaches fully adequate to the task at hand, for the aesthetic problem demanded erosion of the ideological hegemony of the leisure class. So he turned, instead, to a composite of functionality, serviceability, generic beauty, and critical theory. Veblen's attacks on the aesthetics of the leisure class from what appears to be merely a functionalist perspective were, in fact, rooted in an aesthetics that also used fullness of life and intrinsic beauty as reference points. He was not simply castigating leisure-class aesthetics for its reliance on the pecuniary canons of beauty as a means of status enhancement, nor did he consistently endorse waste avoidance as primary evidence of aesthetic value. Rather, he subtly explicated more objective criteria of beauty; and an important source of his aesthetics was Kant, not just Frederick Taylor, as critics have suggested.

Little is said here regarding Veblen's own specific aesthetic tastes. In any case, his encyclopedic knowledge of Western culture and his broad understanding of many non-Western cultures make it impossible to summarize adequately his preferences in art, literature, sculpture, architecture, and music.[26] In conclusion, we must, nevertheless, ask whether his

25. See Tilman and Pasterk, *The Intellectual Legacy of Thorstein Veblen*, ch. 5.
26. Sarah Gregory, a longtime Veblen friend, writes that "Veblen seemed to have read every thing, but he quoted most perhaps, from Chaucer, George Borrow [English author, 1803–1881], William Morris and Cervantes." Gregory to Joseph Dorfman, January 13, 1935, Dorfman Collection.

own artistic tastes were always compatible with the theory of aesthetics extracted here from his writings? To illustrate, according to Becky Veblen, his stepdaughter, one of Veblen's favorite poets was Rudyard Kipling (1865–1936), often regarded as the leading aesthete of British imperialism![27] Yet she also tells us that his favorite Kipling poem was not about the honorific prowess and ferocity of the British army marching to Pretoria to subdue the recalcitrant Boer. Rather it was Kipling's "The Sons of Martha" (1907), which exalts social obligation, devotion to duty, and altruism.[28] Succinctly put, the exploration of whether his own aesthetic tastes were consistent with a noninvidiously meaningful aesthetics culminates in our claim that even his admiration for Kipling rests on commitment to fulfillment of the generic ends of life.

Did Veblen aim at the democratizing and egalitarianizing of an elitist culture by substituting for its present aesthetics based on invidious display, waste, and status enhancement, a naturalistic aesthetics? He never said this explicitly, but it was in keeping with his outlook and intent. Thus the claim that his was an aesthetics for the commoners is ultimately congruent with and a logical outgrowth of his evolutionary naturalism.

27. On Kipling's "Imperialism," see Jay Parini, "Ruddy Good Show," *Chronicle of Higher Education*, February 24, 2006, B5.
28. See Becky Veblen, "Becky's Biography" (Long Version), no date, Thorstein Veblen Collection, Carleton College Archives, Northfield, Minnesota.

Chapter 7

Veblen and the Sociology of Control

This chapter portrays the role of social control in Veblen's analysis of the American version of social existence as it is lived under the thumb of the leisure class. He focused on the conservatizing impact of the invidious structure of everyday life where the underlying populace is under constant pressure to engage in competitive conformity to leisure-class canons of the good and the beautiful. Bill Dugger comments in this regard that:

> With every little physical coercion, the upper class imposes its will on all the lower strata. How can so many knees be bent so easily and so many strong women and men be conquered with so few shots being fired? This conquest is remarkable, on the same level as the Spanish conquest of the New World. Veblen shows how it was done so well, and without force. His exposition is so powerful that it becomes embarrassing. It is the embarrassment of the reader that makes Veblen's most cutting work feel like satire. He is satirical, but more importantly, he is seriously critical. He wants us to change, not laugh. Unfortunately, most of us just laugh.[1]

But Veblen persisted in his assault on the plutocratic, hierarchical, and ideologically legitimizing doctrines and practices of hegemonic elites. Perhaps what he was really trying to do was to change things, not simply be a comedian?

Introduction

Veblen is best known for his theory of status emulation, which is rooted in the practice of imitative rivalry, a process which among other things

1. Bill Dugger to Rick Tilman, October 2005.

increases the sense of self-worth through "unremitting demonstration of ability to pay." However, social scientists and, indeed, many Veblen scholars mostly focus on this aspect of his work on social bonding without adequate explanation of its large role in his sociology of control. This chapter thus aims at elucidating Veblen's sociology of control by drawing insights from contemporary sociologist Jack Gibbs's sociology of control[2] and using these insights to interpret Veblen's theory of status emulation, his theory of nationalism and patriotism, and his myth of the self-made man. The intent is to show that Veblen's work in these areas was a significant contribution to American intellectual history and practice regarding the exercise of hegemony in industrial society. It is to these objectives that I now turn.

Gibbs on Control as Sociology's Central Notion

Despite the fact that Gibbs's metatheoretical underpinnings come from logical positivism and ordinary language analysis that are no longer in vogue in philosophy or the social sciences, his claim that control should be made sociology's central notion is significant and provocative. However, there are several problems with this claim and others he makes that demand further exploration before his sociology of control is used to better understand Veblen.

It is Gibbs's claim that "maximum coherence in any scientific field requires a central notion,"[3] and it is his view that the other social studies should become scientific forms of inquiry. Strongly implied, also, is his conviction that economists, anthropologists, political scientists, and psychologists should be searching for central notions that will enable these disciplines to achieve the maximum of coherence which is the trademark of all scientific fields. Gibbs suggests that the other social and behavioral sciences should find a central common notion that will serve as a unifying agent in each discipline. Some economists have often argued that "equilibrium" or "disequilibrium" serves this function in their field although they cannot agree which of the two should have priority; political scientists relying on Thomas Hobbes, or in a more contemporary vein, Max Weber, see "power" as possessing the potential to play the central

2. See Jack Gibbs, "Control as Sociology's Central Notion," *Social Science Journal* 27 (January 1990): 1–27. Also, see Rick Tilman, "Control as Sociology's Central Notion: An Appraisal," *Social Science Journal* 27 (January 1990): 35–40.

3. Ibid., 23.

role; and, of course, Gibbs now recommends that "control" perform this function in sociology. He writes that:

> A central notion for all of the social and behavioral sciences could be essential for their integration. Hence, one crucial question concerning a central notion candidate for any particular social or behavioral science is this: to what extent could the notion also be central for each of the other social or behavioral sciences?[4]

However, of more immediate concern are the questions he asks regarding the causes and cures of fragmentation in the other social sciences. Probably the only point on which heterodox social scientists, particularly economists, can agree is that their own disciplines are fragmented due to methodological and doctrinal conflicts of various kinds. Given the nature of these conflicts should this fragmentation be judged negatively? Sociologists like Gibbs with a scientistic perspective are more disturbed by this apparent state of disorder than those who view the social sciences as inherently value laden and thus divisive in more than the trivial sense.

The Pervasiveness and Persistence of Values

Gibbs argues that sociology suffers from unprecedented fragmentation brought about by its "indescribably diverse subject matter" and "proliferation of sociological perspectives."[5] But he fails to acknowledge the extent to which this is due to its infection with diverse political and moral attitudes which obstruct his effort to promote unity through the adoption of control as sociology's central notion. To Gibbs's credit, he does concede that individual sociologists can inject whatever values they wish in their use of control as a central notion. For example, he indicates that the exercise of control may result in changes in human behavior that can be judged desirable or undesirable depending on the perspective of the individual sociologist. Nevertheless, he fails to see that the need for the adoption of control as sociology's central notion, so that it can play a unifying role, is itself a consequence of the political, moral, and aesthetic divisions that already exist. It is doubtful that the mere substitution of one central notion such as "power," or "status" for others, will accomplish much by way of bringing liberals, radicals, and conservatives closer together, although it might improve communication between them. Perhaps, in this sense, the

4. Ibid., 1.
5. Ibid., 2.

lasting value of Gibbs's effort will thus be to facilitate dialogue among social scientists who have ceased to communicate with each other.

But the question also remains unanswered as to whether "control" as Gibbs articulates it is politically and morally neutral or whether it is ideologically biased in certain directions. He believes that the unification of sociology cannot occur until there is common agreement on the terms of sociological discourse, which is to suggest that the reunification of sociology awaits a time when claims about values are replaced by claims about control. But this, too, only raises questions about the extent to which the introduction of new common notions will politicize, depoliticize, or facilitate "politically neutral" sociological discourse. Although Gibbs does not face this issue as squarely as he should, he apparently believes he has found a way around the impasse. But the appropriate reaction to his claim may, at worst, be a posture of suspended judgment and, at best, a counterclaim that he has endorsed a common notion for sociology that is politically value laden in more than a trivial sense.

Economics and the Other Social Sciences

As indicated, Gibbs's assertion, which for social scientists in other disciplines may be the linchpin of his argument, is his claim that "maximum coherence in any scientific field requires a central notion."[6] This claim apparently makes it imperative that political scientists, psychologists, anthropologists, and economists also realize that it is incumbent on them to find a central notion if their field is to achieve maximum coherence and thus be regarded as "scientific." Whether Gibbs is correct or not in his claims, the difficulties of achieving this in any of the social sciences are great. His imperative is an important, indeed audacious, one if a common desire for a modicum of unity exists in a field. But at the present time economics, for example, is broken into political, doctrinal, and methodological camps, and its reunification does not seem to have priority in most of them. Political and doctrinal issues and commitments so separate Marxists, institutionalists, social economists, post-Keynesians, neo-Keynesians, public-choice advocates, monetarists, and Austrians that it is hard to imagine the contending camps agreeing on a central notion as a unifying concept to achieve maximum coherence. The different theoretical groupings not only range from far left to far right along the political spectrum, but often have different methodological orientations. To fur-

6. Ibid., 1.

ther complicate Gibbs's vision of a unified discipline, where "central notions" are present at all in the subcultures of economics, they are often not hegemonic, but competitive ones.

Although neoclassical economists usually ignore the competing paradigms that exist, it is nevertheless true that thousands of dissident economists have formed professional organizations separate from the American Economic Association. These have their own journals with a sometimes bewildering variety of political and methodological orientations. With this situation in mind would, as Gibbs suggests, the adoption of a central notion in economics have a unifying effect on the discipline as a whole? Probably it would, but this is very unlikely to happen in the calculable future and, again, the reasons have to do with the divisions that exist. Apparently Gibbs thinks that unification through adoption of control as sociology's central notion is a real possibility in that field. What is it about sociology that makes it more amenable to such a change? One important reason is that, although Gibbs fails to mention it, sociology is more the ideological captive of the liberal-left than is economics. Although it is not a discipline that is doctrinally homogeneous by any means, it is clearly one where the ideological balance point on the political scales is significantly left of center. Among neoclassical economists that balance point may be right of center. Few American sociologists would call themselves politically conservative, but many mainline economists do.

What bearing does this have on the viability of Gibbs's claim, first that sociology has a special need for a central notion and, second, that "control" could be that notion? The answer seems to be that, since sociology is less divided on political and ideological grounds than economics, the chances of a central notion such as control being adopted and playing a unifying role in sociology are better than in economics. However, if the methodological and conceptual divisions that exist in sociology are also taken into account the chances of such changes occurring become slim, indeed.

Methodology, Epistemology, and Central Notions

Methodologically, economists range from crude "barefoot empiricists" with a rigid "metho-quant" orientation at one extreme to a priori rationalists of the Austrian variety at the other extreme. Interspersed at irregular intervals along the methodological spectrum are inductivists and deductivists of a more moderate stripe and conceptualists of various kinds. Traditional epistemologies can be identified including the correspondence theory of British empiricism, the coherence theory of continental rationalism, the consequentialist approach of American pragmatism, and an eclecticism that combines all three in varying ways and proportions. Even a sociology-of-knowledge perspective which defines epistemological and paradigmatic parameters as a function of the system of power relations is commonplace; and it points to hegemonic control in professional organizations as it affects the production and dissemination of concepts and theory in the social sciences. But Gibbs pays little heed to this epistemological fragmentation not only in his own discipline of sociology but in others as well.

As to central notions, what is their present status in economics? Can they be identified at all and, if so, how pervasive are they in the particular subgroups into which the discipline of economics is now fragmented? In neoclassical economics the central notion is equilibrium, but in the several varieties of heterodox economics it is disequilibrium and it has to compete with other notions. In post-Keynesianism, for example, disequilibrium is brought about by pressure exerted by large power blocs; in Austrian economics, disequilibrium is produced by government intervention; in Chicago-type monetarism, equilibrium or disequilibrium are induced by monetary policy, that is, the amount of money in circulation and the velocity with which it circulates. Should the reader now conclude that equilibrium/disequilibrium is the central notion of economics? Hardly, for the central notion in Marxian economics is the alienation of labor and the class conflict this induces; in institutional economics it is the dichotomy between business and industry writ large; in social economics, it is the organic, interconnected nature of the community and the economy. Even the most generic descriptions of economics and economic science such as the "art of social provisioning," or the "choice between alternative uses of scarce resources" are either too general and vague, or too narrow and trivial to be of much discursive value. What Gibbs recommends for sociology is an apparently utopian aspiration for economics because the idea that now, or in the calculable future, economists might agree on a central notion that would provide unity is unrealistic. The view that without a central notion no scientific field can aspire to maximum coherence may be true, but makes little difference to most heterodox economists. For often they have seen "truth" defined in the economics profession by the ideological spokesmen of corporate America, or by mathematical economists and econometricians who, believing in the scientific neutrality of their skills, seek to hone them to higher and higher levels of irrelevance.

Gibbs does not adequately stress the fact that the cognitive modes by which modern sociology interprets human behavior, as well as the perceptive modes by which it observes it, are themselves historical and cultural products. Thus, what seems to underlie his aspirations for a more scientific sociology is the belief that an objective perspective that is no mere cultural or doctrinal artifact relative to time and circumstance is an attainable goal. The residues of positivism are painfully obvious in this assumption and they are not very helpful. It will be tempting for many of his readers to conclude that what he really aims for is the restoration of a strict positivist conception of science as a body of objective, empirically verified or tested propositions designed for the purposes of prediction and control and clearly demarcated from other bodies of knowledge. Sociology is probably no more prone than economics to abuses such as trivialization, ideology mongering, and scientism, but Gibbs will find that many of his readers are particularly sensitive to the deficiencies of the latter.

Social theorists often focus on contemporary legitimation crises whether they concern society at large or internal conflicts within a discipline. Gibbs believes sociology is undergoing a legitimation crisis brought on by its own fragmentation. His recommendation is that sociologists adopt control as a central notion as part of a coping strategy to promote unity in the discipline and thus reduce the severity of the crisis of legitimation. Sociologists who did not find the simplistic optimism of classical positivism persuasive will have more difficulty in rejecting Gibbs's suggestion because the apparent simplicities of an earlier positivism are less pervasive in his work.[7] Nevertheless, the residues of scientism and reductionism that remain provide an obstacle to coping more effectively with the legitimation crisis in sociology. A more realistic approach in sociology, as in economics, will accept the division of these disciplines along conceptual, political, and even aesthetic lines. Gibbs's concern for the future of his discipline is genuine as is his effort to promote unity in it, but such efforts founder on the reef of his own positivistic premises.

However, despite all these caveats it is worth the effort to apply his "control as sociology's central notion" to Veblen in the hopes of elucidating his broader social theory. Indeed, a disproportionate amount of the scholarly writing on Veblen searches for, but cannot adequately identify, the locus of power in his analysis of various societies including the American. Perhaps this is because his sociology of power is better understood as a sociology of control since it does not focus on raw coercive force, authority, or even manipulation so much as on more subtle social psycho-

7. On this problem in the contemporary social sciences, cf. Paul Goldstene, *The Bittersweet Century: Speculations on Modern Science and American Democracy* (Novato: Chandler and Sharp Publishers, 1989), esp. 114–15.

logical insights, which explain social bonding and consensus, if such it be. It is to an exploration of this hypothesis that I now turn.

Veblen on Consumption and Status Emulation as Sociology of Control

Veblen's starting point in the theory of status emulation is "self-esteem." He argues that social factors are driven by self-esteem, that is, "the usual basis of self-respect is the respect accorded by one's neighbors," and he adds that "only individuals with an aberrant temperament can, in the long run, retain their self-esteem in the face of the disesteem of their fellows."[8] Status emulation has long been regarded as an important contribution of Veblen to social theory, and it means that individual and collective character structure evolve in relation to the models available for imitation.[9] Leaders are models of behavior and as such are sources of values.

Veblen argued in his theory of status emulation that social classes competitively emulate the social strata next above them. This is done, wittingly or not, in order to increase their own social status and thus their sense of self-worth and esteem. But emulatory consumption itself is not a value-neutral type of behavior; rather, it is an often intense form of competitive rivalry aimed at status enhancement and, also, at power aggrandizement. In any case, Veblen believed that those classes most able to conspicuously consume, waste, and avoid useful labor are most likely to command social honor and deference from other classes. More recently scholars working in the Veblenian tradition have fleshed out and updated his analysis. Briefly, what they have done is to place consumption into three categories: (1) sybaritic, (2) status enhancing or emulatory, and (3) functional or instrumental.

Sybaritic consumption, named after an ancient people, the Sybarites, was renowned for its emphasis on sensual delights and luxury, but it is

8. Veblen, *The Theory of the Leisure Class*, 30.

9. Phillipe Broda writes: "As far back as 1892, in answer to Spencer, Veblen explained that 'The existing system has not made, and does not tend to make, the industrious poor poorer as measured absolutely in means of livelihood; but it does tend to make them relatively poorer, in their own eyes, as measured in terms of comparative economic importance, and, curious as it may seem at first sight, that is what seems to count.' His judgment was definitive: capitalism could not promote general welfare because each person appraises his own situation only in relativistic terms, compared with the other people who belong to the community." Phillip Broda, "Veblen and Commons on Private Property: An Institutionalist Discussion around a Capitalist Foundation," in *Evolution and Path Dependence in Economic Ideas Past and Present*, ed. Pierre Garrouste and Stavros Ioannides (Cheltenham, UK: Edward Elgar, 2001), 94.

here noted for providing evidence of self-indulgence, not ability to conspicuously consume. Sybaritic consumption may be private or public, but it does not aim at status enhancement, and Veblen thus paid it little heed. The emulatory consumption to which he paid much attention aims at improving one's social position or at least the perception of it by others. The American economist Harvey Leibenstein breaks it into three forms, which he calls (1) "Veblen effects," where price is the main consideration; the higher the price paid for a commodity the more status it gives; (2) "bandwagon effects," where exceeding one's peers in ability to purchase and display conspicuously is not the intent; instead the aim is to achieve rough equality with them, that is, join them on the "bandwagon"; and (3) "snob effects," where emphasis is on the acquisition of culturally esoteric artifacts which only the cultural elite have the expertise or education to appreciate.[10] To this third category Robert Steiner and Joseph Weiss have added a fourth, which they call "counter-snobbery"; it signifies a reversion to more austere and simple tastes and lifestyles.[11] This also aims at status enhancement, but employs a different method for achieving it.

Clearly, Veblen preferred consumption that was functional, that is, instrumental in enhancing the life process of the individual and community. Such consumption aimed at the satisfaction of biological rather than status needs, at human serviceability, not self-indulgence or status display. Veblen was aware that most goods and services achieve some of both, but argued that it was often possible to differentiate the one purpose from the other as the primary function of the commodity. The achievement of the generic ends of life impersonally considered, or "fullness of life," as he sometimes put it, thus depended upon the community's ability to distinguish between functional and status-enhancing consumption.

In his satirical study of the leisure class and the underlying social strata that emulate it, he argued that conspicuous consumption, waste, and ostentatious avoidance of useful work were practices that enhanced social status. In short, Veblen contended that individual utilities could not be understood except in relation to the tastes of others. Individuals were emulating others to strengthen their own sense of self-worth by commanding more social esteem. The assumptions of atomistic individualism and consumer sovereignty, deemed valid by microeconomists, were thus shown to be specious on social-psychological grounds alone.

Although Veblen does not mention the ideas of his contemporary C. Lloyd Morgan's ideas regarding imitative behavior, those familiar with

10. Harvey Leibenstein, "Bandwagon, Snob and Veblen Effects in the Theory of Consumers' Demand," *Quarterly Journal of Economics* 64 (May 1950): 183–207.

11. Robert Steiner and Joseph Weiss, "Veblen Revised in the Light of Counter-Snobbery," *Journal of Aesthetics and Art Criticism* 9 (March 1951): 263–68.

his theory of status emulation will recall that social mimesis plays an important role in it. Indeed, interesting parallels exist between what the two men said regarding imitative traits and to what extent they are learned and conscious versus the degree to which they are innate and unconscious. In this regard Morgan wrote that:

> imitation, or what we are accustomed to regard as such, is an important factor in animal life, especially among gregarious animals, is scarcely open to question. But the biological and psychological conditions are not easy to understand. Some forms of imitation are often spoken of as instinctive; but some are voluntary, and under the guidance of intelligence. It is to the latter that the term "imitation," in its usual acceptation, would seem to be properly applicable. And the exact nature of the connection between this conscious and voluntary imitation and the involuntary instinctive process to which we apply the same term requires careful consideration.[12]

Veblen often wrote of "conspicuous consumption," "conspicuous waste," and "conspicuous exemption" from useful labor. Depending on the context in which he used the terms, "emulation" may be conscious or intentional, or it may involve habitual or instinctual behavior. Morgan thus posed what is an important theoretical issue for Veblen, namely, whether conspicuous consumption, in particular, is intention, motive, or instinct or, perhaps, depending on context, all three.

In Veblen's work status emulation is one of the mainsprings of human conduct, but it should be noted that emulation is not to be confused with mere mechanical imitation. Rather emulatory behavior is by its very nature rivalry, and the emulator aims at equaling or exceeding the emulated through competitive processes. William James penetratingly comments that:

> Emulation or Rivalry, a very intense instinct, especially rife with young children, or at least especially undisguised. Everyone knows it. Nine-tenths of the work of the world is done by it. We know that if we do not do the task someone else will do it and get the credit, so we do it. It has very little connection with sympathy, but rather more with pugnacity, which we proceed in turn to consider.[13]

Veblen, perhaps influenced by James, also links emulation with pugnacity or the "sporting instinct," which is indicative of his view that em-

---

12. C. Lloyd Morgan, *Habit and Instinct* (London: Edward Arnold, 1896), 166.
13. William James, *The Principles of Psychology* (New York: Henry Holt and Company, 1890), 2: 409.

ulation is closely related to the predatory impulses. The view, then, that emulative processes are benign is belied by Veblen's claim that the emulatory strain so characteristic of market economies is, instead, socially pathological. Predation, waste, and futility will be the likely outcome of any cultural process in which the intensification of status emulation is embedded.

Veblen's views on microeconomics, particularly the issue of consumer sovereignty, are perhaps better known than his views on any other subject. Indeed, the theory of status emulation and the vocabulary he developed to explain it are familiar to parts of the American public, who never read a word of *The Theory of the Leisure Class*. Nevertheless, his conservative and liberal critics, while sometimes amused at the satirical qualities of his theory and interested in its implications, ultimately found it an unconvincing explanation of consumer behavior. This is because they attributed more volition and rationality to consumers who are faced with choices between alternatives than Veblen, who, instead, stressed the role of "habits of mind" that induce status-enhancing consumption which provides sustenance to the self-image, particularly the sense of self-worth.

In his satirical study of the leisure class and the underlying social strata which emulate it, he argued that conspicuous consumption, conspicuous waste, and ostentatious avoidance of useful work were practices by which social status was enhanced. *The Theory of the Leisure Class* also contained many of Veblen's ideas regarding the nature of social value, ideas that were fundamentally different from those of the utilitarian tradition that formed the basis of the neoclassical approach to value. Veblen found unconvincing the conventional view of individuals as "globules of desire" attempting to maximize pleasure and minimize pain. He was also caustic in his repudiation of the moral agnosticism that he found pervasive in the neoclassical view of value as subjective preference measurable only by price. This had led to the claim that interpersonal comparisons of utility were impossible or irrelevant since consumer preferences were autonomously rooted in private, subjective states of mind.

Veblen's rejection of neoclassical value theory did not end in nihilism, however, for he outlined an alternative to both neoclassicism and classical Marxism. He wrote of the "generic ends" of life "impersonally considered" and of "fullness of life." In his evolutionary, that is, Darwinian mode of analysis, this implied the existence of some transcultural set of values. He found these values embedded in workmanship, parenthood, and intellectual curiosity, all of which flourish in a properly developing society. Veblen meant that proficiency of craftsmanship, altruism, and critical intelligence were dominant values and processes in communities that were developing in a noninvidious manner. However, in communities that were not so developing, these "instincts" or propensities would be contaminated by their opposites, the pecuniary and sporting traits which were invidious.

Veblen and Other Types of Social Emulation

Veblen's emphasis on the social nature of all human activity reminds us of John Dewey's challenge to a conservative critic to give one example of anything or anyone that exists and acts in isolation from all else. Social mimesis aside, Veblen's emphasis on the social nature of the stock of industrial knowledge and tools, his stress on the organic character of all economies, and his focus on the collective process by which goods and services are produced led him to egalitarian conclusions regarding the distribution of wealth and income. He did not believe it was possible to trace the origin of the value of a commodity to a particular input of land, labor, or capital. This principle of collective wealth is thus fundamental to Veblen's social view of the economy. As he put it:

> Technological knowledge is of the nature of a common stock, held and carried forward collectively by the community, which in this relation is to be conceived as a going concern. The state of the industrial arts is a fact of group life, not of individual or private initiative or innovation. It is an affair of the collectivity not a creative achievement of individuals working self-sufficiently in severalty or in isolation. In the main, the state of the industrial arts is always a heritage out of the past; it is always in process of change, perhaps, but the substantial body of it is knowledge that has come down from earlier generations. New elements of insight and proficiency are continually being added and worked into this common stock by the experience and initiative of the current generation, but such novel elements are always and everywhere slight and inconsequential in comparison with the body of technology that has been carried over from the past.[14]

In some of Veblen's writings on workmanship the contributing elements to social wealth are defined very broadly. In the production of this communal wealth, for instance, he said that, if several local groups are effectively to be represented in a single industrial system, conditions of peace must prevail among them. Community of language seems also to be necessary to the maintenance of such a system. Questions of language, com-

14. Veblen, *The Instinct of Workmanship*, 103.

munication, and cooperation are thus important to his theory of the social production of community wealth. Nor does he neglect "real people" by leaving them out of the process:

> Each successive move in advance, every new wrinkle of novelty, improvement, invention, adaptation, every further detail of workmanlike innovation, is of course made by individuals and comes out of the individual experience and initiative, since the generations of mankind live only as individuals. But each move so made is necessarily made by individuals immersed in the community and exposed to the discipline of group life as it runs in the community, since all life is necessarily group life . . . A new departure is always and necessarily an improvement on or alteration in that state of the industrial arts that is already in the keeping of the group at large.[15]

Veblen went on to say:

> Yet it might be argued that each concrete article of "capital goods" was the product of some one man's labor, and, as such, its productivity, when put to use, was the indirect, ulterior, deferred productiveness of the maker's labor. But the maker's productivity in the case was but a function of the immaterial technological equipment at his command, and that in its turn was the slow distillate of the community's time-long experience and initiative.[16]

The principle of collective social wealth is thus the focal point of his critique of the marginal productivity theory of factors of production and plays a key role in the development of his idea regarding "unearned income." Production is basically a social process, especially where the material product is produced as an integral part of the labor process, but more fundamentally it is due to the collective genesis of knowledge in the industrial arts, and the social origins of knowledge, language, and communication. Unfortunately, the vested interests, Veblen claimed, gain the advantage of the community's social wealth through predation, sexism, war, status emulation, and inheritance.

As Veblen suggests in *The Theory of the Leisure Class*, in the primitive phases of social development the efficiency of the individual mostly enhances the life of the group. What emulation of an economic sort there is between the members of such a group will be largely emulation in industrial serviceability. At the same time the motivation for emulation is not

15. Ibid., 103–4.
16. Veblen, "On the Nature of Capital, I: The Productivity of Capital Goods," *Quarterly Journal of Economics* 22 (August 1908): 339.

strong, nor is the scope for emulation very wide. These provocative comments indicate that social emulatory processes need not be socially or individually destructive. Indeed, Veblen clearly felt that insofar as critical intelligence (idle or not-so-idle curiosity), altruism (parental bent), and proficiency of craftsmanship (instinct of workmanship) could be promoted, intensified, or induced by emulation this was desirable. But throughout much of his career he was pessimistic about projections and prognostications regarding the future of an industrial society that rested on unwarranted assumptions about the plasticity and malleability of social and cultural institutions. The human psyche might contain unlimited capacity for change, yet institutional residuals might drastically limit emancipation of human potential. To illustrate, emulatory processes rooted in pecuniary motives which aim at status enhancement and, ultimately, at strengthening the self-image will contaminate what Veblen called "the instinct of workmanship." This is true because authentic craftsmanship and status emulation represent alternative, and to a large degree mutually exclusive, sources of individual self-esteem.[17]

The Economics of Status Emulation

What precisely are the mechanics of status emulation in Veblen's theory? He describes consumption mimesis as follows:

In a community where class distinctions and class exemptions run chiefly on pecuniary grounds, wasteful conventions spread with great facility through the body of the population by force of the emulative imitation of upper class usage by the lower pecuniary classes.[18]

An important consequence of institutional and economic change can be seen in consumers' behavior. Veblen noted that distinction in consumption or "conspicuous consumption"[19] is a means of gaining reputability which characterizes the majority of wealthy individuals. In the new economic state the "unproductive consumption of goods is honorable, primarily as a mark of prowess and a perquisite of human dignity; secondarily it becomes substantially honorable in itself, especially the consumption of the more desirable things."[20] Veblen attributed conspicuous

17. Veblen, *The Instinct of Workmanship*, ch. 2.
18. Veblen, *Imperial Germany*, 147.
19. Veblen, *The Theory of the Leisure Class*, 75.
20. Ibid., 69.

consumption to institutional as well as technological changes. He argued that conspicuous consumption goods "exceed fifty percent of the current product,"[21] which means entrepreneurial opportunities develop in that "there have grown up an appreciable number of special occupations devoted to the technical needs of reputable spending. The technology of wasteful consumption is large and elaborate and its achievements are among the monuments of human initiative and endeavor."[22]

Veblen stressed that conspicuous consumption is a result of the industrial system where the rate of productivity has increased tremendously as well as the volume of production and wealth.[23] Such behavioral traits are both the effect of emulation and imitation of the wealthiest citizens and countries as well as increased productivity and wealth. The introduction of such "wasteful consumption" thus both extends and enlarges the profit opportunities of entrepreneurs. Thus, as described by Veblen, by increasing production and the rate of wealth in some classes, their proclivities for conspicuous consumption are increased as well as the variety of luxury goods satisfying such preferences.[24] As a result, profitable opportunities in product innovation and entrepreneurial activities are also extended.[25]

Veblen felt that in the modern market system, the role of entrepreneurs is very important for success in the promotion of the various goods and services offered and in the augmentation of their "vendibility." In such a case, advertising and "salesmanship" will bypass "workmanship" requirements in the strategies of the entrepreneur.[26]

The increased profit opportunities in conspicuous consumption goods comes from two consumption preferences: (a) "the superior gratification derived from the use and contemplation of costly and supposedly beautiful products";[27] and (b) the goods must be accompanied by expensiveness as "the canon of expensiveness also affects our tastes in such way as to inextricably blend the marks of expensiveness, in our appreciation, with the beautiful features of the object."[28]

In short, Veblen describes a type of consumption that provides guidance to entrepreneurs. He holds that with a "pecuniary habit of thought"

21. Ibid., 35, 137.
22. Ibid., 35–36.
23. Ibid., 110–11.
24. Ibid., 110–12.
25. Veblen, *Imperial Germany*, 96, 35–36.
26. Ibid., 214.
27. Veblen, *The Theory of the Leisure Class*, 128.
28. Ibid., 130.

and unequal wealth distribution, a stratification of society would take place,[29] where the consumers of the upper and middle classes develop "the habit of disapproving cheap things as being intrinsically dishonorable or unworthy because they are cheap."[30] The study of the economics of status emulation thus lies at the root of the sociology of control along with the role and consequences of nationalism and patriotism to which we now turn.

Patriotism, Nationalism, and the Great War

Despite his intermittent pacifist bias, Veblen's attitude toward American entry into the Great War has not been easy to decipher. His biographer, Joseph Dorfman, ambiguously stated that "on the basis of his books alone, Veblen's position regarding the war might seem rather difficult to determine, but the entrance of the United States on the Entente side seems to have pleased him."[31] Veblen worked for government agencies during the war and even drew up a proposal for an antisubmarine defense system. Yet he was extremely critical of government persecution of the Industrial Workers of the World and of the postwar settlement.

Veblen's thinking during the war may seem ambiguous due to the promotion of his book on *Imperial Germany and the Industrial Revolution* as effective anti-German propaganda at the same time the post office was banning it from the mails as pro-German! However, this ambiguity is more illusory than real when the book is placed within the corpus of Veblen's writings and situated in the context of the development of his thought. Veblen liked to distinguish between republics such as the United States, Great Britain, and France and the dynastic systems of Hohenzollern, Germany; Hapsburg, Austria; and Romanov, Russia. Clearly he preferred the republics to the dynasties, which, no doubt, helps explain his support for the Allies after America entered the war. However, his analysis differed both from revolutionary socialists like Lenin, who conflated both

29. Veblen, *Imperial Germany*, 137.
30. Veblen, *The Theory of the Leisure Class*, 155.
31. Joseph Dorfman, *Thorstein Veblen and His America*, 371–72. Also, see 335. Cf. Carol S. Gruber, *Mars and Minerva: World War I and the Uses of the Higher Learning in America* (Baton Rouge: Louisiana State University Press, 1975), 60, with Isador Lubin, "Recollections of Veblen," in Carlton C. Qualey, ed., *Thorstein Veblen* (New York: Columbia University Press, 1968), 142–43. Bruce Clayton writes that "nearly everyone but Veblen . . . had been engaged in war propaganda." See his *Forgotten Prophet: The Life of Randolph Bourne*, 246.

kinds of political systems and called for their revolutionary overthrow and liberals like George H. Mead who asked their fellow citizens to close ranks and make the world safe for democracy. For Veblen perceived the war itself as caused by the larger system of imperialism of which the republics and dynasties were an integral part; yet he remained cognizant of the important differences between the two. It was evident that he believed the long-term chances for peace depended on dismantling the entire system that brought on the war, not simply on an Allied defeat of the Central Powers.

Veblen was too sophisticated in affairs of state when the Great War began in 1914 to swallow the propaganda of either side with their allegations of primal innocence versus brute aggression. He came to believe, however, as the war progressed that a lasting peace would only be possible if the dynastic states involved in World War I, such as Germany, Austria, Turkey, Japan, and Czarist Russia, were dispossessed of their autocracies and if structural changes also occurred in the liberal democracies. He argued that "the emulative spirit that comes under the head of patriotism"[32] must be subjugated through disallowing the property claims of citizens abroad, eliminating protective tariffs, and ending imperialist control of weaker communities by the stronger in the supposed interest of domestic industry and commerce. Otherwise there would be an inevitable resurgence of militarism, war, and aggression. But radicals, particularly the pacifist left, stressed the moral and political imperatives of international disarmament and the need for a flattened or leveled order to replace the present, highly stratified, one. The radical critique pointed toward a "powerless" international environment where countries would no longer have the will or the means to impose their values on less powerful nations; this critique probably converged with Veblen's own long-run perspective, although not always with his short-term prescriptions.

Generally, radicals have claimed, and Veblen certainly agreed, that liberals and conservatives alike were not cognizant of the degree to which their own values and attitudes were shaped by the dominant classes. Consequently, the radical critique viewed foreign policy options as bound by the dominant class's interests. Much of Veblen's work on politics among nations can best be interpreted as an appeal to his readers, however satirical or subtle, to rid themselves of the hegemonic preconceptions of the ruling classes and their statesmen. Veblen strongly disliked the arms race that produced the Great War, was an acidic critic of the predatory war aims of the bellicose nations, and was a principled opponent of patriotism

32. Veblen, *An Inquiry into the Nature of Peace*, 33.

since he regarded it as a form of chauvinistic nationalism. Yet he clearly believed that it mattered who won the war and favored the Allied cause.

Veblen's *Inquiry into the Nature of Peace* attracted attention from critics, but most of the criticism it received did not focus on his analysis of politics among nations or on his prescriptions in the book for policy change after World War I. Instead, the focus of criticism was his analysis of patriotism. Liberals believed Veblen had distorted the real meaning and significance of patriotism. To both liberals and conservatives, Veblen ignored the mostly positive attributes of the patriot in his eagerness to undermine the ruling class and the support it received for its war aims and duel for empire. Had he perceived patriotism more objectively, they claimed, he might have observed that it was motivated by altruism and other-regardingness, which was a main source of sustenance for whatever degree of human solidarity and community existed. Veblen's focus on the hatred, bigotry, and irrationalism of the patriot was misguided in that it made him ignore the positive qualities of patriotism that was the ultimate social-bonding agent holding the nation together, or so both liberals and conservatives held.

Radicals were more sympathetic to Veblen's perspective because he put in sophisticated conceptual form what many of them had already come to believe; namely, that patriotism was a form of false consciousness that made the common man susceptible to manipulation for warlike aims and imperialistic aggrandizement. However, it is interesting to note that Veblen was criticized by radicals for his failure to understand the liberating force of patriotism as a component of nationalism after World War I. This interpretation of Veblen thus indicted him for ignoring the emancipatory impact of nationalism when it becomes a fulcrum of anti-imperialist radicalism.

Veblen defined patriotism as a "sense of partisan solidarity in respect of prestige" and "a sense of undivided joint interest in a collective body of prestige."[33] However, for Veblen, war only epitomizes the failure of the national community to overcome its self-regarding and destructive "instincts," which are "sport" and "predation." War thus means that the other-regarding "instincts" of parenthood, idle curiosity, and workmanship have become temporarily "contaminated" or at least submerged. Veblen could never forget the primitive tribes that played so prominent a role in his historical anthropology with their ethic of "live and let live." It was this peaceful type-form of humanity and communities constructed in the pacifist image that international relations now lacked.

Obviously, Veblen could not acquiesce in rationalization of war as an

33. Ibid., 31.

essential part of the process of social unification were he confronted with it. To Veblen, war was ordinarily the result of the atavistic thrust of "imbecile institutions." Thus he attributed the militarization of industrial society not simply to the inherent nature of capitalist institutions as such, but also to the survival of archaic feudal attitudes, structures, and ideologies. Such anachronisms remained in Imperial Germany and Japan and in the social structure of England.

The effectiveness with which these advanced industrial societies adapted themselves to war was due, he thought, to the continuing habit of deference to and emulation of a bellicose ruling class. Indeed, Veblen sharply contrasted the parochial politico-military dissension of the age with its cosmopolitan technological knowledge when he wrote that "into this cultural and technological system of the modern world the patriotic spirit fits like dust in the eyes and sand in the bearings."[34]

Veblen had no illusions about the nature of American society including both its alleged democratic nature and the right of dissenters to criticize the involvement in war. Indeed, while his critic Mead proclaimed the achievement of democracy in America and the right of dissenters within limits to criticize the war, Veblen critically observed the persecution of the Industrial Workers of the World (the "Wobblies") and the beginnings of the anticommunist hysteria that was shortly to engulf the country. Nothing could match Veblen's biting indictments of the American mass mind. The war had shocked Americans into a "dementia praecox." The power of organized religion, the "puerile credulities," the persecutions and deportations of advocates of "constructive sedition," the baiting of Wobblies and conscientious objectors, and the criticizing of pacifists for "excessive sanity," all signified mental derangement.[35]

Veblen held that the problems of power in international politics could be more rapidly ameliorated were it not for the vested interests of the governing classes that created them; for weapons systems were often not created to serve real security needs, but to satisfy the interests of militarists and their industrial and financial supporters. Nevertheless, he believed that massive pressure from below might eliminate the whole tangled web of international suspicion and rivalry. This last scenario was made problematic, however, by the hold that the vested interests had on the underlying population in terms of shared values coagulated with patriotic gore. Veblen's view of the darker side of human institutions made it impossi-

34. Ibid., 40. In this same vein, he wrote that "national frontiers are industrial barriers"; ibid., 264.

35. See "Dementia Praecox," in *Essays in Our Changing Order* (New York: Augustus M. Kelley, 1964), 423–36.

ble for him to ignore the likelihood of the persistence of atavistic continuities such as war.

However, Veblen's views on the subject were more complex than this
because there was also a strong vein of pacifism in his thought, often expressed in his admiration for the practice of "live and let live" among
primitive peoples. Consequently, in order to obtain a balanced view of
Veblen's attitudes toward force as a method of conflict resolution, his occasional praise of pacifism must be weighed against his sympathy for the
Russian Revolution. In this last respect, Veblen's outlook bore at least a
family resemblance to neo-Marxist theories of the social dynamics of international relations. In fact, in viewing imperialistic capitalism and its
ideology, Veblen wrote:

> The quest of profits leads to a predatory national policy. The resulting large
> fortunes call for a massive government apparatus to secure the accumula-
> tions, on the one hand, and for large and conspicuous opportunities to
> spend the resulting income, on the other hand; which means a militant, co-
> ercive home administration and something in the way of imperial court
> life—a dynastic foundation of honor and courtly bureau of ceremonial
> amenities. Such an ideal is not simply a moralist's daydream; it is a sound
> business proposition, in that it lies on the line of policy along which the busi-
> ness interests are moving in their behalf.[36]

Thus, as early as 1904, Veblen rejected his Chicago colleague George H.
Mead's liberalism by implicitly denying the efficacy of any conflict-reso-
lution mechanism or policy that fails to recognize a primary source of
modern warfare in capitalist property institutions and the cultural values
they induce.

Of course, if patriotism has indestructible roots in the human psyche,
there is little hope for permanent peace and Veblen's pessimism is fully
justified. On the other hand, if generic man is sufficiently plastic and malleable, then perpetual peace is at least conceivable. There is no clear or
consistent portrayal of human nature in *The Nature of Peace*. Nevertheless,
Veblen's analysis was more far-reaching in its explanatory claims and in
its historical prognosis than Mead's, because he traced the psychological
and moral impetus for war to its roots in economic institutions and the
class system, instead of letting them remain structurally unanchored as
was so often the case with Mead.[37]

36. Veblen, *The Theory of Business Enterprise*, 188–89.
37. Cf. Veblen, *An Inquiry into the Nature of Peace*, and George H. Mead, review of
"An Inquiry into the Nature of Peace," by Thorstein Veblen, *Journal of Political Economy* 26 (June 1918): 752–62.

Veblen's social theory is ultimately rooted in a dichotomous view of class while for Mead considerations of class, power, and authority were assimilated into or submerged under "role." Veblen's view of international politics[38] was thus class based and his social psychology of the common man, including patriotism, linked with the class structure. Veblen's social theory thus forced him to the conclusion that diplomats and statesmen are greatly constrained by the class structure and the dominant ideology, which, no doubt, explained his pessimism regarding the future of mankind and the likelihood of perpetual peace. As he once put it, "history records more frequently and more spectacular instances of the triumph of imbecile institutions than those whose insight saved themselves alive out of a desperately precarious institutional situation, such, for instance, as now faces the people of Christendom."[39] These institutions had, of course, also fostered the myths of radical individualism that flourished on the American frontier.

The Myth of the Self-Made Man

According to the influential historian Frederick Jackson Turner, Veblen's contemporary, this new nation had become an international model for all mankind. The United States, like the frontier from which it emerged, was a study in idealism. In it, mankind could learn about democracy, self-reliance, individualism, progress as a byproduct of a proper mix of nature and technology, and, even more important, a confident mastery of new challenges. Turner's genius was his poetic ability to formulate a mythical view of America that translated a potentially mundane, factual account of American development into a dramatic saga of capitalistic advance and adventure. Turner became this nation's most famous historian because he presented his countrymen with a positive image of themselves that served to mask the growing ugliness of American reality—class conflict, racial animosity, fear of the foreigner, and a growing gulf between the rich and the poor. Paul Tillich asserted that "all historical writing which is to be taken seriously must have in it a mythical ele-

---

38. For analysis of Veblen's view of international relations, see Sondra Herman, *Eleven against War: Studies in American Internationalist Thought, 1898–1921* (Stanford, CA: Hoover Institution Press, 1969), 150–78. Also of interest in this regard are Carl H. Friedrich, *Inevitable Peace* (New York: Greenwood Press, 1969), 219–25; Michael Howard, *The Causes of War and Other Issues* (London: Unwin Paperbacks, 1983), 66–67; Irving Louis Horowitz, *War and Peace in Contemporary Social and Philosophical Theory*, 2d ed. (New York: Humanities Press, 1973), 10–11.

39. Veblen, *The Instinct of Workmanship*, 25.

ment."[40] Turner's history of the United States was a masterpiece of American folklore, rather than an accurate portrayal of American realities.

Veblen stated repeatedly that the incentives of self-help and cupidity of the "English-speaking colonial enterprise" resulted in a lack of any "spirit of community interest in dealing with any of their material concerns."[41] In his view, it was no accident that the farmer placed little emphasis on social solidarity and harmony or that he lacked any organic sense of community. Both the system of exchange relations and the cultural ethos that underlay economic individualism worked against community solidarity. Veblen commented in the same vein that the farmers' "self-help and cupidity have left them at the mercy of any organization that is capable of mass action and a steady purpose."[42] The net result was that their possessive individualism put their own necks under the heel of both local businessmen and large absentee owners. Lurking behind the local front of small businessmen in the country town were those massive business interests whose main function in American society was to get something for nothing. As Veblen put it:

> In a way the country towns have in an appreciable degree fallen into the position of toll-gate keepers for the distribution of goods and collection of customs for the large absentee owners of the business. Grocers, hardware dealers, meat-markets, druggists, shoe shops are more and more extensively falling into the position of local distributors for jobbing houses and manufacturers.[43]

Although Veblen often attacked absentee ownership of commerce and industry, he had less to say regarding absentee landownership than he did about the independent farmer with all of his alleged greed. Nevertheless, the rural population through the sustaining myths of the self-made man played an important role in his sociology of control.

Cultural Lag, Class, and Ideological Control

In contrast to Marx, Veblen gave considerably more credit to the feudal holdovers of subordination and emulation in working-class consciousness. Sidney Plotkin comments that:

40. Paul Tillich, *The Interpretation of History*, 36. Also, see Vernon Mattson and Rick Tilman, "Thorstein Veblen, Frederick Jackson Turner and the American Experience," *Journal of Economic Issues* 21 (March 1987): 219–35.
41. Veblen, *Absentee Ownership*, 132.
42. Ibid., 133.
43. Ibid., 152.

Veblen's emphasis on ideological control is, then, not a peculiar hallmark of his analysis of capitalism anymore than it is the case that predatory systems lived by force alone. Force and ideology combined were politically important long before modern business turned modern advertising and mass media to the purposes of social control. Ruling classes personalized not only their own higher claims to honor; they personalized their relations with the working classes, reinforcing in the latter a steady and assured sense of inferiority and class indignity, identifying industrial work in the minds of its practitioners and dishonor, lower intelligence, and brutishness. The larger social meaning of such notions for the evolution of social and political consciousness cannot be underestimated. Deep inside the social psychological mechanisms of class societies is a tendency to render habitual the idea that the great rule not because they are rich, they are rich because they are great, just as the poor must labor not because they lack income, they lack income because they are weak. In Veblen's notion of class consciousness, the poor, the passive, inert, without requisite personal force in a world that demands predatory aggressiveness as its chief human virtue: contemporary debates about welfare resound with such notions. The permeation of a consciousness of differential status is buried but not dead inside the objective formation of market-based class.[44]

Plotkin concludes that

> Veblen's studies of pre-capitalist societies make clear the dual roles of religion and patriotism as ideological bulwarks of the first importance. Both appeal to the spirit of self-sacrifice, loyalty, and self-abnegation, a willingness to give of one's self—indeed to give one's life—on behalf of higher purpose. Patriotism pulls us toward a deeply sentimental love of country. The sense of place helps to submerge and compensate for indignities of class. An outgrowth of humanity's generic social sense of collective interdependence, patriotism urges us toward intense feelings of "partisan solidarity." Patriotism historicizes social and political consciousness.[45]

Plotkin's astute comments add a radical flavor to the analytics of Jack Gibbs to whose sociology of control we now return for enhancement of our understanding of Veblen.

Conclusion

Both methodologically and conceptually, Veblen and Gibbs are poles apart. Yet the contrast between the two and other leading social theorists

44. Sidney Plotkin, "Illiberal Habits: Veblen's Theory of Power in Pre-Capitalist Cultures" (Paper presented at ITVA meetings, New School, New York, May 2002), 20.
45. Ibid., 26.

of the twentieth century is theoretically illuminating. Gibbs's claim that "control" should be the central notion in sociology may be mistaken, but it cannot be ignored. But why "control" instead of power, class, status, hegemony, or to bring us back to Veblen, emulation, imitation, or social mimesis? In short, what arguments can be advanced in favor of Veblen's theory of status emulation as the central notion of sociology? What advantages or disadvantages might it have versus the other terms mentioned above which other social scientists with good reason might put forward as candidates for the position of central notion?

Although Veblen's social science makes other contributions than his theory of status emulation, surely that is the central focus of his social theory, and any lasting contributions he made to sociology are based on it. Yet it is important not to place more weight on it than it can bear. The British sociologist Colin Campbell has recently written a critique of it as it relates to consumption, arguing that it is badly flawed due to its vagueness and imprecision and the conflicting forms in which it is stated at many points in his work. He claims that it would be difficult or impossible to test it because its deficiencies make it difficult to cast in the form of empirically testable hypotheses. Although Campbell overstates the case against Veblen's theory of status emulation, he cautions against accepting the theory as Veblen originally formulated it more than one hundred years ago, and he insists that it be given more precision before any efforts are made to determine its empirical validity and relevance.[46]

Still, it appears that while status emulation plays a central role in Veblen's work, and should perform a more dominant function in modern sociology than it does (with or without Campbell's suggested refinements), it is incapable of being the central notion itself as, for example, Gibbs's claims for "control." It is more than a minor concept in the social sciences, to be sure, but it is unlikely to develop into a central focus for most academic sociologists or economists. Does this make Veblen a theoretical anachronism or reduce him to a mere figure in modern intellectual history? No, but it suggests a more selective and limited role for his theory of status emulation to play in the modern social sciences, especially economics and sociology.

But to status emulation must be added Veblen's views on the social-bonding role of nationalism and patriotism as control agents. Nor can the social and collective nature of wealth production as explained by him be

46. Colin Campbell, "Conspicuous Confusion? A Critique of Veblen's Theory of Conspicuous Consumption," *Sociological Theory* 13, no. 1 (March 1995): 37–47. Also, see Rick Tilman, "Colin Campbell on Thorstein Veblen on Conspicuous Consumption," *Journal of Economic Issues* 40 (March 2006): 97–112.

ignored along with its egalitarian collectivist thrust, although it was significantly less effective as a control mechanism. Nevertheless, insofar as egalitarian ideology and doctrine had any hold on the American mind (and in Veblen's view it was not very strong), it had to have some basis in existing social reality and institutional configurations. Although these realities were powerful in shaping Veblen's own thinking, he was well aware that the illusions, indeed, the make-believe of the self-made man and the myths of radical individualism, were far stronger in shaping the American psyche and the hegemonic elements in its culture. Thus any viable summary of his sociology of control must emphasize the combination of the dominant elements that shape it, namely, status emulation and the myth of the self-made man that sustains it as well as the social congealants of nationalism and patriotism.[47] Finally, Veblen's analysis suggests that those groups and individuals most likely to support the existing system of power will be sportsmen, gamblers, and conservative religionists.[48] These are essential embellishments of the social theory which he constructed on the philosophical foundations of an evolutionary naturalism.

---

47. Chris Rojek, "Veblen, Leisure and Human Need," *Leisure Studies* 14 (April 1995): 73–86, perceptively comments that:

Veblen rarely refers to the physical power of the leisure class. Instead he typically presents their power as a matter of signification. They influence general social conduct not by coercion but by representation. In Veblen's sociology "the leisure class" figures as the ultimate metaphor for the wastefulness of consumer culture. But the orchestration of emulation in the behavior of the masses is achieved by the mass media. Veblen was of course writing at a time when Taylorism, which recommended regimented, direct physical control systems, was ascendant in management. Given this climate, his sensitivity to the importance of suggestion and the sign economy in shaping leisure practice is remarkable. No less impressive is Veblen's appreciation of the processes of producing consent through preferred presentations of forms of leisure and consumption . . . Veblen in his analysis of "the pecuniary canons of taste," "the belief in luck" and the practice of "devout observances" revealed the machinery for organizing "voluntary" consent to the ruling order. (85)

48. See chapter 5 for a more detailed treatment of this claim.

Chapter 8

Veblen, Darwin, and Sociobiology

Introduction

Few will question the claim that Charles Darwin (1809–1882) was the most important and influential scientist in the nineteenth century. Or to the distress of the Christian Church, that he undermined the idea that life on earth was a result of a master plan authored by divine providence by claiming that life forms were the result of selective pressures of the environment combined with chance variations in individual characteristics. Or that the outcome of evolution was neither bad nor good, simply the result of nonteleological material vectors. This particular view of history thus portrayed science and reason as the potential liberators of mankind; human emancipation was to come about through the efforts of the scientific community of which Veblen considered himself to be part.

He was greatly influenced by Darwin's teaching as well as by pre- and post-Darwinian evolutionary biology.[1] But Veblen could never overcome his belief that the resurgence of atavistic continuities was possible or his belief that it was only a matter of time until the coercive, predatory, and superstitious proclivities of human culture threatened once again to engulf us, or at least encroach on the peaceful, egalitarian, matter-of-fact qualities of modern life in industrial societies with representative and secular political institutions. Perhaps, this is why he attempted to undermine

---

1. See Michael Boyles and Rick Tilman, "Thorstein Veblen, Edward O. Wilson and Sociobiology: An Interpretation," *Journal of Economic Issues* 27 (December 1993): 1195–218; Stephen Edgell and Rick Tilman, "The Intellectual Antecedents of Thorstein Veblen: A Reappraisal," *Journal of Economic Issues* 23 (December 1989): 1003–26; Rick Tilman, "Introduction" to *Social Darwinism and Its Critics*, ed. Frank X. Ryan (Bristol, U.K.: Thoemmes Press, 2001), ix–xxii; and Tilman, *The Intellectual Legacy of Thorstein Veblen*, ch. 2.

social theories that rested on unscientific or, more often, prescientific, that is, pre-Darwinian grounds.

Veblen aimed at discrediting vitalistic, teleological, introspective, animistic, anthropomorphic, metaphysical, and spiritualistic explanations of both the natural and social order. He attacked the "scientific" pretensions and status of neoclassical economics and advocated its replacement with a genuinely evolutionary approach. Veblen thus represented the vanguard of the Darwinian movement in the United States in the social sciences.

Veblen's satirical qualities are at both their best and worst in his use of the Darwinian concepts and legacy, and for this reason it is difficult to interpret his comments. Although it is evident that by claiming neoclassical economics is "pre-Darwinian" he meant "taxonomic," "teleological," and "static," it is less clear as to how Darwinism is actually interwoven in the texture of his own work. Critics of Veblen abound who are puzzled by his use of evolutionary biology. Indeed, a respectable argument can be made that efforts to use Darwin's concept of natural selection as an analogy to processes of economic selection and cultural diffusion or to imitate the causal structure of neo-Darwinian theory in the social sciences has caused more confusion than anything else.

Evolutionism and the Positive Influence of Darwin

In any case, as for its relevance to social science, Darwin's theory of evolution can be reduced to what Stephen Gould has called with great succinctness "two undeniable facts and an inescapable conclusion."[2] These are (1) that species vary enormously; (2) that species tend to reproduce on a scale that precludes the survival of all but the fittest; and (3) that in the ensuing struggle for existence, the variations that are best adapted will survive. In sum, natural selection assures the survival of those species that vary most noticeably in directions favored by the environment. Thus, according to Darwin's materialist conception of evolution, evolution has no purpose and no direction in the sense that it does not inevitably lead from "lower" to "higher" things. In Darwin's own words: "I believe . . . in no law of necessary development."[3]

2. Stephen Jay Gould, *Ever since Darwin: Reflections in Natural History* (Middlesex: Penguin Books, 1980), 11.
3. Charles Darwin, *The Origin of Species*, ed. J. W. Burrow (Middlesex: Penguin Books, 1968), 348.

If one considers what Veblen had to say about the nature of social-scientific inquiry, and if one then considers Veblen's theory of social change in which he effectively substituted the concept "institution" for Darwin's reference to species, it becomes apparent that Veblen was a Darwinian par excellence. He distinguished explicitly between pre- and post-Darwinian science, the former being contrasted with the latter on the basis of its philosophical preconceptions or "point of view" as Veblen called it. Thus, for Veblen, pre-Darwinian science was "essentially classificatory" and concerned with the following questions:

> how things had been in the presumed primordial stable equilibrium out of which they, putatively, had come, and how they should be in the definitive state of settlement into which things were to fall as the outcome of the play of forces which intervened between his primordial and the definitive stable equilibrium.[4]

Thus:

> To the pre-Darwinian taxonomists the centre of interest and attention . . . was the body of natural laws governing phenomena under the rule of causation. These natural laws were of the nature of rules of the game of causation. They formulated the immutable relations in which things "naturally" stood to one another before causal disturbance took place between them. [Post-Darwinian science on the other hand focuses on] . . . the process of consecutive change, which is taken as a sequence of cumulative change, realized to be self-continuing or self-propagating and to have no final term.[5]

Armed with this "modern," post-Darwinian "point of view," Veblen attacked classical economists:

> The ultimate laws and principles which they formulated were laws of the normal or the natural, according to a preconception regarding the ends to which . . . all things tend. In effect, this preconception imputes to things a tendency to work out what the instructed common sense of the time accepts as the adequate or worthy end of human effort. It is a projection of the accepted ideal of conduct.[6]

Veblen suggests that the result of pre-Darwinian methodology is at best "a body of logically consistent propositions concerning the normal rela-

4. Veblen, *The Place of Science in Modern Civilization*, 37.
5. Ibid., 37.
6. Ibid., 65.

tions of things—a system of economic taxonomy" in which the "forces causally at work in the economic life process are neatly avoided." At worst it produces "a body of maxims for the conduct of business."[7]

When Veblen advanced his own substantive theories about society, he attempted to apply Darwinian scientific standards. This can be seen most clearly in his evolutionary theory of change, perhaps the linchpin of his whole macroeconomic sociology.[8] According to Veblen:

> The life of man in society, just like the life of other species, is a struggle for existence, and therefore it is a process of selective adaptation. The evolution of social structure has been a process of natural selection of institutions. The progress which has been and is being made in human institutions and in human character may be set down, broadly, to a natural selection of the fittest habits of thought and to a process of enforced adaptation of individuals to an environment which has progressively changed with the growth of the community and with the changing institutions under which men have lived. Institutions are not only themselves the result of a selective and adaptive process which shapes the prevailing or dominant types of spiritual attitude and aptitudes . . . they are . . . in their turn efficient factors of selection.[9]

More specifically, for Veblen the key institutions are those based upon "workmanship" and "predation," tools and weapons; their growth and influence on individuals and social structure vary in relation to the essentially, though not exclusively, material conditions prevailing in any one historical period. Thus, Veblen suggests that workmanship dominates the savage era, whereas predation dominates the barbarian era and both are a matter of what Veblen called "selective adaptation." Furthermore, although institutions change and develop in two main directions, "the pecuniary and the industrial," Veblen also allows for social regression and cultural borrowing.[10] Hence, for Veblen there is no one single development trend. In other (Darwinian) words, "natural selection" assures the survival of those institutions that vary most noticeably in ways favored by the environment. Consequently, the "notion of a legitimate trend in a course of events is an extra-evolutionary preconception, and lies outside the scope of an inquiry into the causal sequence in any process."[11]

7. Ibid., 67–68.
8. On this point, see Stephen Edgell, "Thorstein Veblen's Theory of Evolutionary Change," *American Journal of Economics and Sociology* 34 (July 1975): 267–80.
9. Thorstein Veblen, *The Theory of the Leisure Class*, 131.
10. See esp. Veblen, *Imperial Germany*, passim.
11. Veblen, *The Place of Science in Modern Civilization*, 76.

The intellectual impact of Darwin on Veblen is thus both clear and unmistakable, unusually so for Veblen. Unsurprisingly, therefore, Veblen can be regarded as a "complete Darwinian" in the sense that he drew heavily upon both his theories and his methods,[12] and often used an identical evolutionary vocabulary.[13] The Darwinian system had three pillars in terms of explaining the evolution of species: natural selection, sexual selection, and the inheritance of acquired characteristics. Although the last two play a decidedly less significant role in evolutionary theory, Darwin believed that the bright plumage of birds or the vivid color of fish during the mating season affect the reproductive pattern of species and thus the character of the offspring. Nor could he abandon his belief that Lamarck's theory of the inheritance of acquired traits contained some scientific validity, although of course not to the extent Lamarck claimed.

To make matters more difficult and complex, evolutionary theory itself was in the process of rapid development and change during this same period. At the outset, Darwinism had to be distinguished from Lamarckianism, which was the earlier creation of the French scientist Jean Baptiste Lamarck (1744–1829). Indeed, Darwin was unable to completely wean himself from this theory so that it existed in uneasy tension in his work with natural selection. In fact, as Peter Bowler has shown it was not until 1940 that natural selection finally carried the day in the biological sciences.[14]

Then, too, the genetic mechanism by which natural selection worked was unknown to Darwin, although the monk-scientist Gregor Mendel explained it as early as 1865 in a paper which was ignored until its "rediscovery" by Hugo de Vries near the turn of the century. Mendel showed that both dominant and recessive genes were inherited by offspring from their parents and this determined the pattern of natural selection that occurred. Also of importance in the development of evolutionary theory was the mutation theory devised by de Vries which explained radical changes in the human genetic structure as "leaps" caused by the earth's natural radiation among other forces. Finally, the work of Auguste Weissman allegedly demonstrated that "germ plasm" in individuals was not affected by changes in the circumstances of daily living—thus acquired characteristics could not be genetically transmitted. It is important to note

12. Cynthia Russett, "Thorstein Veblen: Darwinian, Skeptic, Moralist," in *Darwin in America: The Intellectual Response, 1865–1912* (San Francisco: W. H. Freeman & Co., 1976), 148.

13. Idus Murphree, "Darwinism in Thorstein Veblen's Economics," *Social Research* 26 (June 1959): 311–24.

14. See Peter Bowler, *The Non-Darwinian Revolution*, passim.

that Veblen was familiar with both pre-Darwinian and post-Darwinian evolutionary theory of the sort noted above.

Veblen and Evolutionism

But merely to call Veblen an "evolutionist" or a "Darwinian" is to inadequately summarize or interpret his intellectual pedigree. Indeed, even though many nineteenth-century thinkers were influenced greatly by evolutionary theory, their evolutionism was more Lamarckian than Darwinian.[15] It was Lamarck's theory of evolution by adaptation to the demands of the environment and, in particular, his notion of the inheritance of acquired characteristics that most influenced Victorian intellectuals including Spencer, not Darwin's theory of natural selection.[16] Human modes of sociocultural transmission that preserve species-unique contents are quite different from biological modes of inheritance. It is clear that Veblen regarded himself as a Darwinian since he repeatedly invoked both Darwin's name and the doctrine of natural selection to explain institutional change and claimed that the survival of some institutions and the death of others were evidence of the instrumental adaptive powers of the human community in the process of evolution. Nevertheless, the Lamarckian doctrine of the inheritance of acquired characteristics is a more effective analogy to use in understanding the internal dynamics of institutional change than the theory of natural selection. Through the institutional transmission of values and behavioral traits, acquired characteristics can both direct evolution of the institution and be "inherited." As Bowler puts it: "Darwin himself did not deny a limited role for the inheritance of acquired characters, and he was thus able to admit that the learning of new habits by the animals themselves can play a role."[17]

In Veblen's analysis, the genesis, development, and transmission of individual traits are institutionally induced, and while some institutions may prosper and flourish, others will stagnate and die. The first is best explained by the Lamarckian doctrine of acquired characteristics, while the latter is most aptly explained by Darwinian natural selection. Thus, although Veblen was also a Darwinian, his evolutionary theory of social change, insofar as particular institutions are concerned, may be understood more readily in Lamarckian terms. It is indeed ironic that in spite of

15. Ibid.
16. Peel, *Herbert Spencer*, 134–35.
17. Bowler, *The Non-Darwinian Revolution*, 98.

his castigation of neoclassical economics for being pre-Darwinian, an important part of his own explanation of institutional change is not Darwinian but Lamarckian. Veblen typically failed to acknowledge this point, and textual exegesis of his work is inconclusive.[18] Yet, one may speculate whether in some literal sense, he believed the Darwinian theory applied literally to the theory of evolution of the human species or whether he intended it only as analogy or metaphor. I am inclined toward the former rather than the latter interpretation, but it appears that Veblen, who aimed at discrediting neoclassicists for being pre-Darwinian, also invoked Darwin for analogical and metaphorical purposes.

Reform Darwinism

"Social Darwinism" was often assumed to both glorify and preserve favored individuals, classes, or nations. It was thus understood as a philosophy that exalted competition, power, and violence over cooperation, equality, and pacifism. It exemplified as virtues nationalism, imperialism, and dictatorship as well as militarism and provided ideological undergirdings for cults of the hero, the superman, and the master race.[19] In its conservative doctrinal forms all this was often true, but Veblen was not this kind of Darwinian. Indeed, just the opposite was true of his Reform Darwinism, although he did not explicitly articulate it as such. For Veblen, cooperation and equality were as much a feature of social behavior as competition, conflict, and hierarchy. He thus believed that the political program of conservative social Darwinism failed from a justificatory perspective as did its two leading spokesmen, Herbert Spencer and W. G. Sumner, in their apologetics for it. Indeed, the ideal analogy for human society is not a conflict-filled jungle but a well-cultivated garden. Paul Carter writes:

> Powell, Ward, and the pioneering anthropologist Lewis Henry Morgan attacked Social Darwinism by challenging its premises; the maverick economist and culture-critic Thorstein Veblen deflated it by provisionally accepting those premises and showing that they led into *reduction ad absurdum.* Suppose for the sake of argument that the present ruling elite has risen into

18. The quasi-Lamarckian nature of part of Veblen's work can be found at various points in his writing but see esp. *Imperial Germany,* 125–26.

19. But, see Geoffrey M. Hodgson, "Social Darwinism in Anglophone Academic Journals: A Contribution to the History of the Term," *Journal of Historical Sociology* 17 (December 2004): 429–63.

power by natural selection, Veblen proposed in *The Theory of the Leisure Class* (1899). Protected by wealth and prestige it then proceeded to shelter itself from the very process of evolution which had gotten it there; and the institutions it created, "adapted to past circumstances," could be "therefore never in full accord with the requirements of the present." Such a successful social class, once exempted from the struggle for existence thereafter "will adapt itself more tardily to the altered general situation," and, since evolution presumably does not stop, by virtue of its inherent conservatism this sheltered elite will render itself increasingly and disastrously out-of-date. "The characteristic attitude of the class may be summed up by the maxim 'Whatever is, is right,' whereas the law of natural selection, as applied to human institutions, gives the axiom: 'Whatever is, is wrong.'"[20]

And Carter adds that:

Qué sera sera—dictated not by divine decree, but by pressures of the physical environment, by the inescapable patterns of one's genes, by electrochemical firings within one's skull, or by the compulsion of one's glands.[21]

Carter thus gets down to the basics of human behavior, a view largely held at the most rudimentary level of analysis by both Darwin and Veblen.

In the *Descent of Man*, where Darwin concludes that man is descended from some lower form of life, he applies the ideas of competition and competitive struggle to tribes by arguing the tribes with favorable characteristics will dominate others. He explains:

When two tribes of primeval man, living in the same country, came into competition, if (other circumstances being equal) the one tribe included a great number of courageous, sympathetic and faithful members, who were always ready to warn each other of danger, to aid and defend each other, this tribe would succeed better and conquer the other. Let it be borne in mind how all important, in the never-ceasing wars of savages, fidelity and courage must be. The advantage which disciplined soldiers have over other undisciplined hordes follows chiefly from the confidence which each man feels in his comrades. A tribe rich in the . . . [moral and social] qualities [of obedience and coherence] would spread and be victorious over other tribes; but in the course of time it would, judging from all past history, be in its turn overcome by some other tribe still more highly endowed. Thus the social and moral qualities would tend slowly to advance and be diffused throughout the world . . . Now, if some one man in a tribe, more sagacious

20. Paul Carter, *Revolt against Destiny: An Intellectual History of the United States*, 141.
21. Ibid.

than the others, invented a new snare or weapon, or other means of attack or defense, the plainest self-interest, without the assistance of much reasoning power, would prompt the other members to imitate him; and all would thus profit. The habitual practice of each new art must likewise in some slight degree strengthen the intellect. If the new invention were an important one, the tribe would increase in number, spread, and supplant other tribes.[22]

In addition, Darwin adds, "A tribe including many members who, from possessing in a high degree the spirit of patriotism, fidelity, obedience, courage, and sympathy, were always ready to aid one another, and to sacrifice themselves for the common good, would be victorious over most other tribes; and this would be natural selection."[23]

Clearly, Veblen believed Darwin and Darwinism had unmasked metaphysical delusions and rendered traditional schools of thought obsolete. Since he rejected essentialism in all its forms, it must be recognized that Darwinism was one source of his moral epistemology and a secularized Christian ethic and a Kantian categorical imperative the others. This is more evidence of his eclecticism to be sure, but it is also evidence of his combination of more or less congruent elements into an evolutionary naturalist stew.

Did Veblen find in natural selection a metaphysics, ethics, and political philosophy capable of replacing Christianity? Perhaps, but not quite may be the most appropriate answer because his eclecticism demanded additional support from philosophic naturalism and the other natural and social sciences. In any case, he was not adverse to borrowing from a broad spectrum of intellectual sources including Christianity itself, from which he obtained a lasting respect for the values of altruism and humility.

While Veblen did not always explain and justify the standards of rationality and equity that permitted him to recognize pockets of ceremonialism where they existed, he did accomplish the following: He identified his own particular state of mind, and he explained the form of social discontent which he shared with other left-liberals, especially evolutionary naturalists; and he did focus, if at times iconoclastically, on the inadequacies of social existence from a specific perspective. He was also a keen analyst of the evaluative content of various theories of society and economics. Indeed, Veblen could see no clear line of progress in these theories from the

22. Charles Darwin, *The Descent of Man and Selection in Relation to Sex* (London: John Murray, 1871), Vol. I, 162.
23. Ibid., 166.

earliest to the most recent. This was perhaps still more evidence of the impact of nonteleological Darwinism on him.

Sociobiology

Did Veblen aim at undermining a traditional, deeply entrenched, normative worldview and supplying a substitute for it? Probably he believed that the sense of morality is in part a biological form of social adaptation derived from evolutionary processes. He knew too much about cultural anthropology and social psychology, however, to believe that there is a universal moral sense; for the conscience varies in different races and even ethnic groups because they contain different cultural biases. In short, Veblen did not believe mind including the moral conscience was an imprint on a tabula rasa, but that it was informed by various capacities, constraints, and dispositions that are both innately present and environmentally transmitted.

Nevertheless, the sociobiological claim that individuals ultimately aim for their own individual reproductive benefit and that altruism is thus a byproduct of kin selection was not a claim Veblen would find congruent with his own convictions, nor did he believe the related view that morality is just an aid to survival and reproduction. But neither did he believe that moral standards have a completely independent or autonomous existence external to the evolutionary process of adaptation. Rather morality is the condition that results when rational beings interact socially without undue ceremonial impediments or predatory intent and the "generic ends of life" can flourish. In any case, he saw no deep-rooted tension between the evolutionist's understanding of morality and the ethical claims of Christianity since he accepted altruism and humility as essential to the adaptability and survival of the human species.

To further pursue the topics of ethics and morality in the context of sociobiology, Veblen apparently believed that Darwinism might account not just for self-sacrifice, but for identifiably altruistic ends and the practice of humility. Morality was thus not merely a biological adaptation, yet neither was it an independent variable in cultural outcomes. For clearly the use of force and fraud as means to achieve the social ends of exploitation, subjugation, and self-service were just as common. In short, Veblen perceived the human evolutionary process as one that within limits permitted the preservation and transmission of negative, socially destructive ends as well as what he called the "generic ends of life."

What were Veblen's metaethics anyway? Is morality merely a biological adaptation; does evolutionary naturalism itself deny the independent existence of moral standards? Veblen makes no definitive claims about independent standards beyond and apart from actual existing humans and their cultural evolution that may give us moral rules, but are insufficient for civilized existence.

He did, however, regard the "generic ends of life, impersonally considered" as having powerful transcultural significance along with their invidious and destructive opposites. In short, he knew of no cultures that were devoid of both the instincts of parenthood, idle curiosity, and workmanship which were his peaceful traits and those of acquisition, superstition, and waste which he regarded as the value constants of predation.

Veblen's Darwinian "Radicalism"

Specialists in American intellectual history tend to focus more on conservative social Darwinism than on the liberal, reform version of the Darwinian legacy. It is evident, however, that Veblen is a closer fit with the doctrinal usage of E. A. Ross and L. F. Ward than with that of W. G. Sumner, his professor at Yale. This is to suggest that the "survival of the fittest" understood as a group process, not as a form of individual competition, has a strong egalitarian collectivist significance in Veblen's work. Darwin believed in group selection and Veblen used this to buttress, if not define, his own arguments. These favored cooperation, social instincts, and solidarity not to mention mutualism, altruism, and solidarist values and ethics. In short, a social order organized along collectivist lines with a heavy dose of egalitarianism was more likely to survive and prosper than one characterized by economic individualism, social rivalry, honorific prowess, and emulatory waste and display. This claim can be documented by quoting from both Darwin and Veblen.[24] Yet, in the final analysis, it is Veblen's often unacknowledged extrapolation from Darwinism that is at stake.

The cooperative commonwealth or industrial republic of Veblen's political and social aspirations, in fact, has a social Darwinian basis in that

24. See John Laurent, "Charles Darwin on Human Nature," in *Evolutionary Economics and Human Nature,* ed. John Laurent, preface by Geoffrey M. Hodgson (Cheltenham, UK: Edward Elgar, 2003), 96–113, and Charles Darwin, *The Descent of Man and Selection in Relation to Sex* (Chicago: University of Chicago Press, 1982), in *Great Books of the Western World,* Robert Maynard Hutchins, editor-in-chief, Vol. 49. See esp. chapters 5 and 21 for material that Veblen may have found useful in formulating his egalitarian collectivist version of Darwinism.

he believed a properly developing community that was likely to survive and prosper perhaps in competition with other communities was *ceteris parabis,* one in which the "generic ends of life" flourished. These ends were, of course, altruism, critical intelligence, and proficiency of workmanship and the moral and cultural ethos necessary for their sustenance. It is not force, fraud, and superstition that are at the root of wholesome community growth, not honorific prowess, ostentatious display, or conspicuous waste that provide the fulcrum for the levers of instrumental knowledge that produce noninvidious social orders. It is, in fact, the values and behavioral traits that Darwin identified and articulated in *The Descent of Man* (1871) that exemplify what Veblen used and sanctioned. It is on them that I rely in my assertions regarding the genesis of Veblen's egalitarian collectivism in evolutionary biology and anthropology.

Sociobiology: An Interpretation

The revival of interest in the biological and genetic components, as opposed to the cultural and institutional determinants of human behavior or, more accurately, the resurgence of focus on the interaction between both, suggests a new interpretation of the ideas of leading figures in the history of heterodox economics. One such figure is Veblen, whose relationship to what is now known as "sociobiology" is only now being systematically explored.

The American historian Carl Degler has recently argued that the idea of a biological root to human nature was widely accepted by the end of the Victorian era, when it all but disappeared from social theory, only to reappear in the aftermath of the Second World War.[25] Degler also demonstrates that Victorian biology provided ideological legitimation in the form of conservative social Darwinism for racism, sexism, and imperialism. Offshoots of Victorian science, or pseudoscience, included intelligence testing, the eugenics movement, and the involuntary sterilization of criminals.

Aside from different interpretations of Darwin's major writings, especially *The Origin of Species* and *The Descent of Man* (1871), it is evident that what separates the social Darwinists from their critics are assumptions of a metatheoretical nature. Metatheory itself has three parts: (1) epistemol-

---

25. See Carl Degler, *In Search of Human Nature: The Decline and Revival of Darwinism in American Social Thought* (New York: Oxford University Press, 1991), and Mary Maxwell, ed., *The Sociobiological Imagination* (Albany: SUNY Press, 1991).

ogy, or the study of the origin, nature, methods, and limits of knowledge, particularly the meaning of "truth"; (2) ontology, the study of being (reality); and (3) axiology or the study of value, value inquiry so to speak. Even though the social Darwinists and their doctrinal opponents overlap at times on the meaning and relevance of these metatheoretical issues, metatheory often separated them into distinct camps. To illustrate this, the system of W. G. Sumner and his critics is briefly analyzed below.

It is fair to say that the system of Sumner, Veblen's professor at Yale, was accused by its critics of being built on (1) a deductive epistemology in which truth claims are made by deducing conclusions from general principles not from factual evidence; (2) a radically individualist ontology which asserted that society is composed of individuals and nothing more, that is, individuals are more real than society and thus have ontological primacy over the collective; and (3) axiomatically speaking, individual "freedom" is the highest value available to humans during their earthly existence, especially freedom of exchange and contract and the right to unlimited acquisition of wealth and property. With these metatheoretical assumptions in mind, although his critics sometimes exaggerated their role in his thought, it is easier to understand Sumner's political and social philosophy and its opponents.

Sumner's views of the proper role of the state can be summarized under three headings: (1) administration of justice, that is, provision of courts, police, and prisons for the punishment of criminals and enforcement of the obligation of contract; (2) national defense, that is, provision of an army and navy to deter and punish aggression by foreign powers; and (3) provision of public works when goods and services absolutely essential to the welfare of the community are not forthcoming because it is not profitable to provide them. Sumner's philosophy regarding the proper role of government is not too different from that of other classical liberals, including Adam Smith, whatever their disagreements in terms of policy application. Although Sumner was to modify his views in later life, particularly with regard to the use of state power to curb monopoly, most of his critics have focused on his prescriptions for strict laissez-faire expressed in *What Social Classes Owe to Each Other* (1883). It is these which are used here to characterize his philosophy for it was these with which his student Veblen had an intimate familiarity.

Sumner claims in effect that what social classes owe to each other is nothing except mutual acquiescence in the structures of unlimited capitalism. In a free market, each owner of a factor of production, whether it be land, labor, or capital, will be paid what his contribution is worth provided that neither unions, government, nor monopolists interfere with the free movement of the forces of supply and demand. Indeed, such interference will likely help the unfit to survive, multiply, and transmit their inferior genes to future generations, which will only slow down the evolutionary processes that effectively cull out the weak and unfit. In short, unlimited capitalism contains and gives rise to the most efficient social institutions for the processes of natural selection and adaptation to work toward their ends; and these ends are the transmission of only those traits conducive to the survival of the fittest. This is how Sumner links free-market processes with evolutionary biology, and his critics were quick to seize on the callousness of his theory and its apparent denial of social justice and social conscience. Sumner, however, was shrewd enough to distinguish between the struggles that went on in nature and the conflicts that ensued between human beings in the social order and he did not consistently extend the logic of his argument from the natural to the social order. If, ultimately, he drew back from the Darwinian portrayal of "nature red in fang and claw" as the model for human existence, he continued to sanction and praise competitive processes as the highest form of human existence that evolution could bring at the existing level of social development.

However, there were also those who were skeptical of any significant role for biology and, in fact, thought that culture shaped human nature. Some of those scientists who believed this included Franz Boas, John B. Watson, and, later, Ruth Benedict and Margaret Mead. What lay behind this change in scientific thinking was a basic paradigm shift as these intellectuals sought a rationale for their own political agendas and struggled against racism and sexism. They found it in the concept of culture broadly interpreted as environmental determinism.

In more recent years, however, there has been a revival of interest in the instinctual and biological nature of humankind led first by ethologists such as Karl Von Frisch, Nikolas Tinbergen, Konrad Lorenz, and Jane Goodall, who argue that direct parallels exist between animal and human behavior. They have been followed in varying degrees by Melvin Konner, Alice Rossi, Jerome Hagen, and many other social scientists. However, the best-known and most influential of contemporary sociobiologists is the Harvard entomologist Edward O. Wilson (born in 1929, the year Veblen died), whose ideas provide the basis for a comparison with Veblen. First, however, let us continue exploring Veblen's relationship to the biological sciences.

Biological Aspects of Veblen's Work

In developing the ideas that are presented particularly in *The Theory of the Leisure Class* and *The Instinct of Workmanship*, Veblen utilized many biological concepts. Principles of genetics and evolution are firmly entrenched in Veblen's analyses, which was unusual at the time for a social scientist because both of these fields of biology were relatively young and what was known about them was a small fraction of what is known today. Two chapters in *The Theory of the Leisure Class* are particularly noteworthy. In chapter 8, "Industrial Exemption and Conservatism," Veblen explains the relationship between economic institutions and human character. He states that the continual change of our institutions is selective for those individuals who can best adapt to the changes. In other words, only those individuals who can modify their lives to the circumstances created by new institutions have any hope of being successful. He also notes, however, that people are creatures of habit and that changes in their scheme of life are made reluctantly and only in response to dire economic exigencies. This partly explains the conservatism of the leisure class. Their wealth shelters them from economic pressures, so there is no need for change, and the class retains its characteristic habits of life. What Veblen has described is not far removed from what might be read in a textbook on evolution.[26] One aspect of the theory of evolution is that organisms change through time, and this change is driven by environmental pressures that make the survival of some forms more likely than the survival of others, which, of course, is the process known as natural selection.

In chapter 9, "The Conservation of Archaic Traits," Veblen describes traits and their expression at different stages of socioeconomic development. He begins by saying that people are hybrids of different ethnic types. Long ago, these ethnic types, or races, existed in pure form, and each had its own unique temperament. As a result of hybridization, most humans are descended from several different races, and the type of temperament that they express is determined by social and cultural forces. For example, in the stage of savagery—the earliest stage of social development considered by Veblen—individuals tend to be community oriented. Technology is so primitive that communities must struggle just to provide what is necessary for subsistence. Cooperation is necessary for survival. Such a culture promotes peace, honesty, and goodwill in its citizens. On the other hand, in the predatory or barbarian stage, technology is advanced enough to create a surplus of goods, and the differential distribution of these goods creates a class system, with the classes being distinguished on the basis of wealth. Thus barbarism promotes ferocity, selfishness, and invidious emulation.

26. See, for example, John Alcock, *Animal Behavior* (Sunderland: Sinauer Associates, Inc., 1989).

Veblen articulates a theory of natural selection in describing how different environments select for different traits in the individuals that live in them. However, there are other ideas in this chapter that also have biological significance, one of which is his notion of hybrids. Hybrids are the offspring that result from the mating of genetically unlike strains of organism. For example, a mule is a hybrid produced by crossing a mare with a jackass. Analogously, the intermixing of dolicho-blonds and dolicho-brunettes would also produce hybrids. Veblen's understanding of the ramifications of such a process is quite impressive. After all, the results of Mendel's primitive experiments on hybrids were not published until 1865 and were largely ignored for many years afterward. These experiments were conducted to determine the mechanisms of heredity, not the significance of hybridization. Nevertheless, Veblen explains in detail the importance of the hybridization of human cultures. Since individuals now contain elements of several racial temperaments, their behavior can evolve in a way that complements the evolution of economic institutions. In the peaceful state of savagery, the traits of the Mediterranean and dolicho-brunettes will dominate, while in barbarism the predatory temperament of the dolicho-blonds will be most prevalent.

Many contemporary social scientists attribute the differences between groups to different cultural regimes rather than to genetic variability and selection. Nevertheless, what Veblen is saying is that the interbreeding of different groups has led to an increase in the gene pool of different communities. "Gene pool" is a collective term referring to all of the genes in a population. Populations with larger gene pools exhibit greater variability and have a better ability to adapt to their environment. Variability is the key to evolution. Since hybrids must necessarily have greater variability in their genes than either of the two strains from which they were derived, they are often the organisms best able to cope with their environment. Biologists refer to this as "hybrid vigor." Genes that code for advantageous traits will increase in frequency in the population, and the traits will become more common. In the hybridized communities that Veblen discusses, there is the potential for any number of traits and temperaments to be expressed. However, the cooperation necessary for survival in savagery tends to increase the frequency of peaceful traits, while the competitiveness necessary in barbarism increases the frequency of predatory traits.

Veblen's insight into population genetics does not end here. He notes that although a particular phase of social development may last for a protracted period of time, perhaps thousands of years, those traits that are selected against are not completely eliminated from the population. Thus, when one temperament yields to another, there is always the potential for the first temperament to regain its position of dominance, and according to Veblen, such reversions are seen. For example, when Veblen describes how the exigencies of the predatory culture tended to weaken the peaceful traits in society, he makes sure to note "the almost ubiquitous fashion in which they [the peaceful traits] assert themselves whenever the pressure of special exigencies is relieved."[27] The traits "seem to have persisted by force of the tenacity of transmission that belongs to an hereditary trait. . . . Such a generic feature is not readily eliminated, even under a process of selection so severe and protracted as that to which the traits here under discussion were subjected during the predatory and quasipeaceable stages."[28]

There is biological significance in those statements. First, they reflect Veblen's understanding of the difficulty with which genes are eliminated from a population. Even genes that code for deleterious traits tend to persist in populations because of the complicated mechanisms of genetics. A particular gene may be present in the genetic makeup of an organism even though the particular trait for which it codes is not expressed by the organism. To use the proper biological vocabulary, the gene is part of the organism's genotype, but it is not expressed in its phenotype. Such genes are not eliminated from the gene pool because the traits for which they code, however undesirable, are not expressed and therefore are not subjected to the selective forces of the environment. Thus, as Veblen says, even the peaceful traits were conserved when they were not useful and reemerged when once again the environment became favorable to them. The conservation of a wide range of characteristics allows human behavior to continually change in response to continually changing socioeconomic institutions. That Veblen was aware of these kinds of complexities reflects both the completeness of his arguments and the evolutionary nature of his social theory.

The other significant aspect of Veblen's statements is that they related to a controversial discipline—sociobiology. Strictly speaking, sociobiology is the study of the ultimate adaptive significance of social behavior in animals, using an evolutionary approach. However, the term has come to refer to the study of human behavior from a biological perspective. In other words, it is often thought of as the intersection of biology and sociology.[29] In both *The Theory of the Leisure Class* and *The Instinct of Workmanship,*

27. Veblen, *The Theory of the Leisure Class,* 150.
28. Ibid., 151.
29.
The disagreement between sociobiologists and cultural materialists on the issue of human nature is a matter of the contraction versus the expansion of the postulated substance of human nature. Cultural materialists pursue a strategy that seeks to re-

Veblen describes the way behavior has changed throughout history. The essence of his claim is that economic institutions shape human actions. However, to say that the institutions themselves are directly responsible for changing human behavior is an oversimplification of Veblen's point. More accurately, Veblen states that it is institutional influence on hereditary makeup that determines human behavior.

At this point, a brief recapitulation of Veblen's theory of social change, as described in *The Instinct of Workmanship,* is in order. Veblen hypothesized four distinct eras in human history: savagery, the predatory culture, the era of handicraft, and the machine age. Each era can be defined in terms of the state of the industrial arts and the state of human nature found within it. In the state of savagery, humans are already identifiable as tool-using animals, but the tools have not advanced enough to allow for the production of anything more than what is necessary for survival. Communities are so concerned with fulfilling essential physical needs that they do not have the time or the energy for violent attacks on others; nor is there any benefit in attacking another person or group because no one produces enough goods to make the raid valuable or worthwhile. Such conditions select for humans who are peaceful, hardworking, and cooperative because only individuals with these traits can hope to be successful under these conditions.

The situation changes in the predatory culture. Here, the advancement of technology allows for the production of a surplus of goods, and it becomes possible for an individual to have more than he or she needs. Success can now be measured in individual terms, rather than community terms. Successful individuals are those select few who are able to appropriate the surplus of goods. Now, for two reasons, aggression is more common. First, certain individuals need to devote less time and energy to survival and thus have more time for acquisition. Second, aggressive behavior is now beneficial because there is a surplus of goods that can be obtained by aggressive measures. As a result, those individuals who are genetically "programmed" to be physically stronger, more cunning, and more competitive have a decisive advantage, and human behavior evolves accordingly.

---

duce the list of hypothetical drives, instincts, and genetically determined response alternatives to the smallest possible number of items compatible with the construction of an effective corpus of sociocultural theory. Sociobiologists, on the other hand, show far less restraint and actively seek to expand the list of genetically determined traits whenever a plausible opportunity to do so presents itself. From the cultural materialist perspective, the proliferation of hypothetical genes of human behavioral specialties is empirically as well as strategically unsound.

Marvin Harris, *Cultural Materialism: The Struggle for a Science of Culture* (New York: Random House, 1979), 127 and ch. 5.

The handicraft and machine eras have tended to be somewhat more peaceful than the predatory culture, although now the motive is different. The development of ownership puts emphasis on pecuniary gain, so people tend to behave in ways that will increase their profits. There is still an element of competitiveness, but a "quasi-peaceable" state is necessary because this is conductive to industry, and industry (i.e., the control of industry) is the means through which profits are made. Once again, the conditions of society select for the proper genetic type, which in turn codes for the proper behavior.

Sociobiology and Its Critics

Despite criticisms to the contrary, sociobiologists do not assume that organisms, human or otherwise, need to be consciously aware of the reasons for their activities. Just as birds that migrate to tropical locations during the winter are not consciously aware that their behavior increases their chance of survival, humankind need not know that their behavior has evolved to maximize their success in society. Veblen makes this clear in his discussion of emulatory processes. Not only are people unaware that their competitiveness and ferocity are traits that have evolved to promote success in a culture that rewards wealth and conspicuous waste, they are often unaware that they are engaging in conspicuous waste. As Veblen puts it, "It frequently happens that an element of the standard of living which set out with being primarily wasteful, ends with becoming, in the apprehension of the consumer, a necessity of life; and it may in this way become as indispensable as any other item of the consumer's habitual expenditure."[30] In short, a good part of emulatory consumption is not intentionally emulative; rather, it is a result of habits created by society. Nevertheless, it is behavior that is adaptive to the environment.

Another criticism of sociobiology is that it is merely a means of justifying social injustice and inequality. Critics argue that saying a trait is adaptive is equivalent to saying that it is genetically determined and therefore good and cannot and should not be changed. This interpretation of sociobiology was probably the biggest factor contributing to the controversy over Edward O. Wilson's work. Many social scientists believed that Wilson developed a theory that would allow those in power to rationalize the maintenance of the status quo and the preservation of existing social injustices and inequalities. This, in Wilson's view, is a misconception

30. Veblen, *The Theory of the Leisure Class*, 79.

of sociobiology that attempts only to explain why a behavior exists, not to justify it.

Veblen obviously was not justifying the behavior he described. He was opposed to much social convention, and his work can be interpreted as an attempt to erode it. His point is that human behavior is adapted to the existing social institutions. At the time that Veblen wrote, institutions were producing undesirable behavior. It was Veblen's hope that institutions could be reformed so that human behavior could become more acceptable. He knew that the only way to make humans act more productively was to change the institutions that influenced them.

Furthermore, Veblen realized that a discussion of adaptive behavior does not necessarily imply that a characteristic is determined absolutely by genes. Sociobiology is validated as long as there is significant correlation between genotype and behavior. Certain behaviors may only develop under the proper combination of genetic makeup and environmental conditions. Such behaviors will not develop in individuals with the right genotype and the wrong environment, or in those with the right environment and the wrong genotype. As an example, there is evidence indicating that phenylketonuria, a disorder that can lead to mental retardation if left untreated, is a result of both genetic and environmental influences.[31] With advances in sociobiology, awareness of the many facets of human existence that are formed this way is increased. The famous "nature versus nurture" debate is becoming less relevant as more is learned about the interplay of the two sides.

Although it has taken years of research to advance knowledge to this point, Veblen preconceived much of it. He believed that the reason for the aggressive behavior exhibited by certain cultures at certain periods of development was not that the individuals had inherited an "aggressive gene" from their ancestors. Almost certainly, there is no such gene, and it is only the interaction of a number of genes in combination that predispose an individual to aggression under the right social conditions, which is what Veblen describes. Each developmental era of society exerts a unique influence on the genotype, creating certain types of behavior, which thereby become characteristic of the era. Thus, we have the peacefulness of savagery, the aggression and force of barbarism, and the competitive, quasi-peacefulness of the handicraft and machine ages.

There are points at which Veblen tends toward a more complete environmentalism than is characteristic of sociobiology. For example, in chapter 2 of *The Instinct of Workmanship*, titled "Contamination of the Instincts

---

31. Alcock, *Animal Behavior*, 512.

in Primitive Technology," he bluntly states that "all instinctive behavior is subject to development and hence to modification by habit."[32] For Veblen, the term "habit" approximates what some anthropologists call "cultural traditions," that is, behavior influenced by what has been done in the past.

Even as he so transforms his usage of the phrase "instinct of workmanship," Veblen further reduces its biological implications. Linking it with what he terms an instinctive "parental bent," he argues these two instincts cannot be treated apart from each other.

> The only other instinctive factor of human nature that could with any likelihood dispute [the] primacy [of the sense of workmanship] would be the parental bent. Indeed the two have much in common. They spend themselves on much the same concrete objective ends, and the mutual furtherance of each by the other is indeed so broad and intimate as to leave it a matter of extreme difficulty to draw a line between them. Any discussion of either, therefore, must unavoidably draw the other into the inquiry to a greater or less extent, and a characterization of one will involve some dealing with the other.[33]

This parental bent is not only solicitude for children, but a "solicitude in mankind [that] has a much wider bearing than simply the welfare of one's own children."[34]

As Dorothy Ross has argued, Veblen's instinct is not a single heritable trait, but at most a stable disposition shaped and passed on by cultural experience.[35] It appears that in his later writing what Veblen calls "instincts" are mostly learned behavior and are the consequence of environmental conditioning, whereas in Veblen's earlier writing, "instinct" often carried hereditarian implications.

For Veblen, human nature was expressed through culture but could not be reduced entirely to it. He also believed that the genetic endowment is relevant for a scientific understanding of human sociality. Veblen did not go so far as to claim that human beings are nothing but the form into which their particular culture molds them, and yet he placed considerable emphasis on plasticity of behavior, social amelioration, and the great impact of environmental variation as a determinant of behavior. Whatever

---

32. Veblen, *The Instinct of Workmanship*, 38.
33. Ibid., 25–26.
34. Ibid., 26.
35. Dorothy Ross, *The Origins of American Social Science* (Cambridge: Cambridge University, 1991), 384.

man's instinctual endowment, and Veblen was sometimes vague about it,[36] this endowment primarily affected behavior in competition with environmental determinants. He rarely went so far as to portray the human organism as a tabula rasa, but neither did he set any rigid limits on human plasticity. Veblen was a proper Darwinian and, in that perspective, biology and culture are both adaptive. He knew better than to speak authoritatively on the nature versus nurture controversy, and he did not claim the influence of heredity and culture as a question of one or the other. Genetic predispositions were not rigidly deterministic, because they permitted flexible adjustment to a wide range of socioeconomic circumstances. In short, he believed genes were coded for a broad range of forms under a wide array of social-ecological conditions.

Clearly, there are points of both convergence and divergence in Veblen's relationship to sociobiology that are traceable to his own intellectual odyssey and development as a theoretician. That part of Veblen's work that best lends itself to a sociobiological interpretation is his writing up through the publication of *The Theory of the Leisure Class* in 1899. It is then that his use of evolutionary theory and genetics is heaviest, and it is also during this period that the influence of racialism is most pronounced. The term "racialism" is used rather than "racism" for two reasons: first, because different racial and ethnic groups were thought to have different psychological traits and temperaments, and this was commonly supposed to be due more to heredity than to environment. Veblen was not immune to this trend of thought, which brings us to the second point, which is that he did not think these groups were better or worse than one another; he only believed they were different, and these differences were, in part, genetic.[37] In his later works, particularly *The Instinct of Workmanship*, the racialism has mostly disappeared in favor of environmental and culturological explanations, and he is more careful to distinguish "in-

---

36. Clarence Ayres once wrote that:

I met Veblen once or twice at the home of Walter Stewart, in Amherst; and once (I believe in the spring of 1920) when one of his step-daughters was in a class of mine at Chicago, she invited me to dinner with Veblen, and to my great embarrassment there were no other guests. Also, after dinner the family left us alone together. This was my only conversation with Veblen. It was on this occasion that a conversational exchange occurred which I believe often repeated to students apropos V's [Veblen's] conception of "instincts." He asked me point blank if I recalled exactly how he had defined instincts. I was flabbergasted, and after a long moment's soul-searching I replied that I couldn't recall any exact definition. He beamed, and said, "No, you can't, because I never did!" [Ayres to Louis Junker, July 18, 1966]

37. See Veblen's attack on German claims to racial superiority in *Imperial Germany*, ch. 1.

stinctive" behavior, which is learned, from "tropismatic" behavior, which is innate. Thus, by 1914, Veblen's social theory rested on a more radical environmental determinism than it had earlier when he was prone to use explanations more congruent with contemporary sociobiology. Apparently, between 1899 when he published *The Theory of the Leisure Class* and 1914 when *The Instinct of Workmanship* appeared, there was a shift in his thought from viewing instincts both as genetically and culturally induced to a view of them as more largely the result of environmental forces.

Enter Edward O. Wilson

In 1975, Harvard professor and avowed secular humanist Edward O. Wilson published *Sociobiology: The New Synthesis,* for which he was criticized for taking a biological approach to the study of human behavior. Interestingly, human sociobiology was the topic of only one of the book's twenty-seven chapters, whereas the same theme permeates entire books written by Veblen.

Although Wilson's work on human societies is at times vague and ambiguous, his assertions are theoretically interesting because like Veblen he employs an evolutionary approach that is traceable to Darwin, and his writing is studded with both biological and sociocultural concepts and definitions. Also, like Veblen, part of his work is explicitly political and aimed at discrediting radical environmentalism of the sort articulated by Veblen himself. Although Veblen is not an explicit target, it is evident that Wilson's attacks on Marxists, radical environmentalists, and extreme behaviorists are of a political nature and point at the "utopian" essence of socialism and anarchism. According to Wilson, schemes for radical reconstruction of society are doomed to failure because intransigent human nature will not permit their realization without the destruction of freedom and individuality. These are specifically political claims, of course, and it is important to note that Wilson's political agenda compelled him to use a glossary in which several key terms are defined in politically loaded ways. Take, for example, these definitions drawn from the glossary in *On Human Nature,* which he uses often and consistently in his sociobiological writings. The prejudicial way in which Wilson defines "human nature," "sociobiology," "instinct," and especially, the different forms of "altruism" biases his case against radicalism from the outset. To illustrate, "human nature" is defined as "the full set of innate behavioral predispositions that characterize the human species; and in the narrower sense, those predispositions that affect social behavior."[38] This definition rests on the dubious assumptions that these "predispositions" are identifiable premises that preclude applicability to a broad range of human activities at least as yet. Wilson identifies "sociobiology" as "the scientific study of the biological basis of all forms of social behavior in all kinds of organisms, including man."[39] This definition too, rests on the same problematic assumptions of identification and separability. Interestingly, Wilson identifies "instinct" in a manner quite different from Veblen when he describes it as:

> Behavior that is relatively stereotyped, more complex than simple reflexes such as salivation and eye blinking, and usually directed at particular objects in the environment. Learning may or may not be involved in the development of instinctive behavior; the important point is that the behavior develops toward a comparatively narrow, predictable end product.[40]

For Veblen, on the contrary, "instinctual behavior" usually involves the acquisition of knowledge or skill and the end product often is neither narrow nor necessarily predictable because of cultural dilution and deflection and institutional channeling. Consequently, critics have often rightly complained that Veblen's use of the term "instinct" is part of a vocabulary in transition and a manifestation of archaic usage, though in its later form, the term becomes less useful in explaining human behavior.

Finally, and perhaps most prejudicial in terms of Wilson's definitional strategy, he defines genuine or idealistic "altruism" in the following manner:

> Self-destructive behavior performed for the benefit of others. Altruism may be entirely rational, or automatic and unconscious, or conscious but guided by innate emotional responses.[41]

This politically loaded definition is then employed by Wilson to undermine any schemes, ideologies, doctrines, or programs that presuppose that human beings can be conditioned to behave in a consistently "altruistic" manner. But the extremist nature of the definition itself should be

38. Edward O. Wilson, *On Human Nature* (Cambridge: Harvard University Press, 1978), 217–18.
39. Ibid., 222.
40. Ibid., 218.
41. Ibid., 213.

noted, for Wilson has defined "altruism" as "self-destructive behavior performed for the benefit of others." He may be correct in assuming that no society could exist for long based only on such a moral code without destroying freedom and individuality. However, he fails to identify the secular political creed or ideology today held in the West to the effect that a viable social order can rest on such an ethical system. Wilson has thus selected a doctrinal caricature for demolition.

Much of what is commonly understood as "altruism" is neither consciously self-destructive nor based on what Wilson calls "reciprocity" in which individuals behave altruistically because they anticipate that other individuals will do the same for them. Instead, it is rooted in the Kantian categorical imperative that endorses treating people as ends having value in themselves rather than as means, that is, as mere instruments of our own pleasure and its Christian equivalent, which is "to love thy neighbor as thyself." Kantians, Christians, and secular humanists, no doubt, have believed that in extreme cases, altruists should sacrifice themselves on behalf of others, but they do not claim that altruism itself must be continuously self-destructive as Wilson suggests.

Wilson and Political Ideology

Another of the most explicitly political statements Wilson makes is his comment in *On Human Nature* to this effect:

Marxism is sociobiology without biology. The strongest opposition to the scientific study of human nature has come from a small number of Marxist biologists and anthropologists who are committed to the view that human behavior arises from a very few unstructured drives. They believe that nothing exists in the untrained human mind that cannot be readily channeled to the purpose of the revolutionary socialist state. When faced with the evidence of greater structure, their response has been to declare human nature off limits to further scientific investigation ... anxiety about the health of Marxism as a theory and a belief system is justified. Although Marxism was formulated as the enemy of ignorance and superstition, to the extent that it has become dogmatic it has faltered in that commitment and is now mortally threatened by the discoveries of human sociobiology.[42]

Given the crude similarity that exists between Veblen and Marx's conceptions of human nature, it is likely that Wilson would apply much of

42. Ibid., 191.

the same to Veblen except, of course, that the latter was able to use evolutionary biology and genetics more effectively than Marx because they were more scientifically mature by Veblen's time.

Wilson makes still another revealing comment when he writes that:

> As our knowledge of human nature grows, and we start to elect a system of values on a more objective basis, and our minds at least align with our hearts, the set of trajectories will narrow still more. We already know, to take two extreme and opposite examples, that the worlds of William Graham Sumner, the absolute Social Darwinist, and Mikhail Bakunin, the anarchist, are biologically impossible. As the social sciences mature into predictive disciplines, the permissible trajectories will not only diminish in number but our descendants will be able to sight farther along them.[43]

At this point, it is evident that Wilson's sociobiology has become a species of political ideology. Certain political and ideological objectives are said to be impossible to achieve because the human biological and genetic structure will not permit their attainment. The two examples given above are apparently supposed to represent anarchism of the right and left. Wilson does not elaborate at this point on the impossibility of achieving still other political goals with which he disagrees. Nevertheless, by extrapolation from what he says, it must be biologically impossible to achieve either the noninvidious, nonemulatory industrial republic from which Veblen's dreams were made or the classless society in which the state has withered away favored by Marx.

Genes on a Leash

Wilson's value commitments are often evident in his skepticism regarding the possibility of any large-scale social transformation rooted in moral ideals. In one of the most famous and often quoted passages in his work, he asks:

> Can the cultural evolution of higher ethical values gain a direction and momentum of its own and completely replace genetic evolution? I think not. The genes hold culture on a leash. The leash is very long, but inevitably values will be constrained in accordance with their effects on the human gene pool. The brain is a product of evolution. Human behavior—like the deepest capacities for emotional response which drive and guide it—is the cir-

43. Ibid., 208.

cuitous technique by which human genetic material has been and will be kept intact. Morality has no other demonstrable ultimate function.[44]

In still another passage, the strong element of genetic determinism and biological reductionism that is pervasive in much of Wilson's writing manifests itself:

> Genetic determination narrows the avenue along which further cultural evolution will occur. There is no way at present to guess how far that evolution will proceed. But its past course can be more deeply interpreted and perhaps, with luck and skill, its approximate future direction can be charted. The psychology of individuals will form a key part of this analysis. Despite the imposing holistic traditions of Durkheim in sociology and Radcliffe-Brown in anthropology, cultures are not super-organisms that evolve by their own dynamics. Rather, cultural change is the statistical product of the separate behavioral responses of a large number of human beings who cope as best they can with social existence.[45]

The list of human traits that Wilson alleges have a hereditarian basis is long and specific. It includes aggression, the capacities to select particular aesthetic judgments and religious beliefs, incest taboos, taboos in general, xenophobia, the dichotomization of objects into the sacred and profane, hierarchical dominance systems, susceptibility to indoctrination, intense attention toward leaders, charisma, trophyism, and susceptibility to trance-induction. As Wilson puts it:

> All of these processes act to circumscribe a social group and bind its members together in unquestioning allegiance. Our hypothesis requires that such constraints exist, that they have a physiological basis, and that the

44. Ibid., 16. Sociobiologist Joseph Lopreato cites Veblen in his book *Human Nature and Biocultural Evolution* and then argues that possession of wealth is the criterion necessary for dominance. The expression of dominance takes the form of conspicuous consumption. When individuals put their wealth on display, they are showing their "superiority" through their ability to acquire a disproportionate amount of the community wealth, even though their contribution to the community good may be no greater than, or possibly less than, some of the other, more "inferior" individuals. See Lopreato, *Human Nature and Biocultural Evolution* (Boston: Allen and Unwin, 1984), 35, 111–12, 163, 164. Lopreato displays no awareness, however, of the ways in which Veblen diverges from sociobiology. Unlike Lopreato, William Dugger does not believe that the arguments of Veblen and Wilson parallel each other. See his "Do Genes Hold Culture on a Leash?" *Social Science Quarterly* 62 (June 1981): 243–46, and "Sociobiology for Social Scientists: A Critical Introduction to E. O. Wilson's Evolutionary Paradigm," *Social Science Quarterly* 62 (June 1981): 221–33.
45. Wilson, *On Human Nature*, 16.

physiological basis in turn has a genetic origin. It implies that ecclesiastical choices are influenced by the chain of events that lead from the genes through physiology to constrained learning during single lifetimes.[46]

Although at times Wilson is careful to qualify his stress on heredity by alluding to the influence of culture, his emphasis on genetic inheritance overall is marked:

Although the genes have given away most of their sovereignty, they maintain a certain amount of influence in at least the behavioral qualities that underlie variations between cultures. Moderately high heritability has been documented in introversion-extroversion measures, personal tempo, psychomotor and sports activities, neuroticism, dominance, depression, and the tendency toward certain forms of mental illness such as schizophrenia. Even a small portion of this variance invested in population differences might predispose societies toward cultural differences. At the very least, we should try to measure this amount. It is not valid to point to the absence of a behavioral trait in one or a few societies as conclusive evidence that the trait is environmentally induced and has no genetic disposition in man. The very opposite is true.[47]

Wilson's hereditarian stress reaches its apex in his claims that:

It is possible, and in my judgment even probable, that the positions of genes having indirect effects on the most complex forms of behavior will soon be mapped on the human chromosomes. These genes are unlikely to prescribe particular patterns of behavior; there will be no mutations for a particular sexual practice or mode of dress. The behavioral genes more probably influence the ranges of the form and intensity of emotional responses, the thresholds of arousals, the readiness of learning certain stimuli as opposed to others, and the pattern of sensitivity to additional environmental factors that point cultural evolution in one direction as opposed to another.[48]

Not surprisingly, Wilson suggests that there may be some correlation between genes and worldly success, that is, that power, class, and status may have a hereditarian basis. He admits, however, that this is an unsettled issue when he comments that "a key question of human biology is whether there exists a genetic predisposition to enter certain classes and to play certain roles [in them]."[49]

46. Ibid., 78.
47. Ibid., 177.
48. Ibid., 550.
49. Ibid., 47.

It is clearly implied, although not always explicitly stated by sociobiologists including Wilson and his collaborators, that disparities in class, status, and power have biological roots. However, they have presented little empirical evidence to prove this. If, and when, sociobiologists are able to demonstrate that social aggregates are divided in particular unequal ways along lines of class, status, and power, due to the interaction of genes with social environment, then social scientists will find the claims of sociobiology more compelling.

To most social scientists, questions regarding the origin and genesis of disparities of power, status, and wealth are enormously important. Sociobiology a la Wilson claims either that it can explain or at least has the potential to explain why some social aggregates achieve social superiority and dominance while others remain in subservient positions. Why, then, a full academic generation after the revival of sociobiology is it still unable to explain the relationship between class, status, and power in classical Weberian terms or in any other conceptual framework acceptable to social scientists? Sociobiology may have much to contribute to the dialogue between ideological antagonists and conflicting schools of thought in the social sciences. But up to now, its contribution consists largely of explanations regarding why some individuals differ from other individuals. Indeed, with the possible exception of gender differences, sociobiologists are presently unable to offer much illumination on the genesis of inequalities of power, class, and status.

In his efforts to be more precise regarding the ways in which the human mind is genetically geared and driven, Wilson writes:

> So the human mind is not a tabula rasa, a clean slate on which experience draws intricate pictures with lines and dots. It is more accurately described as an autonomous decision-making instrument, an alert scanner of the environment that approaches certain kinds of choices and not others in the first place, then innately leans toward one option as opposed to others and urges the body into action according to a flexible schedule that shifts automatically and gradually from infancy into old age. The accumulation of old choices, the memory of them, the reflection on those to come, the re-experiencing of emotions by which they were engendered, all constitute the mind. Particularities in decision making distinguish one human being from another. But the rules followed are tight enough to produce a broad overlap in the decisions taken by all individuals and hence a convergence powerful enough to be labeled human nature.[50]

50. Ibid., 67.

Veblen, of course, does not describe the mind in any such fashion, because he views it primarily as containing layer after layer of material deposited from the environment; "habits of mind," as he calls them, are the consequences of the cultural and institutional situation and, except possibly for his Kantian assumptions regarding the tendency of the brain to view time, space, and motion in particular ways, he does not delineate the actual psychological mechanisms by which culture is induced. The reason for this is that Veblen is more influenced by radical environmentalism and behaviorism than Wilson and, although he does not deny the role of genes in determining the evolution and behavior of the species, the human psyche in his view is more a product of the cultural environment.

Conclusion

Anglo-American evolutionary psychologists and physiologists such as George Romanes, William James, C. Lloyd Morgan, William McDougall, and Jacques Loeb all worked more or less consciously in the Darwinian tradition. All used Darwin's argument that human nature, including our moral and emotional life, evolved from animals just as human anatomy and physiology did, and all emphasized the role of instinct in human behavior, which was then viewed as a fundamental insight of psychology. Veblen approvingly cites parts of the body of literature produced by these evolutionists and was perhaps more influenced by them than by any other group of thinkers in formulating the psychological aspects of his social theory. He proposed the development of an evolutionary economics in the Darwinian sense and, in fact, was more swayed by Darwin's evolutionary theory than by any other single ideational source.[51] Since sociobiology, too, traces much of its pedigree to Darwinism, it and Veblen share a common intellectual heritage.

However, Veblen was also influenced by some radical environmental determinists including John Dewey and Edward Bellamy[52] to name two. He was also very familiar with the work of social and cultural anthropologists such as Franz Boas,[53] James Frazier, Edward Tylor, and Lewis Hen-

51. See esp. Veblen, *The Instinct of Workmanship*, ch. 1.
52. See Stephen Edgell and Rick Tilman, "The Intellectual Antecedents of Thorstein Veblen: A Reappraisal," 1003–26.
53. Veblen was personally acquainted with Boas, and correspondence between the two indicates that Veblen was active in attempting to secure an academic post for Alexander Goldenweiser, a radical environmental determinist much influenced by Boas, at the New School for Social Research in New York City. See Franz Boas to

ry Morgan and with their emphasis on the malleability and plasticity of the human species. Thus, tracing his intellectual antecedents, as we have just done, will not resolve the issue of his relationship to contemporary sociobiology.

The overall Darwinian framework in which Veblen's analysis is set has never been in doubt, although the consistency of his application of Darwinian principles has been challenged. Nevertheless, a distinction must be made between the selection arguments Veblen makes in *The Theory of the Leisure Class* and the arguments for adaptation through habituation often found in his later work. The material quoted here from the former deals primarily with the institutional framework while, for example, Veblen's 1913 essays on "The Mutation Theory and the Blond Race" and "The Blond Race and the Aryan Culture" focus on adaptation through changes of habit.[54] Even so, it is a matter of emphasis that differentiates the two phases of Veblen's work from each other.

Veblen explains the origin of the malignant and destructive tendencies that he claims emerged during the second or predatory phase of cultural evolution by suggesting that they were due to mutations. But he qualifies his claim by asserting that hybrids, such as Western peoples, tend to breed "true" to the original type of the peaceful, savage stage when cultural evolution began.[55]

Veblen is treated here neither as a sociobiologist nor as an opponent of sociobiology. Rather, an effort has been made to demonstrate how in certain respects his theories converge with it, yet in other respects diverge from it. However, Wilson's version of sociobiology is not compatible with Veblen's institutional economics or that brand of evolutionary economics that currently labels itself "radical institutionalism."[56] Ultimately, what

---

Thorstein Veblen, March 12, 1919, and Veblen to Boas, March 12, 1919, Franz Boas Collection, American Philosophical Society, Philadelphia.

54. Veblen, *The Place of Science in Modern Civilization*, 457–96.

55. Veblen, *The Theory of the Leisure Class*, ch. 9.

56. No discussion of this sort on sociobiology would be complete without citing Stephen Jay Gould, one of the main protagonists in the debate. "Genes influence many aspects of human behavior, but we cannot say that such behavior is caused by genes in any direct way. We cannot even claim that a given behavior is, say, 40 percent genetic and 60 percent environmental, and thereby defend at least a partial old-fashioned genetic determinism. Genes and environment interact in a non-additive way, yielding emergent features in the resulting anatomies, physiologies, and behaviors." See, for example, the radical egalitarian and environmental determinism that informs the essays in Dugger and compare with E. O. Wilson as exemplified in his interview with Claude Fischler. William Dugger, ed., *Radical Institutionalism: Contemporary Voices* (Westport, CT: Greenwood Press, 1989); Claude Fischler, Interview with Edward O. Wilson, in *Le Monde*, February 24, 1980, 15; and Stephen J. Gould, "The Confusion over Evolution," *New York Review of Books* 39, no. 19 (November 19, 1992).

separates Veblen as institutional economist from the sociobiologist Wilson is the former's belief in the human potentiality for social transformation and the latter's pessimism, indeed, at times, cynicism regarding the transformative political potentiality of the species.[57] This, too, is what separates sociobiology from contemporary evolutionary economics. However, as we have tried to demonstrate, biological concepts and methods are too deeply embedded in heterodox economics to warrant exclusion from its evolving social theory. In any case, no analysis of Veblen's evolutionary naturalism would be complete without considering the sociobiological elements in his thought.[58]

57. Although Wilson further refined his arguments, there was no significant change in their political import. See, for example, Charles Lunsden and Edward O. Wilson, "The Relation between Biological and Cultural Evolution," *Journal of Social and Biological Structures* 8 (1985): 343–59; "Genes, Mind and Ideology," *The Sciences* 21 (November 1981): 6–9; *Genes, Mind and Culture: The Coevolutionary Process* (Cambridge: Harvard University Press, 1981); *Nature Revealed: Selected Writings, 1949–2006* (Baltimore: Johns Hopkins University Press, 2006), 167–70; and Michael Ruse and Edward O. Wilson, "Moral Philosophy as Applied Science," *Philosophy* 61 (1986): 173–99.

58. However, a leading Veblen scholar writes:

> If Social Darwinism simply means the application of Darwinian ideas to social phenomena, then Wilson stands condemned, along with Kropotkin, Ritchie, Veblen and many modern writers who have also applied Darwinian principles of variation, selection and inheritance to socio-economic change. In contrast, if Social Darwinism means the use of Darwinism to justify individualist, conservative or racist views, then Wilson must be acquitted. Alternatively, if Wilson is charged with claiming to explain human social phenomena *entirely* in biological terms, then he must also be acquitted, partly on the grounds of his explicit and repeated claims to the contrary. But if Wilson is charged with exaggerating the possibility of using biology to explain human behavior, then there still remains a strong case against him to be answered. It all depends on the precise charge. The imprecise accusation of "Social Darwinism" is of little help.

Hodgson, "Social Darwinism in Anglophone Academic Journals," 448.

Chapter 9

Veblen and the Sociologists
of Knowledge

Introduction

During the interwar and immediate postwar eras, the sociology of knowledge became a formal part of the discipline of sociology. Some of the best-known sociologists of the period who made notable contributions to its development included Karl Mannheim, Robert Merton, C. Wright Mills, and, of course, Max Weber. To this list must be added the Italian communist activist and theoretician Antonio Gramsci, who was not a professional sociologist, and the philosopher and historian of science Thomas Kuhn. Our aim is to use some of their theoretical contributions to better understand Veblen's sociology of knowledge.[1]

The early professional sociology of knowledge dealt with the influence of social phenomena upon knowledge. At its most effective, it gave no theoretical immunity from sociological analyses to any special class of cognitive phenomena. Indeed, as in Veblen's own analysis, most cognitive elements can be linked to social determinants, although so linking them does not necessarily relieve the social scientist of the obligation of ascertaining their empirical validity or logical consistency. Understanding the cultural genesis and social determinants of ideas may, however, further an understanding of what Thelma Lavine calls "directional norms," that is:

---

1. On Veblen's sociology of knowledge, cf. Walter P. Metzger, "Ideology and the Intellectual: A Study of Thorstein Veblen," *Philosophy of Science* 16 (April 1949): 125–33; Frank J. Weed, "Thorstein Veblen's Sociology of Knowledge," *Review of Social Theory* 1, no. 1 (September 1972): 1–11; Warren J. Samuels, "The Self-Referentiability of Thorstein Veblen's Theory of the Preconceptions of Economic Science," *Journal of Economic Issues* 24, no. 3 (September 1990): 695–718; and Rick Tilman, "The Frankfurt School and the Problem of Social Rationality in Thorstein Veblen," *History of the Human Sciences* 12, no. 1 (February 1999): 91–109. These are reprinted in Rick Tilman, ed., *The Legacy of Thorstein Veblen*, Vol. I (Cheltenham: Edward Elgar Publishing, 2003).

The principle of cultural uniqueness, entailing the demonstrable unique-ness of cognition as a culturally responsive phenomenon; the principle of the historical and social continuity between primitive and contemporary forms of thought; the concept of religion as a symbolic formulation of social structure; the concept that the variant of established religion consists in re-sponsiveness to a specific social demand; the principle of the multiple re-sponsiveness, intellectual, social, and political, of technological innovation; the concept of technological innovation as responsiveness to the concrete historical stage of class conflict; the concept that objective political thought is ideally responsive to a particular social class at a particular historical stage of the class conflict; the concept that the fundamental characteristics of thought are responsive to an alternating fixed polarity of epochal social qualities; the concept that canons of logic, basic categories, criteria, and pro-cedures developed historically in connection with the special sciences are responsive to a multiplicity of social demands.[2]

Veblen scholars will readily grasp the relationships between the direc-tional norms articulated by Lavine and key aspects of Veblen's thought. There is, for example, the historical and social continuity between primi-tive and contemporary forms of thought which plays such an important role in his theory of status emulation. Moreover, conspicuous consump-tion, waste, and exemption from useful labor can all be linked with both the social mores and cultural practices of early modernity and the thought of the contemporary leisure class of industrial society in his theorizing. Many analyses of these aspects of Veblen's work exist, but they do not em-ploy the paradigms of the particular sociologists of knowledge used here. In short, no systematic attempt exists to apply the analytic apparatus of the leading sociologists of knowledge of the early- to mid-twentieth cen-tury to his ideas. Veblen was a highly perceptive analyst of the social and cultural determinants of ideas, and it seems appropriate to turn the tables on his mockery and satire by subjecting him to his own theory and those of others.

I shall examine Kuhn's "scientific paradigm" and "paradigmatic change" and Gramsci's notions of "false consciousness," "ideological he-gemony," and "organic intellectuals" to better understand their relation-ships to their equivalents in Veblen's sociology of knowledge. Mann-heim's concepts of "ideology" and "utopia," "relationism," "rationality," and the "floating strata of socially unattached intellectuals" will be ana-

---

2. Thelma Lavine, "Naturalism and the Sociological Analysis of Knowledge," in *Naturalism and the Human Spirit,* ed. Yervant H. Krikorian (New York and London: Co-lumbia University Press, 1944), 201–2.

lyzed with the same purpose in mind. Weber's notion of "rationality," Merton's "latent and manifest function of ideas," and Mills's views on rationality will also be given due consideration in terms of finding the equivalents in Veblen, or in using these devices to reinterpret the meaning of his work. In short, this chapter attempts to use the literature on the sociology of knowledge, particularly its leading exemplars in the interwar and early postwar eras, to better understand Veblen's own contribution to this perspective.

It is important to note that several of these sociologists of knowledge were quite knowledgeable about Veblen's work, including his own sociology of knowledge. Weber, Mannheim, Merton and, especially, Mills knew his work and made comments on or use of it. However, due to Veblen's status as an economist, it is essential to look briefly at the doctrinal divisions that now exist in that discipline before the analytic tools of the sociology of knowledge are used to dissect his work. I will briefly focus on the exclusionary nature and lack of social reality in mainline (neoclassical) economics before returning to an analysis of Veblen through the theoretical apparatus provided by leading figures in the sociology of knowledge.

Ideological Bias, Epistemological Privilege, and Intellectual Conflict in Political Economy

The claim of Nobel laureate Milton Friedman that only neoclassical economics can lay claim to the continuing status of an empirical science is interesting primarily because it is simply more evidence of the obliviousness of many economists to the impact of postmodernism on the study of society. Contemporary philosophy of science, history of science as paradigmatic shift, sociology of knowledge, and social epistemology all point to a more sophisticated and penetrating view of what it means to claim freedom from ideological bias as well as to claim a position of epistemological privilege. Although Friedman's contributions to modern economics do not require his recent ruminations for their legitimation, who can look at the continued proliferation of both paradigms and doctrinal groupings in economics and political economy, and share his confidence that his school of thought, and his alone, is "scientific"?[3]

3. Recently, at a conference in London before an audience that included Margaret Thatcher, Arthur Schlesinger Jr. commented that "because Stalin has been proved wrong does not make Milton Friedman right." Friedman's old professor at the Uni-

Friedman's fellow Nobel Prize winner and former student Gary Becker informs us that the "combined assumptions of maximizing behavior, market equilibrium, and stable preferences, used relentlessly and unflinchingly, form the heart of the economic approach."[4] No doubt, in response to such claims, Douglas Dowd asserts that:

> By assumption and by focus, the neo-classical economist is led to ignore (1) the persistent clash of interests in society, as reflected in the realm of politics and social movements, and (2) the fundamental and unsettling force of technology and technological change. In taking institutions as given, the neo-classical economist (3) is led to define as an improvement in welfare only those changes that involve neither social conflict nor institutional change, and (4) is inhibited from examining the social process at all, except as something "outside" the theory.[5]

To this Dowd adds:

> In contrast to the neo-classical economists—and many conventional economists still today—Veblen took as the central problem of the economist, (1) the restlessness and imperatives of technology, (2) the institutions shaping human behavior, (3) the social psychology of man, and (4) the interaction of all these as determining the quality of a society at any time and as giving rise to social change and conflict.[6]

---

versity of Chicago Frank Knight, himself a strong critic of Veblen, reacted as follows to the challenge to neoclassicism:

> To no one's surprise, Knight launched into a passionate defense of the old order's superior methodology and of the uncompromising claim that neoclassical economists, unlike the vulgar Keynesians, had membership in the natural sciences. All proper theory, he told his listeners, was based to begin with in axioms and "*eternal and immutable laws.*" Then, in a vivid premonition of the later "Rational Expectations" and "New Classical" schools that emerged from Chicago in the 1970s, Knight said flatly and summarily that economics required assumptions of "rational and errorless choice, presupposing perfect foresight" and of "foreknowledge free from uncertainty," based on the "assumption of 'atomic' units negligible in size and hence continuous variability of all magnitudes." With these and a few other assumptions based on an idealized model of perfect competition, the essence of economic theory was complete. "*There is no possibility,*" he imperiously concluded, "*that new laws will be discovered comparable in generality and importance with the basic principles long recognized.*"

Quoted from Richard Parker, *John Kenneth Galbraith: His Life, His Politics, His Economics* (New York: Farrar, Straus and Giroux, 2005), 197.

4. As quoted in Victor Fuchs, *How We Live* (Cambridge, MA: Harvard University Press, 1983), 1.

5. Douglas Dowd, "The Strengths and Weaknesses of Veblen," 22.

6. Ibid., 24.

Interestingly, Dowd also comments:

> The economist alone has the professional function to examine and under-
> stand the economic interests of society as a whole; when he abdicates that
> function, whether by conscious choice or by the focus and method of his
> analysis, he leaves the public unarmed to face the continuing and self-seeking
> thrusts of "the vested interests"—in business, in labor, in agriculture, in pol-
> itics, or wherever they may exist.[7]

Clearly, Veblen, like Dowd, did not permit the neoclassical economists of
his time to interpret and articulate the economic interests of society as a
whole without objection.

Consider, too, the formation of new associations within the discipline
Friedman claims to exemplify, most of which meet regularly, have hun-
dreds of professional economists as members, and publish their own journals. Doctrinally and politically these organizations can be grouped from
left to center across the ideological spectrum and include several of substantial size.[8]

Claims of freedom *from* politico-ideological bias are usually linked with
claims *for* a position of epistemological privilege; but skepticism regarding such claims need not necessarily lead to intellectual deadlock, nihilism, or radical relativism. As Mannheim argued long ago in *Ideology and
Utopia* (1936) the emergence of a "floating strata" of socially detached intellectuals does not mean that the "truth-claims" of social-scientific work
must be divorced from the demands of either logical consistency or em-

---

7. Ibid., 23.

8. O'Hara and Sherman offer this brief historical overview:

In the 1960s and 1970s there was a revival of non-neoclassical political economy, or
heterodox economics, in many nations which has continued to this day. Original institutionalists started their own organization, the Association for Evolutionary Economics (AFEE) in 1965, followed by the *Journal of Economic Issues* in 1967. Other radicals and Marxists organized the Union for Radical Political Economics (URPE) in
1968 and the *Review of Radical Political Economics* the following year. In many nations
similar organizations and journals were instituted in the 1970s. Post Keynesians
started to publish the *Cambridge Journal of Economics* in England and the *Journal of
Post Keynesian Economics,* inaugurated by Sidney Weintraub and Paul Davidson from
the USA. The Association for Social Economics started to become more inclusive and
participatory by including many heterodox themes in its *Review of Social Economy.* It
wasn't until the late 1980s that the European Association for Evolutionary Political
Economy was formed and the early 1990s when feminists became active in organizing their own International Association for Feminist Economics along with their
journal, *Feminist Economics.*

P. A. O'Hara and H. J. Sherman, "Veblen and Sweezy on Monopoly Capital, Crises,
Conflict and the State," *Journal of Economic Issues* 38 (December 2004): 969.

pirical validation. He did suggest, however, that such claims cannot be fully or adequately understood apart from their social genesis, historical and paradigmatic development, and the political and social consequences of their adoption and implementation.

Of particular interest to those hoping to escape the intellectual malaise that infects the change-resistant elements in the economics profession are the ways and means by which economic inquiry is inescapably value-impregnated. This does not mean merely that they are skeptical toward or reject the positivist version of the fact-value dichotomy in epistemology or the positive-normative distinction in economics. It also signifies a conviction that there is little of significance in contemporary social science inquiry that is not permeated in one way or another with moral, political, and aesthetic values. Economists, for example, make value judgments when they choose topics of inquiry, select methods to use in this inquiry, and decide how their findings are relevant to policy analysis. So how and why is it that neoclassical economists like Friedman claim both freedom from politico-ideological bias and, for themselves, a position of epistemological privilege? Upon what articulated criteria can such claims be based other than *hubris?*

To illustrate, perhaps the least flattering estimate of Veblen in recent years comes from the economist Thomas Sowell, himself a Veblen scholar. This is not unexpected in view of Sowell's own political conservatism and his affiliation with the strongly procorporate and ideologically conservative Hoover Institution. In an "authoritative" entry in *The New Palgrave: A Dictionary of Economics,* Sowell finds no real merit or validity in Veblen's work and concludes that "it is difficult to see how economics as it exists today is any different from what it would have been had there been no Thorstein Veblen."[9] Were this statement applied to the discipline of economics as a whole, its inaccuracy would be evident. However, it is likely from the context in which Sowell wrote that he did not regard heterodox forms of economics such as Marxism, institutionalism, social economics, and, perhaps, post-Keynesianism as "objective" kinds of economic inquiry. Sowell's statement was apparently intended to apply to an economics profession that is so entirely neoclassical in orientation that dissenters such as Veblen are banished to the hinterlands of sociology and social philosophy. Although Sowell's statement rests on paradigmatic assumptions that fundamentally distort what actually exists in the politically fragmented discipline of economics, it is important in that it exemplifies a mindset that has long existed in American neoclassical econom-

9. Thomas Sowell, "Veblen, Thorstein," 800.

ics. Indeed, it shows that the doctrinal biases Veblen fought against in his own time remain both tenacious and influential. Their pervasiveness can only lend legitimacy to efforts to account for the persistence of such interpretations of the corpus of Veblen's work. In 1973 Joseph Dorfman wrote in summation of the controversy over Veblen that:

> as to the merits of his work, opinions differ more widely and more fervently than on any other writer of equal prominence. He is rated among the great economists of history, or as no economist at all; as a great original pioneer or as a critic and satirist without constructive talent or achievement. And he was, one might almost say, all of these things; from different standpoints and by different criteria, each of which it is possible to understand and even to appreciate. One thing at least can be said. If he chose to paint after a futurist technique of his own devising, it was not for lack of capacity to master the academic canons.[10]

Veblen, a man of massive erudition, had few peers when it came to a knowledge of the social sciences, philosophy, and languages. His mastery of the academic canons is thus not in doubt. It is his utilization of them to achieve political and moral ends not shared by his critics that is the real point of the controversy.

Veblen and Hubris

Veblen warns against such hubris in several ways. First, and perhaps most important, he cautions against the use of knowledge, scientific or not, on behalf of vested interests whether they are class interests, corporate interests, ideological, racial, cultural, and religious identities, or otherwise; in short, in opposition to hierarchical or antiegalitarian claims, values, and processes. One such set of vested interests were those of neoclassical economics and its practitioners; the political, moral, and ideological functions of their ideas were obvious to Veblen, and he spent considerable time in exposing the underpinnings of these doctrines. Nevertheless, the possibility of obtaining access to "objective knowledge," whatever that might mean, still appeals to many social scientists. But, at least partly under Veblen's influence, they have come to recognize the difficulty of attempting to achieve this. Indeed, particularly in the social sciences, the sociology of knowledge points to the role of history, cul-

10. Joseph Dorfman, "New Light on Veblen," in *Essays, Reviews and Reports* (New York: Augustus M. Kelley, 1973), 596.

ture, social structure, moral and political conflict, and power relations in thwarting such efforts. Veblen's work in these realms can be treated as a running commentary on the cultivation and exposure of such bias.

More specifically, in Veblen's time, perhaps five epistemological stances outside the realm of religion and revelation had emerged. These were (1) the correspondence theory of British empiricism, (2) the coherence theory of Continental rationalism, (3) the instrumentalist theory of the American pragmatists, (4) an eclectic theory that combines the above in varying ways and proportions, and (5) the sociology of knowledge. Although Veblen did not adhere unequivocally to any of these, he clearly used aspects of all. His methods of reasoning, inquiry, and justification show that he (1) sought out the correspondence between images in the mind, sense data that imposed them and external objects that produced the sense data; (2) preferred coherent reasoning, that is, a systematic consistent explanation of all the facts of experience; (3) looked favorably at theories that "worked," that is, facilitated causal explanations of events or accurate forecasts of future occurrences; (4) used eclectic combinations of all five theories in his own work and criticized some of the existing eclecticism that he encountered for its paste-pot composition, that is, its undiscriminating lumping together of an endless diversity of ideas into a shapeless form; and, (5) focused on the social determinants of knowledge such as class, power, status, authority, and tradition.

Examples in Veblen's work of each of the five different epistemologies can be found in the methodological approaches he used during his long scholarly career, which lasted from 1884 to 1925. To illustrate, the correspondence theory cannot be disaggregated from the quantitative-statistical method that he used in his study of American wheat prices.[11] Veblen's use of the coherence theory can be found in his exchange with John Cummings where he rebutted the latter's attack on *The Theory of the Leisure Class* by criticizing Cummings's use of the logical principle of the excluded middle which led him to view its subject matter in terms of "exclusive alternative."[12] Veblen's employment of pragmatic-instrumentalist epistemology to explain the past and future behavior of consumers relies on his theory of status emulation which predicts emulatory consumption to enhance status through "unremitting demonstration of ability to pay" as a

---

11. See Veblen, "The Price of Wheat since 1867," *Journal of Political Economy* 1 (December 1892): 68–103, and "The Food Supply and the Price of Wheat," *Journal of Political Economy* 2 (June 1893): 365–79.

12. See Veblen, "Mr. Cummings' Strictures on 'The Theory of the Leisure Class,'" in *Essays in Our Changing Order*, ed. Leon Ardzrooni (New York: Augustus M. Kelley, 1964), 16–31.

continuing and persistent kind of social behavior. His own utilization of a sociology-of-knowledge approach is evident in the methods he employed to explain the "make-believe" of animism, natural law, and vitalism,[13] nor can his amalgam of theory and insight be withdrawn from cultural lag and the data of cultural anthropology.

But Veblen's eclecticism should not mislead the reader. In his view eclecticism should *not* be a pluralism that gives an equal weight to all factors. Neither should it be an accommodationism from which no transcending synthesis emerges, nor a balancing process within which conflicting elements are not modified or purged but remain to produce a disharmonious whole. Veblen believed much "liberal" scholarship in the social studies suffered from such faults, and that a paste-pot eclecticism was not the way to overcome them.

But what were his intent and emphasis as an eclectic? Apparently, it was to use social theory and criticism to unmask the predatory intentions and functions of classes or elites bent on the exercise of power, wealth, and status through the manipulation of invidious knowledge. In short, his eclecticism aimed at stopping or inhibiting the re-creation of invidious social orders through the use of knowledge which is noninstrumental; and, conversely, as Marc Tool suggests, at the re-creation of noninvidious community through the use of instrumentally warranted knowledge.[14]

Objectivity in Veblen's Social Science

Veblen disliked both *formalist* and *providentialist* theoretical perspectives and thus embedded in his socioeconomic analyses a skeptical emphasis on wholism and contingency (uncertainty and indeterminacy are thus important factors in his work). He also believed that findings in biology, psychology, and philosophy naturalized the human conception both of the world and of themselves. In particular, after Darwin's monumental findings were publicized, no intellectual rationale for the existing forms of society could claim permanent ontological status or anything more than temporary epistemological and axiomatic status. For these reasons among others Veblen rejected redemptive philosophies of history, static epistemologies, and ontologies and rigidly formal as well as entire-

13. See Charles Rasmussen and Rick Tilman, *Jacques Loeb: His Science and Social Activism and Their Philosophical Foundations* (Philadelphia: American Philosophical Society, 1998), ch. 5, for analysis of Veblen's (and Loeb's) views on animism and vitalism.

14. See Marc Tool, *The Discretionary Economy: A Normative Theory of Political Economy* (Santa Monica: Good Year Publishing Company, 1979), 300–312.

ly subjective kinds of ethical systems in favor of a naturalistic theory of humankind.

The meaning of "objectivity" is often debated and with good reason and, in any case, critical social scientists more or less continuously ponder its meaning. The political theorist Wilson Carey McWilliams writes:

> Modern social scientists have abandoned any hope for total "objectivity," conceding that it is impossible to eliminate the influence of emotions and sentiments on perception. In fact, if there is one generally accepted principle in social science, it is that individual needs and satisfactions influence what an individual sees and is willing to learn; and because groups play a vital role in satisfying needs, they are recognized as influencing and structuring knowledge and perception. Social scientists, however, continue to treat objectivity as desirable if unattainable, a fact which causes them to center on the distortions of reality produced by group influences. Rarely if ever has social science concerned itself with the positive role which groups or emotions can play in individual perception.[15]

But the point of this brief commentary on "objectivity" in the sociology of knowledge is not to focus on Veblen's personal eccentricities as a thinker so much as to locate and define what he means when he uses synonyms for it. For Veblen "objectivity" and "dispassion" are the concomitants of the scientific laboratory and the machine process as well as the accoutrements of scholarship and learning. However, in a unique passage in *The Higher Learning*, Veblen writes:

> The modern technology of an impersonal, matter-of-fact character in an unexampled degree, and the accountancy of modern business management is also of an extremely dispassionate and impartially exacting nature. It results that the modern learning is of a similarly matter-of-fact, mechanistic complexion, and that it similarly leans on statistically dispassionate tests and formulations. Whereas it may fairly be said that the personal equation once—in the days of scholastic learning—was the central and decisive factor in the systematization of knowledge, it is equally fair to say that in latter time no effort is spared to eliminate all bias of personality from the technique or the results of science or scholarship. It is the "dry light of science" that is always in request, and great pains are taken to exclude all color of sentimentality.[16]

15. Wilson Carey McWilliams, *The Idea of Fraternity in America* (Berkeley: University of California Press, 1973), 56.

16. Thorstein Veblen, *The Higher Learning in America: A Memorandum on the Conduct of Universities by Business Men* (New York: Augustus M. Kelley, 1965), 7. J. Patrick Raines and Charles Leathers, *The Economic Institutions of Higher Education: Economic*

The novelty of this passage in his writing resides in the fact that the reader is told that "objectivity" can be found in "modern technology," "the accountancy of modern business management," and "modern learning," that is, "science and scholarship." It is significant also in that this signifies extreme dispassion and precision, "matter-of-fact, mechanistic complexion," "elimination of all bias of personality," and "exclusion of all color of sentimentality." It would be difficult to find a more succinct summation in Veblen's writing of what "objectivity" apparently meant to him. However, in the larger context of his thought it also explains the dangers of such "objectivity" when it is at the command or the disposal of the captains of industry and finance not to mention the militarists to which Veblen fails to allude, although he commonly mentioned them elsewhere in his writings on war and dynasticism.[17]

Shifts and Transitions in Habits of Thought

There is, of course, much emphasis in Veblen on the psychological impact of the machine process. This explains in part cultural lag as well as cultural progression in his theorizing:

> The continuity comprised in the concept of process as applied to conduct is consequently a spiritual teleological continuity: whereas the concept of process under the second head, the non-teleological sequence, comprises a continuity of a quantitative, causal kind, substantially the conservation of energy. In its turn the growing resort to categories of process in the formulation of knowledge is probably due to the epistemological discipline of modern mechanical industry, the technological exigencies of which enforce a constant recourse to the apprehension of phenomena in terms of process, different therein from the earlier forms of industry, which neither obtruded visible mechanical process too constantly upon the apprehension nor so imperatively demanded an articulate recognition of continuity in the processes actually involved. The contrast in this respect is still more pronounced between the discipline of modern life in an industrial community and the discipline of life under the conventions of status and exploit that formerly prevailed.[18]

Theories of University Behavior (Cheltenham: Edward Elgar, 2003), write that: "The nature, character, and content of knowledge are subjective to the forces of institutional evolutionary change over time," 93.

17. See Veblen, *An Inquiry into the Nature of Peace* (New York: Macmillan, 1917), ch. 2 and 7, and *Imperial Germany*, ch. 5 and 8.

18. Veblen, *The Place of Science in Modern Civilization*, 158–59.

The key phrase in this quotation is "epistemological discipline of modern mechanical industry," but Veblen is careful to point out that often this does not work to its logical end in any rigorous fashion because circumstances will not permit it to:

> Life in an advanced industrial community does not tolerate a neglect of mechanical fact; for the mechanical sequences through which men, at an appreciable degree of culture, work out their livelihood, are no respecters of persons or of will-power. Still, on all but the higher industrial stages, the coercive discipline of industrial life, and of the scheme of life that inculcates regard for the mechanical facts of industry, is greatly mitigated by the largely haphazard character of industry, and by the great extent to which man continues to be the prime mover in industry. So long as industrial efficiency is chiefly a matter of the handicraftsman's skill, dexterity and diligence, the attention of men in looking to the industrial process is met by the figure of the workman, as the chief and characteristic factor; and thereby it comes to run on the personal element in industry.[19]

On the whole then, Veblen believes that at least since the Industrial Revolution got under way in various countries on a large scale the epistemological discipline of the machine process linked with the spread of the ethos of science are the main causes of the shifts and transitions in habits of thought. That these shifts and transitions are incomplete and that the resulting habits of thought are not fully modern nor rigorously consistent with the long-run impact of science and technology is due to inadequate exposure to them on the part of the labor force and to certain forms of institutional and cultural resistance. It is to the latter that we now turn.

Institutional and Cultural Resistance to Change under Conditions Approaching Modernity

Veblen's stages of social development—savagery, barbarism, handicraft, and machine technology under corporate hegemony and, often, absentee ownership—are well known to his readers. Our treatment of his theory of cultural lag is primarily reserved to analysis of the last two stages, particularly the stage of the machine process and corporate domination. Yet the intent of his theory is sufficiently generic to cover all four stages of social development as this quotation indicates:

19. Ibid., 104.

With the advent of the predatory stage of life there comes a change in the requirements of the successful human character. Men's habits of life are required to adapt themselves to new exigencies under a new scheme of human relations. The same unfolding of energy, which had previously found expression in the traits of savage life recited above, is now required to find expression along a new line of action, in a new group of habitual responses to altered stimuli. The methods which, as counted in terms of facility of life, answered measurably under the earlier conditions, are no longer adequate under the new conditions. The earlier situation was characterized by a relative absence of antagonism or differentiation of interests, the later situation by an emulation constantly increasing in intensity and narrowing in scope. The traits which characterize the predatory and subsequent stages of culture, and which indicate the types of man best fitted to survive under the regime of status, are (in their primary expression) ferocity, self-seeking, clannishness, and disingenuousness—a free resort to force and fraud.[20]

But let us turn once again to the final stage of his scheme where he attempts to explain institutional and cultural resistance to modernity. On this occasion, he is explaining why the retardation of the business community vitiates full use of the existing industrial plant. As he puts it:

A somewhat thorough review of the pertinent facts would probably persuade any impartial observer that, one year with another, such businesslike enforced idleness of plant and personnel lowers the actual output of the country's industry by something nearer fifty percent of its ordinary capacity when fully employed. To many, such an assertion may seem extravagant, but with further reflection on the well-known facts in the case it will seem less so in proportion as the unfamiliarity of it wears off. However, the point of attention in the case is not the precise, nor the approximate, percentages of this arrest and retardation, this partial neutralization of modern improvements in the industrial arts; it is only the notorious fact that such arrest occurs, systematically and advisedly, under the rule of business exigencies, and that there is no corrective to be found for it that will comport with those fundamental articles of the democratic faith on which the businessmen necessarily proceed. Any effectual corrective would break the framework of democratic law and order, since it would have to traverse the inalienable right of men who are born free and equal, each freely to deal or not to deal in any pecuniary conjuncture that arises.[21]

A page earlier, he wrote that:

20. Veblen, *The Theory of the Leisure Class*, 152.
21. Veblen, *The Nature of Peace*, 173.

It is not that the captains of industry are at fault in so failing, or refusing, to supply the needs of the community under these circumstances, but only that they are helpless under the exigencies of business. They cannot supply the goods except for a price, indeed not except for a remunerative price, a price which will add something to the capital values which they are venturing in their various enterprise. So long as the exigencies of price and of pecuniary gain rule the case, there is manifestly no escaping this enforced idleness of the country's productive forces. It may not be out of place also to remark, by way of parenthesis, that this highly productive state of the industrial arts, which is embodied in the industrial plant and processes that so are systematically and advisedly retarded or arrested under the rule of business, is at the same time the particular pride of civilized men and the most tangible achievement of the civilized world.[22]

Much of the "law and the prophecy" in Veblen's sociology of knowledge are thus embedded in his indictment not only of the business community itself; but in the structural obstacles to change which prevent it from maximizing the full productive capacity of the industrial system itself.

Historically and anthropologically, Veblen's sociology of knowledge has deep roots, first, in his critique of conventional economics and, second, in his stages theory of social and economic development. I have dealt with the ideological structure of classical and neoclassical economics elsewhere, but to briefly recapitulate them as Veblen perceived them, they were viewed as a form of apologetics for the activities of the business community and in a larger sense the rationale for the ownership and acquisitiveness of the leisure class. Utility, taxonomy, teleology, and hedonism combined with unrealistic assumptions about reality ("normality" and "natural rights") to distort the actual structure and workings of the economy and legitimize unearned income and existing power relations.

More important, however, is Veblen's historicist sociology of knowledge. Werner Stark sardonically comments:

Unspoilt primitive communities, Veblen imagines, are dominated by the "instinct of workmanship." This so-called instinct will express itself in certain mental modes appropriate to it: men, for instance, will experience as beautiful what is purposive or adapted to its end. They will find a cow or a cart-horse more beautiful (because it is more useful) than a war-horse or a race-horse, a natural, flower-studded meadow more satisfactory to look at than an artificial, weeded lawn, a woman with large hands made for work and wide hips good for child-birth more comely than one with long, pale, anaemic fingers and a wasp-waist. But as soon as society alters its charac-

22. Ibid., 172.

ter, when property develops and, with it, "predaciousness," aesthetic conceptions are turned inside out. The race-horse is ranked higher in the scale of beauty than the cart-horse because all that has to do with work is despised, all that has to do with leisure exalted. Meadows are transformed into lawns, women locked up in crippling corsets, their hands made to end in long brittle fingernails painted red which prove beyond the shadow of a doubt that their owners need not, and do not, descend to the degrading level of domestic drudgery, and so on, and so forth . . . even religion is transformed from an adoration of the creative forces of nature, from an agriculture rite, into the adulation of an authoritarian personal god modeled on the warrior-noble, the man of power. All these changes of culture come about because the course of history replaces one mind-determining factor by another: first it is a spontaneous drive, the instinct of workmanship; later it is a factitious arrangement of human relationships, the emergence and existence of a leisured master class, and its associated "pecuniary" canons of thought and taste.[23]

This brief summation of Veblen's sociology of thought is casually but critically drawn from *The Theory of the Leisure Class* by the author of a comparatively early study (1958) of the sociology of knowledge. But it illustrates how humans throughout the ages are caught in the persistence and resurgence of atavistic continuities—that is, exploitation, predation, invidious distinction, wasteful emulation, and religious superstition. For Veblen, although not most of his critics like Stark, the study of the genesis and use of ideas is a device for unmasking those social values and practices that defeat the generic ends of life.

Robert Merton's Atypical Analysis of Veblen

It is to both a liberal critique and employment of Veblen's social theory that we now turn because its author was one of postwar sociology's most penetrating thinkers. Trained at Harvard by Talcott Parsons, Robert Merton (1910–2004) served for many years as chairman of the Department of Sociology at Columbia and became a leading proponent of structural-functional theory. Merton was also an advocate of "presentism," a view in which social-scientific writings of the past are seen in terms of contemporary sociological preoccupations and concerns. The development of social-scientific ideas was, he thought, to proceed incrementally, with the

23. Werner Stark, *The Sociology of Knowledge* (London: Routledge and Kegan Paul, 1958), 234.

current state of knowledge representing the culmination of this process.[24] According to Merton, the "systematics" of sociology, in the form of the empirically validated residues of earlier theories, ought to be distinguished from the "history" of the field, as embodied in "the false starts, the now archaic doctrines, and both the fruitless and fruitful errors of the past."[25]

Talcott Parsons deplored the unfortunate fact "that at present few economists and sociologists have even a modicum of interest or competence in each other's subject matter."[26] On the same page, he makes reference to the "great synthetic minds" of such luminaries as Alfred Marshall and Vilfredo Pareto, but no mention is made of Karl Marx and Thorstein Veblen.[27] It is evident that while Parsons neglected, or paid only peripheral attention to, seminal thinkers such as Simmel, Mannheim, Marx, and Veblen, his contemporary and former student Merton took a greater array of ideas from a broader variety of sources. Merton's skillful and judicious use of Veblenian concepts vividly demonstrates Veblen's potential for incorporation into a continually evolving and usefully eclectic social theory.

Merton's use of Veblen is most evident in regard to the distinction between manifest and latent functions that Merton popularized in American sociology. Indeed, Merton argued that the Veblenian analysis has entered so completely into popular thought, that its latent functions were now widely understood. Merton's analysis in this respect bears further elaboration, for it led to perhaps the most astute employment of Veblenian concepts in the literature of mainline American sociology. His most important use of Veblenian ideas thus resided in his skillful explanation of status emulation from a functionalist perspective. In Merton's words:

> however, says Veblen in effect, as sociologists we must go on to consider the latent functions of acquisition, accumulation and consumption, and these latent functions are remote indeed from its naïve meaning (i.e., manifest function) that the consumption of goods can be said to afford the incentive from which accumulation invariably proceeds. And among these latent functions, which help explain the persistence and the social location of the pattern of conspicuous consumption, is a symbolization of "pecuniary strength and so of gaining or retaining a good name." The exercise of "punctilious discrimination" in the excellence of "food, drink, shelter, service, or-

24. William Buxton, *Talcott Parsons and the Capitalist Nation-State* (Toronto: University of Toronto Press, 1985), 5.

25. Robert K. Merton, *Social Theory and Social Structure* (New York: Free Press, 1956), 309.

26. Talcott Parsons and Neil Smelser, *Economy and Society* (Glencoe, IL: Free Press, 1956), 309.

27. Ibid.

naments, appeal, amusements" results not merely in direct gratifications derived from the consumption of "superior" to "inferior" articles, but also, and Veblen argues, more importantly, it results in a *heightening or reaffirmation of social status.*

The Veblen paradox is that people buy expensive goods not so much because they are superior but because they are expensive. For it is the latent equation ("costliness—mark of higher social status") which he singles out in his functional analysis, rather than the manifest equation ("costliness—excellence of the goods").[28]

Notice Merton's careful analysis of Veblen's famous concept of conspicuous consumption. Merton removed it from its "utopian" context and put it to work. He refined the notion, gave it greater precision, and applied it to a number of situations—with the felicitous outcome that social theory has been enriched.

It is often remarked that "necessity is the mother of invention." But Merton approved of Veblen's inversion of this idea and its role in both manifest and latent conceptualizations of function. He wrote that "it is more often the case, as Veblen has remarked, that invention is the mother of necessity. The ulterior consequences of the more important mechanical inventions have been neither foreseen nor intended, though they have commonly demanded a whole series of institutional and technical adjustments."[29] In another context, Merton also suggested that scientific data may have had only a latent function in ordinary, everyday activity, but a manifest function when part of scientific discourse. He credited Veblen with this insight.[30]

However, by treating Veblen as a functionalist, Merton allowed himself to ignore the political significance of his work. Thus Veblen's radical critique of capitalism was simply absorbed into mainline sociology. Merton's emphasis on function rather than structure became a device for stripping Veblen of radical import, ultimately rendering his work conservative. While a sympathetic interpreter of Veblen, such as Merton, may praise his originality and utilize his insights, Merton nonetheless deradicalized Veblen's most important ideas and, perhaps inadvertently, became his critic. But his contemporary Mills, to whom we now turn, made different use of Veblen.[31]

28. Robert Merton, *Social Theory and Social Structure,* 69.
29. Robert Merton, *Science, Technology and Society in Seventeenth-Century England* (New York: Howard Fertig, 1970), 158.
30. Ibid.
31. See Mills, intro. to Veblen's *Theory of the Leisure Class.*

Mills, Veblen, and the Sociology of Knowledge

Within the context of exploring the possibilities of the human potential, C. Wright Mills (1916–1962) searched for a more adequate theory of value. His eclectic approach to this problem is evident in his combination of Mannheim's distinction between functional and substantial rationality, in his use of Marx's dichotomy between false and true consciousness, and Veblen's separation of ceremonial from instrumental values.[32] Mills never succeeded in fully integrating the core value systems or the analytic modes that underlie these concepts; however, he often used them as equivalencies that can substitute for each other in measuring the performance of social institutions and processes. For example, he attributes to Veblen "the realization of this false consciousness all around him."[33] Actually, Veblen never used the term "false consciousness," but his idea of "ceremonialism" roughly parallels it. Mills put it this way:

> For Veblen, technology, widely construed, stands opposite irrational (ceremonial) institutions. And for both, in whatever other respects they may differ, the rational, the technical pole of history will come through; it will increase to dominate the social life of the West . . . The irrational is identified with "pecuniary institutions."[34]

Mills learned from Veblen, and from his disciple Clarence Ayres, that prices do not measure real values, but only quantify judgments made prior to price transactions. Whether those judgments are really good or bad should be determined not by the price system, but by their relation to the technological continuum. Ayres, who was Mills's professor at Texas, distinguished between "price" values and "real" or "technological" values. In his view price values are frequently ceremonial values, and often reflect antiquated mores, such as the power of money and class distinctions. In modern capitalism, use value is what has utility for an individual. But, as Ayres explained, this is not determined by some mystical "inner nature" of mankind demanding a particular good or service. On the contrary, what is useful or valuable for an individual is determined by the cultural environment in which he lives. Too often, this value-determining cultural environment comprises a backward-looking and change-resistant institutional complex based on traditional mores, beliefs, and attitudes

32. See Tilman, *C. Wright Mills,* 53–59.
33. Ibid.
34. Mills, *Power, Politics and People,* ed. with intro. I. L. Horowitz (New York: Oxford University Press, 1963), 54.

Table 4. Veblen's Taxonomy of Habits of Thought

Positive	Negative
Scientific	Animistic
Matter-of-Fact	Make Believe
Dynamic	Static
Evolutionary	Pre-Darwinian
Sheer Undirected Process	Melioristic-Progressive Process
Impersonal Causal Sequence	Preordained Ends
Materialism-Naturalism	Spiritualistic-Providential
Secular Trend	Teleological Order–Final Causes
Natural Uniformities	Fortuitous Necessity (Luck)
Natural Order	Divine Necessity
Noninvidious Objectives	Invidious Distinctions and Honorific
Instrumental Means	Prowess
	Emulatory and Honorific Aims
Functional Observances	Ceremonial Practices
Common Man	Vested Interests
Peaceful Type	Predatory Type

Sources: Thorstein Veblen, *The Theory of the Leisure Class, The Place of Science in Modern Civilization and Other Essays, The Instinct of Workmanship;* and Joseph Dorfman, *Thorstein Veblen and His America* (New York: Viking Press, 1966), 155–56.

rather than the technological or instrumental qualities that enhance the life process. Mills agreed with much of this system of evaluation, and yet he doubted the ultimate potential of the technological continuum as the basis for development of a fully adequate value theory.

He was also critical of Veblen's claim that technology would foster a new rationality in the working class and, by implication, in industrial managers. Nor did he think, as Veblen did, that this would also happen to scientists and make them capable of critically analyzing basic social institutions and values. Extrapolating from this, Mills argued that Veblen mistakenly assumed that the development of a "functional rationality" would be accompanied by "substantial rationality." Veblen tended in his early works to hold that the industrial labor force would combine its insights into the nature of the industrial process with a broader understanding of complex social relationships because an understanding of technical means would be fused with critical intelligence:

What the discipline of the machine industry inculcates therefore, in the habits of life and of thought of the workman, is regularity of sequence and mechanical precision; and the intellectual outcome is an habitual resort to terms of measurable cause and effect, together with a relative neglect and disparagement of such exercise of the intellectual faculties as does not run on these lines. . . . The machine throws out anthropomorphic habits of thought. It compels the adaptation of the workman to his work, rather than adaptation of the work to the workman. The machine technology rests on a knowledge of impersonal, material cause and effect, not on the dexterity, diligence, or personal force of the workman, still less on the habits and propensities of the workman's superiors.[35]

Mills did not think that the increasing rationality of technology would make the individuals who operated the machines more rational, except perhaps in the functional sense. In fact, he argued that the judgment of engineers and technicians, combined with their capacity for substantive rationality in social and political affairs, was often no better than that of businessmen. Mills judged Veblen's work from the perspective of the distinction between substantial and functional rationality. He found it strong in its claims that the machine process fostered the latter, but weak in its assertion that it encouraged the former. He warns against assuming that an understanding of means-ends congruence can be equated with an ability to comprehend complex sets of social relationships.

But analysis of Veblen's view of the psyche of industrial workers indicates that he portrayed it largely in negative terms: in terms of disbelief and distrust of convention and precedent. Veblen believed blue-collar workers were skeptical about the "natural rights" of property, freedom of contract, customary authority, and traditional religious outlooks. Perceiving this skepticism is not the same as claiming a positive causal relationship between interaction with the machine process and the development of substantive rationality. Veblen's claims would have to be more extreme than they were for Mills's critique to be fully justified. However, it is to Weber that we now go for a more complex discussion of rationality as it pertains to Veblen.

35. Veblen, *The Theory of Business Enterprise*, ch. 9.

Enter Max Weber

As a taxonomist and sociologist of knowledge, Max Weber (1863–1920) had insights that help illuminate Veblen's views on the genesis and nature of social rationality. Before turning to them, however, a few cautionary words are in order. Unlike Weber, Veblen did not believe that capitalism epitomizes rationality; indeed, his dislike for business irrationality caused him to articulate one of his most flamboyant oxymorons, to describe the "astute mismanagement" of industry by business.[36] His insistence on the irrationality of modern business should not, however, be disregarded as mere satire or mockery, for his position complements Weber's influential theory of the motivation of capitalists set forth in *The Protestant Ethic and the Spirit of Capitalism*.

Weber's claim that the "spirit of capitalism" reflects the Calvinist idea of a predestined "calling" for the elect leads to his view that modern business is characterized by its rationalization. This is to say that Weber contrasted premodern business as "irrational" work "directed to acquisition by force, above all the acquisition of booty," with the penchant of modern business for rational "calculation of capital in terms of money. . . . [Wherein] everything is done in terms of balances."[37] Veblen's argument regarding the irrationality of modern business does not simply contradict Weber's view, for what Weber meant by "rationality" has more to do with what Veblen called "pecuniary accountancy," or the flattening of all evaluation to numbers, than with irrationality itself. Veblen also suggested that what Weber perceives as the rational calculation of interest actually cloaks a profound irrationalism; this is because, in Veblen's view, other standards, such as serviceability, productivity, and usefulness to the larger community, offer better ways to measure value than does the pecuniary calculus. In short, Veblen suggested that the allegedly rational reduction of value to numerical calculations is, in fact, "businesslike imbecility."[38] His characterization of modern business as destructive and predatory is thus the reverse of Weber's claim regarding the peaceful, rational nature of modern business.

In Weber's sociology there are four types of social action, and two of these contribute most directly to clarification of Veblen's ideas regarding social rationality. For Weber, *zweckrational* was basically expedient ratio-

36. Veblen, *Imperial Germany and the Industrial Revolution* (New York: B. W. Huebsch, 1915), 320.
37. Weber, *The Protestant Ethic and the Spirit of Capitalism*, 20, 18.
38. Veblen, *Absentee Ownership*, 360. For a comparison of Weber and Veblen on several of these points, see Diggins, *The Bard of Savagery*, 113–19 and 133–38.

nality that indicated a system of action whereby the actor weighed the alternative ends and means open to them in terms of his or her goals and chose the course of action most appropriate for him or her. A system of particular ends existed for the actor, but before action was taken to achieve them, the likely costs and consequences must be taken into consideration. In Weber's analysis, *wertrational* was distinguishable from *zweckrational* because the incorporation of an "absolute value" eliminated the possibility of the actor's choosing alternative ends, and ultimately, therefore, also barred the possible choice of certain means. The only important consideration of the actor was the realization of the cherished value whether it is ethical, aesthetic, religious, or otherwise.[39]

Weber Continued: "Transfer Effects" or Not?

Altogether, Weber used four types of rationality that enhance the meaning of Veblen's machine-induced rationality. First, his "practical rationality" accepts the realities of everyday existence and calculates the most expedient means of coping with the difficulties it offers. It thus suggests a purely adaptive subordination of individuals to circumstances and a tendency to avoid behavior that transcends daily routine. Clearly, the imperatives of the industrial economy and its corollary, machine-induced rationality, rest on such rationality if the system is to function effectively. Second, Weber's "theoretical rationality" is also relevant to understanding the impact of the machine process on the psyche of the worker. Here Weber's focus is on such cognitive processes as logical deduction and induction, the attribution of causality, and the genesis of symbolic meanings. Third, his "formal rationality" (zweckrational) indicates means-ends rational calculation with reference to applied rules, laws, or regulations. Its meaning in Veblen's time in a partly or largely bureaucratized work setting is too obvious to need further elaboration.

Weber's fourth category of "substantive rationality" (*wertrational*) di-

39. The other two are "(3) in terms of affectual orientation (*affektuell*), especially emotional, determined by the specific affects and states of feeling of the actor; (4) as traditionally oriented (*traditional*) through the habituation of long practice." Max Weber, *The Theory of Social and Economic Organization*, 115. Also, cf. Ernest M. Manasse, "Moral Principles and Alternatives in Max Weber and John Dewey," I and II, *Journal of Philosophy* 41 (January 20 and February 3, 1944): 29–48, 57–68; and Donald N. Levine, "Rationality and Freedom: Weber and Beyond," *Sociological Inquiry* 51, no. 1 (1981): 5–25. For an analysis of culture-bound definitions of "rationality," see Anne Mayhew, "Contrasting Origins of the Two Institutionalisms: The Social Science Context," *Review of Political Economy* 1 (November 1989): 319–33.

rectly orders action into patterns. But it does so on the basis of some value postulate. Such a postulate is not simply a single value; on the contrary, it signifies entire clusters of values. Veblen's machine-conditioned proletarian is clearly an exemplar of one such cluster—for he was undevout, matter-of-fact, increasingly unpatriotic, inclined toward collectivist-egalitarian values, and oriented toward the practice of workmanship. Or so Veblen claimed. Yet he also detected strong countervailing tendencies. It should be noted, however, that these values and their behavioral correlates are time-specific and have a particular cultural locus, whereas for Weber the exercise of substantive rationality may extrude values and value clusters that are enormously varied and culturally transcendent. Nevertheless, Veblen's machine-conditioned worker does exemplify a certain form of substantive rationality in the Weberian sense.

Weber's typology of rationality also subsumes five important aspects of rationality which include (1) inductive inference, (2) causal attribution, (3) symbolic abstraction, (4) systematization of belief, and (5) rules of conduct, all of which play a significant role in Veblen's machine-induced rationality. In fact, the machine process cannot function effectively without inculcating the cognitive skills so essential to these aspects of rationality, or so he claims. They also play a key role here in clarification of the meaning of Veblen's machine-conditioned rationality that presupposes it will incorporate at least a modicum of each. To illustrate, in Veblen's analysis industrial workers when interacting with machines use induction to grasp the interactive role of the machine. That is, the worker learns to both generalize and then particularize about the nature of machinery through protracted experience. Inductive logic thus becomes reflexively embedded in the worker's psyche and this can be traced to the machine process. But it does not follow that there is any direct or immediate transfer effect outside the realm of industry, a fact that Veblen sometimes failed to adequately stress.[40]

Causal attribution also resulted from the worker's observance of both the performance of individual machines and their concatenation with other machines, both singly and in the aggregate. At the very least, the operator comes to understand the role of boilers, pistons, and flywheels in animating heavy machines, as well as the functional relationship between and among machines. Still, causal attributive abilities developed in the mechanical sphere may not carry over directly into the realm of social causation, or so Veblen's critics claimed.

40. But, cf. Veblen's discussion of "borrowing," "crossing," and "grafting" in *The Instinct of Workmanship; Imperial Germany*, ch. 1, 2, 6, and *The Theory of Business Enterprise*, ch. 9.

They also argued that it was more doubtful yet that machines induced the ability to manipulate abstract symbols except insofar as these symbolized aspects of the machine process are closely related facets of industrial experience. Still, as Veblen pointed out in *The Theory of Business Enterprise*, the socially valuable skills of accounting and inventorying also required symbolic understanding and manipulation, and it was difficult to imagine an industrial process without them.

It is questionable whether the machine process helped workers systematize their belief system about anything except industrial machinery, its immediate environment, and the work habits essential to their maintenance. In short, it is difficult to see why Veblen thought that important transfer effects to other cultural and institutional realms would occur in terms of formulating and systematizing their more general belief systems. However, he did argue that both the scientist and the worker would be transformed into "finikin skeptics"; that is, agnostic, secular materialists. And the formation and application of rules of conduct, insofar as these involve the imperatives of industrial work, do have some likelihood of transfer effects. It is difficult to deny that the rules also have some impact on the need for punctuality, efficiency, and harmonious interaction in other walks of life. Schools, churches, and places of consumption, service outlets, and transportation facilities all require formally or informally defined rules of conduct if they are to function effectively, and these rules converge with those of the scientific laboratory and machine shop.

While critics were skeptical about Veblen's views regarding transferability to other cultural realms of inductive inference and causal attribution, they were even less sanguine regarding the impact of the machine process on the development of proletarian abilities to engage in symbolic abstraction, systemization of belief, and formulation and manipulation of rules of conduct. The transfer effects of industrial technology alleged by Veblen to have a transformative impact on the worker's psyche were said by his critics to have a cultural and ideational genesis instead. Probably Weber, who is the inspiration for most of this analysis, would agree with them. Veblen clearly failed with regard to transfer effects to move from hypothesizing about authenticity to demonstrations of matter-of-fact and empirical testing. Nevertheless, his views are highly suggestive for hypothesis creation and testing.

Enter Karl Mannheim

Karl Mannheim (1891–1947) offered further explanation of how European social theory can be used to enhance the meaning of Veblen's ideas regarding social rationality. Mannheim distinguished between "function-al" and "substantial" rationality. In regard to "rational" in the functional or instrumental sense, there are two criteria:

> (a) Functional organization with reference to a definite goal; and (b) a con-sequent calculability when viewed from the standpoint of an observer or a third person seeking to adjust himself to it. . . . [E]ach act is functionally ra-tional [if] (1) it is organized with reference to a definite goal, and (2) [if] one can adjust oneself to it in calculating one's own actions.[41]

He also defined "substantial" rationality in a way that is helpful to social scientists attempting to better understand Veblen's point of view: "We un-derstand as substantially rational an act of thought which reveals intelli-gent insight into the interrelations of events in a given situation."[42] Sub-stantial rationality thus referred to the perceptions individuals obtain of a situation, which may permit them to control it in keeping with their con-scious purposes. Mannheim thus distinguished between the ability to ad-just means to ends when the ends are given, or fixed, which he called "functional" rationality, and designating certain ends as superior to or worse than alternative ends, which he called "substantial rationality," based on what he asserts is the latter's superior development of human insight into matters of value.[43]

However, Mannheim was more given to conceptualization and analy-sis than he was to prescription; indeed, although hopeful that intellectu-al "floating strata" might obtain privileged epistemological access to su-perior forms of social rationality, he was uncertain of this eventuality.[44]

41. Karl Mannheim, *Man and Society in an Age of Reconstruction*, 53–54. "Instru-mental" or "functional" rationality is open to the charge that it can be used to shore up any kind of political regime. As Eduard Heimann said in a letter to Mannheim, "Your rationalism is not opposed to fascism, but rather, because of its reduction to the social-technical, politically neutral . . . therefore supportive of the prevailing power." Eduard Heimann to Karl Mannheim, January 31, 1935, in Mannheim, *Sociology as Po-litical Education*, David Kettler and Colin Loader, eds. (New Brunswick, NJ: Transac-tion Publishers, 2001), 179.

42. Ibid., 53.

43. See the brief but penetrating discussion of Mannheim in this regard in Russell Jacoby, *The End of Utopia Politics and Culture in an Age of Apathy* (New York: Basic Books, 1999), 110–11, 123. Also, see Jacoby's analysis of intellectuals, ideology, and policy in the same study.

44. Warren Samuels, "The Self-Referentiability of Veblen's Theory," *Journal of Eco-nomic Issues* 24 (September 1990): 714–15. Assuming that Veblen was often an advo-cate and a practitioner of instrumental valuation in the American pragmatic tradition, his generic ends of life are not and cannot be ends-in-themselves. Instead, although more value-specific than John Dewey's ends-in-view, they must also take their place in the continuum of means-ends. Ends, however important they may be, are also

Like Weber, he did not think social scientists should engage in political advocacy without explicit acknowledgment of their own political and moral orientation. Veblen, too, recognized the degree to which any kind of social science is value laden and reflective of the social position and professional aspirations of its practitioners. Even his tongue-in-cheek claims that he is dispassionately and objectively analyzing social values, institutions, and actors cannot disguise the satirical mockery in which he so often engaged.

Nevertheless, by extrapolating from Mannheim, it is possible to interpret "rationality" in three ways that are relevant to Veblen. The first, functional or instrumental rationality, can be interpreted as means-ends congruence; the second, substantive rationality, signifies an understanding of the relationship between and among alternative ends; and the third involves the substance of ends themselves as understood by Veblen. It is these ends to which I now turn.

Mannheim, Veblen, and the Generic Ends of Life

Veblen claimed that the human community had "generic ends of life," or as he also put it, "fullness of life," to pursue. Do these contain institutionally dependent value constants? Or are they of transcultural moral and social significance for the human species throughout history and prehistory? If Veblen is properly understood, "idle curiosity," "workmanship," and the "parental bent" are not passing fancies but values embedded in the social matrix of a properly developing (meaning generically human) community. The exercise of "social rationality" properly construed must focus on the revalidation of these values as well as their implementation. Social scientists, no doubt, will continue to debate whether or not these values and the social processes linked to them are superior to others. But they will find an effective explication of them in Veblen's work. In this vein, Warren Samuels wrote:

> One can read *The Leisure Class* as a satire of the practices of conspicuous consumption and status emulation by the socio-economic elite, as Veblen's denigration of those practices for what they are, namely, situation dependent practices having no independent existence or value, despite how much they mean to certain practitioners.[45]

---

means to other ends whose existence may not as yet even be recognized. See Veblen, *The Instinct of Workmanship*, 48.

45. Warren Samuels, "The Self-Referentiability of Thorstein Veblen's Theory of the Preconceptions of Economic Science."

Veblen's way of valuing, then, drew on various sources, but there were limits to its diversity. This was because the utilization of certain values would obstruct his method of valuation itself. Thus *The Theory of the Leisure Class* was highly suggestive of the limits that must be placed on social values and evaluative processes if they were to remain conducive to "fullness of life." However adequate his formulation of ends may or may not be judged to be, he articulated value constants, though relative to time and circumstances, which provided standards by which to judge existing social values and processes. Whether these were congruent with European conceptualizations of substantive rationality or not, he was clearly more explicit than Mannheim and Weber regarding the nature and significance of values and the reasons for sanctioning certain of them.

In any event, it is with the explication of what Veblen called the "generic ends of life, impersonally considered," that I am concerned because this is the most explicit articulation of value constants in his work. As indicated, these value constants included "idle curiosity," which in industrial society signified critical inquiry; the "parental bent," which meant altruism; and the "instinct of workmanship," which denoted taking pride in and obtaining gratification from the craftsmanlike performance of work. Veblen was thus using transcultural standards of judgment which found their locus in his abbreviated discussions of the "fuller unfolding" of the life process.[46]

An important statement regarding the theoretical relationship between Mannheim and Veblen which leads to a clearer understanding of the latter's views on social rationality is provided by Mannheim's comments to this effect:

if in analyzing the changes of recent years, people had kept in mind the distinction between various types of rationality, they would have seen clearly that industrial rationalization served to increase functional rationality but that it offered far less scope for the development of substantial rationality in the sense of the capacity for independent judgment. Moreover, if the distinction between the two types of rationality which emerges from this explanation has been thought out, people would have been forced to the conclusion that functional rationalization is, in its very nature, bound to deprive the average individual of thought, insight, and responsibility

46. See Veblen, *The Theory of the Leisure Class*, ch. 13. Also, cf. Veblen's *The Instinct of Workmanship*, ch. 1–2, and Michael Sheehan and Rick Tilman, "A Clarification of the Concept of 'Instrumental Valuation' in Neoinstitutional Economics," *Journal of Economic Issues* 26 (March 1992): 731–44.

and to transfer these capacities to the individuals who direct the process of rationalization.[47]

Clearly, Mannheim believed that functional rationalization does not necessarily enhance substantial rationality which he uses to explain the frailty of Veblen's sociology of knowledge as it pertains to human rationality:

we see that the social source of rationalization can be clearly determined and that indeed the force which creates in our society the various forms of rationality springs from industrialization as a specific form of social organization. Increasing industrialization, to be sure, implies functional rationality, i.e. the organization of the activity of the members of society with reference to objective ends. It does not to the same extent promote "substantial rationality," i.e. the capacity to act intelligently in a given situation on the basis of one's own insight into the interrelations of events. Whoever predicted that the further industrialization of society would raise the average capacity for independent judgment must have learned his mistake from the events of the past few years. The violent shocks of crises and revolutions have uncovered a tendency which has hitherto been working under the surface, namely the paralyzing effect of functional rationalization on the capacity for rational judgment.[48]

Although Mannheim did not agree with Veblen's conflation of functional and substantial rationality his comments are helpful in understanding why. The contrast between the two was amplified by his claim that disparities of power and limited access to information are likely to strip the industrial labor force of the opportunity to develop substantial rationality, a prospect which he felt Veblen overlooked. Although Veblen did not actually use the terms "functional" or "substantive" rationality, it was evident that what the Europeans call "substantive rationality" was, in his analysis, partly machine induced; at the very least it was influenced by the machine process. "Substantive rationality" was thus synonymous or at least closely linked with these particular ends. In his eyes, it was not simply as Mannheim wrote, an act of thought which reveals intelligent insight into the interrelations of events in a given situation, as important as this

47. *Man and Society in an Age of Reconstruction*, 58. Mannheim referred to Veblen's *The Vested Interests and the State of the Industrial Arts* (1919), the title under which it was originally published. And, in a footnote, Mannheim concluded:
Cf. Veblen, Th. B., *The Vested Interests and the 'Common Man'* . . . for an *exposition of a divergent interpretation* of the influence of industrialization on the possibilities of substantial rationality (author's emphasis).
48. Ibid., 58.

may be for liberal intellectuals like Mannheim, whose focus was on cognitive processes. Rather, for Veblen it was a transcultural value orientation embedded in a social existence which eschewed emulatory consumption, waste, and exemption from useful labor. This existence prescribed a system of social arrangements and material provisioning which enhanced the life process, and this life process was characterized by an abundance of critical intelligence, altruism, and a proficiency of workmanship.

Mannheim, Veblen, and the Intellectuals as "Floating Strata"

Perhaps the most provocative, enduring, and influential claim Mannheim made was to the effect that intellectuals could or would become socially detached, culturally uprooted, "floating strata" without determinative economic interests, and that this background and existential presence would give them an objectivity about politics and the public interest denied to others. In short, the uprooted intellectuals with loose or no class affiliations would be the least influenced by the ideology of their groups and would thus make the most competent social scientists. Mannheim, himself, had qualms and misgivings about his own assertion, and it was undoubtedly the source of much of the volatility of the debate over his work. But what would Veblen say about this proposition were it presented to him?

As an ideal state of affairs for intellectuals, it might well have appealed to him. Still, he had few illusions regarding the social detachment and cognitive objectivity of any social group, and few men of his time knew more about the intellectual predispositions and political values of their contemporaries than Veblen. However, merely to be socially detached and culturally uprooted as Veblen was may induce intellectual and scholarly iconoclasm. But this does not necessarily make one more "objective," whatever this might mean.

Mannheim's free-floating intelligentsia are an important part of his sociology of knowledge. Does Veblen have an equivalent in his own social theory? No definitive answer can be given to this question, and even a tentative provisional hypothesis is difficult to construct without further elaboration of the meaning of "floating strata" in Mannheim's theory. Suffice it to say that his socially detached intellectuals are potentially capable of understanding the public interest in objective ways denied to much of the rest of the population. Veblen, on the other hand, relies on the habit-forming effects of the machine process to reshape the thinking and values of blue-collar workers and the growth of a scientific aptitude and reason among the intelligentsia that would work in favor of social reconstruction.

The political theorist Eugene Meehan claims that:

> Mannheim provided a methodological base from which social criticism could proceed more or less unhindered by the requirements of scientific cogency. The arguments on which he based his position bear close examination. Three points are of prime significance: (1) the claim that all knowledge is relative to social position, and particularly class status; (2) the tendency to concentrate on the source of knowledge or means of acquiring knowledge rather than verification procedures; and (3) the close relationship that is assumed between social criticism and active participation in social life.[49]

Whether this is an accurate portrayal of Mannheim's thought or not is beside the point, since Meehan is interested primarily in assessing the frailties of his sociology of knowledge, while we are focused primarily on using Mannheim's theoretical apparatus for better understanding Veblen. However, the three points Meehan makes can all be applied to Veblen as well as Mannheim, yet they leave something to be desired as a definitive interpretation of Veblen's position.

As to the claim that "all knowledge is relative to social position," Veblen clearly believed, to the contrary, that matter-of-fact knowledge, as opposed to make-believe, is epistemologically superior, perhaps even privileged. Like Mannheim, Veblen had a tendency to concentrate on the source of knowledge and the meaning of acquiring it rather than on how to verify it. This is commonly referred to as his "critical genetic" method; yet he did not object in principle to the use of quantitative-statistical methods of verification. Finally, he was well aware of the close relationship between social criticism and active participation in social life even if he often disguised his own normative commitments behind massive barrages of satire, mockery, and irony. In short, there is considerable resemblance between Mannheim and Veblen in all three of the realms mentioned by Meehan.

To them must be added the distinction Mannheim made between "ideology" and "utopia." *Ideology* denotes doctrinal justification and rationalization of particular institutional configurations and the material privileges emanating from them to certain groups and classes; *utopia* suggests idealistic visions of social and political reconstruction often lacking in realism. Much of Veblen's work can be interpreted as an effort to distinguish between ideology and the utopia in the existing social and political orders and in the thought of other intellectuals.

---

49. Eugene Meehan, *Contemporary Political Thought: A Critical Study* (Homewood, IL: Dorsey Press, 1967), 90.

About Antonio Gramsci

Antonio Gramsci (1891–1937), Marxist theoretician and cofounder of the Italian Communist Party, has attracted massive attention from a generation focused on the genesis and role of power and control. His most important insights dovetail with that part of the sociology of knowledge which establishes the basis for ideological hegemony, political dominance, class power, and the development of false consciousness among the masses. His theory to the extent that it parallels or sheds light on Veblen is as follows:

> A third moment is that in which one becomes aware that one's own corporate interests, in their present and future development, transcend the corporate limits of the purely economic class, and can and must become the interests of subordinate groups too. This is the most purely political phase, and marks the decisive passage from the structure to the sphere of the complex superstructures; it is the phase in which previously germinated ideologies become "party," come into confrontation and conflict, until only one of them, or at least a single combination of them, tends to prevail, to gain the upper hand, to propagate itself throughout society—bringing about not only a unison of economic and political aims, but also intellectual and moral unity, posing all the questions around which the struggle rages not on a corporate but on a "universal" plane, and thus creating the hegemony of a fundamental social group over a series of subordinate groups.[50]

In Gramsci's view, the superstructural apparatus of capitalism, that is, its prevailing values and culture and the acquiescent, if not supportive, role of its organic intellectuals gave its upper class and their social satellites "ideological hegemony" and thus political control. An older and more conventional Marxist variant of American exceptionalism stresses the existence of "false consciousness" among the masses, that is, lack of awareness of objective self-interest fostered by the social and cultural apparatus of hegemonic capitalism. In short, people are not necessarily interested in what is to their interest.

Gramsci's focus is on "Americanism" among the working classes as a form of false consciousness. But, contrary to Veblen, it assumes that proletarian militancy harnessed by an intellectual vanguard to revolutionary objectives is somehow the norm in industrial societies with the single exception of the United States. In any case, Veblen places more emphasis than Gramsci on status emulation—that is, conspicuous consumption,

---

50. Antonio Gramsci, *Selections from the Prison Notebooks*, ed. and trans. with intro. Quintin Hoare and Geoffrey Nowell Smith (London, 1971), 181–82.

conspicuous waste, and conspicuous avoidance of useful labor as power-ful social bonding agents which mitigate class conflict and ideological pol-itics. Also, as John Diggins notes, "In treating cultural ideas as the uncon-scious foundations of social life," Veblen anticipated the Gramscian notion of "hegemony . . . a phenomenon made all the more perplexing be-cause it involves man's subjugation to ideas rather than to power and co-ercion."[51] Diggins's formulation can be enhanced by insisting that for Veblen, "subjugation to ideas" is precisely a form of power.

In *The Theory of the Leisure Class,* Veblen sets forth his central charge against the leisure class: it is hampering cultural advancement by imped-ing the full adjustment of society to a contemporary industrial economy. It does this, he suggests, through the conservatism of the class itself, for it has a material interest in leaving things be, through rendering an exam-ple of conspicuous waste and conservatism, which become honorable in the eyes of the other classes even though it causes deprivation in the low-er ones. The British literary critic Michael Spindler concludes that:

> we can thus see that, like Antonio Gramsci, Veblen is concerned with the question of hegemony: how a ruling class manages to maintain its domi-nance in the face of the numerical and technical superiority of the subordi-nate classes, and how it manages (to borrow a phrase from Noam Chom-sky) to "manufacture consent." The answer is that it maintains political supremacy through command of the ideological field, and the latter half of the book [TLC] sets out to demonstrate how leisure-class canons of con-spicuous consumption, status, and invidious distinction ramify through such diverse surface phenomena as manners, sports, religion, and univer-sity education. This pecuniary culture, promulgated throughout society and participated in by the middle class, is thus demystified by Veblen and exposed as masking class interests and ensuring the continuation of the rul-ing class' dominance.[52]

In summation, then, can it be said that Veblen has rough equivalents to Gramsci's ideas regarding "ideological hegemony" and "false conscious-ness"? Several scholars have seen close parallels between Gramsci's "ide-ological hegemony" and Veblen's views on the ideological and cultural dominance of the upper classes. Perhaps the main differences between the two men in this regard lie not so much in substantive theoretical or con-ceptual terms as in the former's idiosyncratic Marxism, which makes his

51. John P. Diggins, *Thorstein Veblen, Theorist of the Leisure Class* (Princeton: Prince-ton University Press, 1999), 108.

52. Michael Spindler, *Veblen: Modern America Revolutionary Iconoclast* (London: Plu-to Press, 2002), 38.

analysis more difficult to compare with the latter's sometimes convoluted English and satire; although Veblen undoubtedly thought that the class structure of capitalism was more complex and differentiated and its ideational patterns more complex than Gramsci.

Nevertheless, both men thought that the upper classes, particularly large businesses and their owners, exercise a massively disproportionate influence on the making of public policy because of their hold on the minds and values of these social strata beneath them. This is accomplished at all levels of the socialization and communication process, from schools, families, and churches to news media and places of employment. And, of course, it is reinforced through the central government by the influence of corporations and the military which manipulate mass patriotism and nationalism for purposes of self-aggrandizement.

However, the problem of the "organic intellectuals" in the thought of Gramsci and, conceivably Veblen, is more difficult to explain for the latter does not use the term or really have such a concept. Gramsci, of course, believed in the possibility, indeed, the likelihood of proletarian cultural hegemony through domination of the work process; Veblen, too, thought, at least in 1904 when he published *The Theory of Business Enterprise*, that interaction with the industrial process was promoting an egalitarian collectivist mentality among blue-collar workers, although he did not go so far as to predict proletarian cultural hegemony. However, Gramsci distinguished between organic intellectuals of the working class and traditional intellectuals from outside it. As Quinton Hoare and Geoffrey Smith put it:

> The central argument of Gramsci's essay on the formation of the intellectuals is simple. The notion of "the intellectuals" as a distinct social category independent of class is a myth. All men are potentially intellectuals in the sense of having an intellect and using it, but not all are intellectuals by social function. Intellectuals in the functional sense fall into two groups. In the first place there are the "traditional" professional intellectuals, literary, scientific and so on, whose position in the interstices of society has a certain inter-class aura about it but derives ultimately from past and present class relations and conceals an attachment to various historical class formations. Secondly, there are the "organic" intellectuals, the thinking and organizing element of a particular fundamental social class. These organic intellectuals are distinguished less by their profession, which may be any job characteristics of their class, than by their function in directing the ideas and aspirations of the class to which they organically belong.[53]

53. Gramsci, *Selections from the Prison Notebooks*, 3.

Veblen would undoubtedly recognize the distinction between "organic" and "traditional" intellectuals. But he would not attribute to it the ideological and political significance that Gramsci does. In this vein, Gramsci writes:

> The working class, like the bourgeoisie before it, is capable of developing from within its ranks its own organic intellectuals, and the function of the political party, whether mass or vanguard, is that of channeling the activity of these organic intellectuals and providing a link between the class and certain sections of the traditional intelligentsia. The organic intellectuals of the working class are defined on the one hand by their role in production and in the organization of work and on the other by their "directive" political role, focused on the Party. It is through this assumption of conscious responsibility, aided by absorption of ideas and personnel from the more advanced bourgeois intellectual strata, that the proletariat can escape from defensive corporatism and economism and advance towards hegemony.[54]

Veblen with his focus on the stresses and strains, that is, cross-cutting cleavage, to which blue-collar workers are subjected in capitalist-market economies minimized such potential. In fact, he stressed the role of emulatory consumption and nationalism and patriotism as social bonding agents and denied that America was likely to be governed by a militant proletariat in the future. The mere presence of organic and traditional intellectuals would not, in Veblen's view, lead to either ideological coagulation or concerted proletarian political action. Indeed, Gramsci himself came to have misgivings about the role of the working class and its intellectuals in the United States when he finally grasped the impact of "Americanism."

As for "false consciousness," that is, lack of awareness of objective self-interest, Veblen recognized its existence and its class basis long before Gramsci began to theorize. But the social and cultural determinants of knowledge a la Lenin and Gramsci did not exist in the United States on such a scale as to facilitate the kind of "genuine" or "true" consciousness (awareness of objective class interest) it would take to bring about a genuine proletarian upheaval. The myth of the self-made man, the economic aspirations and real material achievements, shared emulatory bonding patterns, and the congealing effects of nationalism and patriotism on the underlying population all saw to that. Gramsci's frailties of analysis help to clarify Veblen's sociology of knowledge in a broad political sense, but they also explain why Veblen was not a Marxist.

54. Ibid., 4.

Veblen and Paradigm Shifts

Thomas S. Kuhn's *Structure of Scientific Revolution* (1962) provided critics with new analytic tools both for analyzing Veblen's own sociology of knowledge and for making his life and work a focal point of the sociology of knowledge itself.[55] Perhaps of most interest to economists was the question of whether or not Veblen's critique of mainline economics required a paradigm shift and if some form of Veblenian economics was to be the new paradigm. Although it is impossible to definitively respond to these questions here, I will attempt a brief exploration of the issues involved, starting with a brief restatement of Kuhn's basic thesis.

First, consider the rather rudimentary definition by Kuhn of what he means by "paradigms." As he put it: "these I take to be universally recognized scientific achievements that for a time provide model problems and solutions to a community of practitioners."[56] Of course, it can be argued that the discipline of economics never really achieved the status of a "paradigm" in the Kuhnian sense, which was Veblen's point when he suggested that it was pre-Darwinian. Even today, many economists question the scientific status of their own discipline. Kuhn even suggests that competing paradigms may exist in the same areas of science for considerable periods of time and that the preparadigmatic phase of its existence may persist indefinitely. This may be the present situation in economics.

But Kuhn believes that at some point practitioners of science in whatever realm will change their "perception and evaluation of familiar data," in part because of "the manner in which anomalies, or violations of expectation, attract the increasing attention of a scientific community."[57] This leads to a Kuhnian explanation that parallels Veblen's own thinking on the subject:

> I have said nothing about the role of technological advance or of external social, economic, and intellectual conditions in the development of the sciences. One need, however, look no further than Copernicus and the calendar to discover that external conditions may help to transform a mere anomaly into a source of acute crisis. The same example would illustrate the way in which conditions outside the sciences may influence the range of alternatives available to the man who seeks to end a crisis by proposing one or another revolutionary reform.[58]

55. Thomas S. Kuhn, *The Structure of Scientific Revolutions*, 2d ed., enlarged (Chicago: University of Chicago Press, 1970).
56. Ibid., viii.
57. Ibid., ix.
58. Ibid., x.

Clearly, Veblen believed that conditions external to the study of economics had invalidated much of its claim to scientific status, not to mention
the pernicious impact of cultural lag intrinsic to the doctrines of the discipline itself. As Kuhn put it:

> history of science becomes the discipline that chronicles both these succes-
> sive increments and the obstacles that have inhibited their accumulation.
> Concerned with scientific development, the historian then appears to have
> two main tasks. On the one hand, he must determine by what man and at
> what point in time each contemporary scientific fact, law, and theory was
> discovered or invented. On the other, he must describe and explain the con-
> geries of error, myth, and superstition that have inhibited the more rapid
> accumulation of the constituents of the modern science text.[59]

To Veblen, of course, economics was still mired down in the "congeries of
error, myth, and superstition" that acted as a drag on the development of
a truly evolutionary economic science. Kuhn breaks the problem(s) down
still further:

> These three classes of problems—determination of significant fact, match-
> ing of facts with theory, and articulation of theory—exhaust, I think, the lit-
> erature of normal science, both empirical and theoretical.[60]

Probably, Veblen would agree with Kuhn that ascertaining the relevant
facts, matching them with theory and articulating the theory itself are part
of the development of most scientific paradigms. At least nothing he
wrote contradicts this. And, beyond this, Kuhn makes a point which has
particular application to Veblen's sociology of knowledge:

> The invention of other new theories regularly, and appropriately, evokes the
> same response from some of the specialists on whose area of special com-
> petence they impinge. For these men the new theory implies a change in the
> rules governing the prior practice of normal science. Inevitably, therefore, it
> reflects upon much scientific work they have already successfully complet-
> ed. That is why a new theory, however special its range of application, is
> seldom or never just an increment to what is already known. Its assimila-
> tion requires the reconstruction of prior theory and the re-evaluation of pri-
> or fact, an intrinsically revolutionary process that is seldom completed by a
> single man and never overnight. No wonder historians have had difficulty

59. Ibid.
60. Ibid., 34.

in dating precisely this extended process that their vocabulary impels them to view as an isolated event.[61]

Veblen, like Kuhn, was more than skeptical about the alleged "scientific" nature of the social sciences.[62] Indeed, he mentioned economics and political science by name as unworthy of the name "science." Infected by religious teleology, natural law, or Hegelian residues and the metaphysics of normality, the two disciplines lacked the essentials of scientific objectivity. Veblen held up the evolving model of the biological sciences and, more specifically, the Darwinian revolution as the most promising direction in which to proceed. A Kuhnian gloss thus illuminates Veblen's sociology of knowledge by focusing on the progress of the human mind, or lack thereof, in both the natural and the human sciences.

For Kuhn, the reality for most scientists is "normal science," which means working away at problems set within an accepted overall background or "paradigm." It appears that one essential component of a Kuhnian paradigm is habits of mind; in the case of scientists this means central habits of mind. A "paradigm shift" in Veblenian terms would be a special sort of change in habits of mind. Yet habits are socially shared in that while they facilitate communication, they also constrain novel thinking and innovative cognition. Yet, as Howard Margolis points out they play no essential role in Kuhnian paradigms.[63] Veblen's focus at times on core habits of mind thus gives his perspective on scientific change a different flavor from those of Kuhn. Facilitating habits must somehow overcome entrenched habits if the new is to triumph over the old; but in Veblen's analysis habits are difficult to change and much repetition is usually required. The barriers to change are often rigid, and iconoclastic; that is to say, radically unconventional ideas are slow to emerge. Indeed, multiple barriers may exist in the cultural and disciplinary matrix of inquiry and breakthrough effects rare in coming. "'Economy' and 'comfort' as properties favored by the endemic cognitive propensities . . . ultimately govern the formation and hence also the breaking of habits of mind."[64]

Few Veblen scholars doubt the powerful role that "habit" and "habits of mind" play in his social psychology, or for that matter, in his explana-

61. Ibid., 7.
62. Ibid., 15.
63. See Howard Margolis, *Paradigms and Barriers: How Habits of Mind Govern Scientific Beliefs* (Chicago: University of Chicago Press, 1993), 31 and ch. 1–3. Margolis does not use or cite Veblen, but he illuminates much of what Veblen wrote about "habit" and "habits of mind."
64. Ibid., 42.

tion of the sociology of science. What we now call "paradigm shifts," including the genetic endowment that is prior to, yet part of habit itself, as well as the sociocultural conditioning essential to it, can lead to radical restructuring of well-entrenched beliefs. The underlying cognitive propensities that shape habits of mind, constrain intuition, and promote inertia and conceptual obstacles to change are also a major focus in Veblen's thought. In short, as Margolis also puts it, "'habits of mind' suggests entrenched responses that ordinarily occur without conscious attention, and that even if noticed are hard to change."[65] Or so Margolis and Veblen would have us believe.

Although Veblen never had the opportunity to read Darwin's early (and in the former's lifetime mostly unpublished) writings, his own use of Darwin and Lamarck is plagued by the same uncertainties and ambiguities that perplexed Darwin himself.[66] Darwin never rid himself entirely of the Lamarckian tendencies in his work nor did Veblen, although in Veblen's case a more defensible rationale exists for his continued use of Lamarck; in that functional use and disuse, adaptation and nonadaptation can be found within their specific historical and cultural anchorage. However, while he believed that institutions have features that change in the direction of adaptive efficiency that enable them to perform better than the unsuccessful, apparently that did not mean an adaptive role for every function. In Veblen's at times Darwinian social analysis, there are evidently cultural and societal furniture and artifacts floating around that may provide clutter, but do not directly aid or hinder adaptability or survival. Again, however, the formation and development of habit plays a key role in Veblen's analysis even in the genesis of science and the scientist. As he put it in this salient passage in his writings on the scientific community:

> This question of a scientific point of view, of a particular attitude and animus in matters of knowledge, is a question of the formation of habits of thought; and habits of thought are an outcome of habits of life. A scientific point of view is a consensus of habits of thought current in the community and the scientist is constrained to believe that this consensus is formed in response to a more or less consistent discipline of habituation to which the community is subjected, and that the consensus can extend only so far and maintain its force only so long as the discipline of habituation exercised by the circumstances of life enforces it and backs it up.[67]

65. Ibid., 7.
66. Cf. Charles Darwin, *Metaphysics, Materialism and the Evolution of Mind* (Chicago: University of Chicago Press, 1980), 21–78, 114–18, on habit(s).
67. Veblen, *The Place of Science in Modern Civilization*, 38–39.

Veblen's emphasis on habit thus can play an important corrective role in understanding and embellishing a Kuhnian paradigm shift.

Sociology of Knowledge: Epilogue

The ways and the extent to which truth claims are culturally relative and socially constructed is still a matter of debate among philosophers and social scientists. To illustrate, the British sociologist Steven Lukes claims that:

> (1) there are no good reasons for supposing that all criteria of truth and validity are (as many have been tempted to suppose) context-dependent and variable; (2) there are good reasons for maintaining that some are not, that these are universal and fundamental, and that those criteria which *are* context-dependent are parasitic upon them; (3) it is only by assuming such universal and fundamental criteria that a number of crucial sociological questions about beliefs can be asked, among them questions about differences between "traditional" and "modern" or "prescientific" and "scientific" modes of thought; and therefore (4) despite many possible difficulties and pitfalls, the sociologist or anthropologist need not prohibit, indeed he should be ready to make, cognitive and logical judgments (however provisional) with respect to the beliefs he studies.[68]

How would Veblen's views on the sociology of knowledge fit into Lukes's interpretative framework, or would they? How would he react to Luke's claim that not all criteria of truth and validity are context-dependent and variable particularly in the hard sciences. Individual and social values may penetrate the realms of physics, chemistry, and biology in certain respects, but they intrude more in the social sciences. But in the social sciences of his time, particularly economics and political science, much of the inquiry was heavily value laden at least in his view. Thus the criteria of truth and validity were context dependent and variable, so pervasive was the doctrinal influence of the Enlightenment and the Age of Romanticism.

Veblen's claims regarding the value bias and ideological skewing of the received economics of his day need no reiteration here. Of course, by Lukes's time the social sciences had by consensus, perhaps, made headway toward achieving a more robust scientific rigor. Still, the latter over-

---

68. Stephen Lukes, *Essays in Social Theory* (New York: Columbia University Press, 1977), 138.

states the likelihood of the existence of universality and fundamentalism of objective criteria even if he recognizes the pragmatic necessity, indeed, inescapability of making cognitive and logical judgments with regard to what scholars study. Nevertheless, Lukes's caveats are valuable in bringing coherence to the study of the sociology of knowledge as Veblen perceived it.

Although Veblen suggests that scientific method is more likely to produce desirable results in the social sciences than other approaches, he also recognizes its moral and cultural limitations:

> But while the scientist's spirit and his achievements stir an unqualified admiration in modern men, and while his discoveries carry conviction as nothing else does, it does not follow that the manner of man which this quest of knowledges produces or requires comes near answering to the current ideal of manhood, or that his conclusions are felt to be as good and beautiful as they are true. The ideal man, and the ideal of human life, even in the apprehension of those who most rejoice in the advances of science is neither the finikin skeptic in the laboratory nor the animated slide-rule. The quest of science is relatively new . . . . The normal man, such as his inheritance has made him, has therefore good cause to be restive under its dominion.[69]

Veblen thus saw considerable lag time between the advent of the ethos of science and its assimilation by the common person; or, for that matter, political scientists.

Nor is this to suggest that he thought it impossible to predict the course of future political events including revolution itself; rather, it was that he viewed science as having made only slight inroads into the study of politics. His view of the future of the politics of industrial societies was, therefore, several sided and, in the final analysis, open ended. This simply reinforces my earlier claims that Veblen did not have a scientistic or a positivistic view of the social sciences.[70] In accord with the logic of evolutionary naturalism, his sociology of knowledge precluded such dogmatism.

Conclusion

What conclusions may be drawn from this brief venture into the sociology of knowledge using early- and mid-twentieth-century social theorists to better understand Veblen's own sociology of knowledge?

69. Veblen, *The Place of Science in Modern Civilization*, 30–31.
70. See Rick Tilman, *Thorstein Veblen, John Dewey, C. Wright Mills and the Generic Ends of Life*, 233–37.

First, placing Veblen in their midst gives historical as well as theoretical perspective to his views on the social and cultural determinants of ideas. Not only his own role in the unfolding of American intellectual history, but its relationship to contemporary theoretical developments in the social sciences is evident. Veblen was not a Marxist, structural functionalist, nor a consensus historian, much less a political scientist advocating pluralism as an explanation of the American power system, and it should now be more evident as to why.

Secondly, his epistemological stance in favor of "matter of fact" as opposed to "make-believe" has far-reaching implications for personal and individual cosmology to be sure. More important, for our purposes, is its significance for social-scientific inquiry where it becomes imperative to make precise and accurate epistemological judgments, and where it is plain that animistic, natural law, and vitalistic explanations of the social and natural order are worse than useless.

As a social ontologist, Veblen was critical of both radical individualism and extreme collectivism. On the one hand, like Dewey his work challenges individualists to give examples of anything or anyone that acted in complete isolation from everything else. On the other, he refused to strip individuals completely of will or volition even though he thought habits or habits of mind were past-binding and, often, retardants providing a foundation for cultural lag and drag.

As an axiologist, that is, inquirer into the nature of value, Veblen believed he had located those values that are most likely to enhance life in a properly developing community or social order. He referred to these as the "generic ends of life, impersonally considered," and they included altruism, workmanship, and idle curiosity. But he believed these ends could not flourish, or in the case of idle curiosity (critical intelligence) even become publicly visible, without the appropriate social determinants and cultural sustenance.

Epistemology, ontology, and axiology thus form the basis for a metatheory, and Veblen's sociology of knowledge is linked with his role as a metatheoretician. Whether or not the lens of the sociology of knowledge is used to interpret his life and work or examine his own employment of the social determinants of thought, an examination of the metatheoretical basis of his thinking is a precondition for understanding his evolutionary naturalism and the possibility of its enrichment.

Chapter 10

Veblen's Anthropology and Structural-Functional Theory

Introduction

It is commonplace for Veblen scholars to acknowledge his use of the literature and data of modern anthropology. Interesting and helpful taxonomies of the influence of late-nineteenth- and early-twentieth-century anthropologists on him are available, and, of course, there is always the massive study of intellectual influences on him by Joseph Dorfman to fall back on when all else fails, however reliable one judges it to be.[1] In any case, we now recognize Veblen's sources and use of the stages theory of historical and anthropological evolution; the theory of cultural lag and institutional retardation and the ramifications of the corollary doctrine of cultural survivals; the hybridization and racial amalgamation of the human species particularly in Northern Europe; the plasticity and malleability of humankind as well as the biological constants in its composition, and the role of Darwin and Lamarck and post-Darwinian evolutionary theory in explaining adaptation, variation, and natural selection in the Veblen corpus.

Veblen's use of the literature and data of anthropology has long inter-

1. Anthropologists have paid Veblen little heed, at least in print. The only piece on Veblen that has ever appeared in a major journal in anthropology is Melville J. Herskovits, "The Significance of Thorstein Veblen for Anthropology," *American Anthropologist* 38 (April–June 1936): 351–52. The reason for this may be that anthropologists take for granted what is new or revelatory to other social scientists particularly neoclassical economists. However, see John P. Diggins, *The Bard of Savagery: Thorstein Veblen and Modern Social Theory* (New York: Seabury Press, 1978), 97–108; "Animism and the Origins of Alienation: The Anthropological Perspective of Thorstein Veblen," *History and Theory* 16 (May 1977): 113–36. Hugh Dawson, "E. B. Tylor's Theory of Survivals and Veblen's Social Criticism," *Journal of the History of Ideas* 54 (July 1993): 489–504; Dorothy Ross, *The Origins of American Social Science* (Cambridge: Cambridge University Press, 1991), 207–14.

ested social scientists, yet no systematic study of his broader employment of it exists. However, more to the point here is the analysis of a salient aspect of it, namely, criticism of his work using structural-functional theory. Talcott Parsons and two other sociologists trained or influenced by him, Arthur Davis and Bernard Rosenberg, used the theory in various ways to interpret Veblen. Interestingly, however, they failed to recognize his own critical use of aspects of structural-functionalism to attack the various social processes and practices they used the theory to defend. Veblen was unimpressed by the demands of functional unity, indispensability, and universalism of the theory and his critics missed the point of part of his theorizing by failing to understand this.

Veblen's understanding of anthropology mostly stems from his reading of Lewis Henry Morgan, Edward B. Tylor, Franz Boas, Herbert Spencer, William Graham Sumner, and their contemporaries. He was Sumner's student at Yale and corresponded with Boas, apparently when Veblen lived in New York City after the First World War, and he was in close reading contact with Morgan, Tylor, and Spencer. What familiarity he later had with the work of Bronislaw Malinowski[2] is documented, while his knowledge of A. R. Radcliffe-Brown was probably modest. It is interesting to note that none of the above published anything of any consequence about Veblen. But what is significant is that although the latter two belonged primarily to a later generation of theorists than Veblen, it is they who were primarily responsible for refining and formalizing structural-functional theory as it came to be understood in the English-speaking world. And Veblen critics Talcott Parsons, Arthur Davis, and Bernard Rosenberg came to view society in part through a lens the Britishers ground in their publications from shortly after the turn of the century until the end of the interwar era. The generic version of structural-functional theory used to interpret Veblen is as much the product of Malinowski and Radcliffe-Brown, whatever their differences, as it is of earlier or later theorists.

This chapter is thus organized around the following themes, analyzed here in numerical order: (1) Veblen's relationship to functionalism, historically speaking, that is, his intellectual pedigree; (2) criticism of Veblen's social theory by structural-functional theorists and evaluation of

2. Dorfman writes that Veblen "expressed a good-natured ridicule for the anthropologist B. Malinowski because "he claims to be reporting only facts, whereas actually on every line he builds up his own charming theories." Dorfman, *Thorstein Veblen and His America*, 499. An important analysis of the literature, data, and personalities of anthropology when Veblen was at the University of Chicago is George W. Stocking Jr., *Anthropology at Chicago* (Chicago: University Chicago Press, 1979).

these criticisms by the author; and (3) a more detailed and systematic effort than any existing to show the implications of his social theory for structural-functional analysis under three headings; (a) the functional indispensability of norms and practices, that is, the existence of functional prerequisites as well as substitutes or alternatives to them; (b) universal functionality; the view that all persisting forms of culture are inevitably functional; and (c) the assumption of functional unity, that is, the full integration of all societies. It will be shown that Veblen's social theory, while it can be analyzed through this paradigm, can be used more effectively to illuminate the paradigm's own political and ideological flavor. It also suggests the chimerical, that is, utopian nature of any system that is claimed to be self-equilibrating or self-maintaining, using Veblen's theory of cultural lag as the basis for comparison. The likely resurgence of atavistic continuities such as religion, sport, and economic predation, in his view, does not serve human well-being on the whole. For their persistence does not always signify the normative value of functional indispensability, universal functionality, or functional unity, the three prerequisites of structural-functional theory; in fact, quite the opposite or so Veblen believed.

Functionalism

"Functionalism," as a concept in philosophy and the social sciences, has a long pedigree. As long ago as 1931, Horace Kallen wrote that:

> Functionalism is a term which came into the foreground of philosophic discourse in the last quarter of the nineteenth century and has maintained an increasingly strong position there ever since. It sums up and designates the most general of the many consequences of the impact of Darwinism upon the sciences of man and nature. This was to shift the conception of "scientific thinking" into a temporal perspective; to stress relations and activities as against terms and substances, genesis and development as against intrinsic character, transformation as against continuing form, dynamic pattern as against static organization, processes of conflict and integration as against formal composition out of unchanging elements. In short, the shift was from "structure" to "function" as the principal tool of scientific explanation and interpretation.[3]

3. Horace M. Kallen, "Functionalism," *Encyclopedia of the Social Sciences*, ed. E. R. A. Seligman and Alvin Johnson, Vol. 6 (New York: Macmillan Company, 1931), 523–26. The quotation is on 523.

Veblen was recognized early on as a proponent of functionalism in economics. One of the first scholars to attach this label to Veblen's economics was A. B. Wolfe in 1924.[4] Kallen followed Wolfe's lead, writing that:

In economics functionalism has taken mainly the form of institutionalism as embodied in the point of view of Thorstein Veblen and his followers. The hates and enthusiasms of Veblen often deviate him into teleology, but on the whole he appears as the most uncompromising functionalist among social philosophers of the last generation. Malinowski appears to be aiming at an equally thoroughgoing functionalism in anthropology, insisting that questions of origins, stages and laws of development in culture are inferential and secondary and must wait upon the discernment of functions. Functions are events going on, operations of bodily needs and the instrumental uses of objects which constitute their cultural character. They are contents of direct experience, susceptible to observation and analysis. Seen functionally, religion, the arts and sciences become reduced to specific habits, materials, meanings, activities, within the context of a cultural situation, and the forms and structures of such cultural objects become derivatives, concretions or deposits of the dynamic relations in play. What ceases to function, ceases to be.[5]

Kallen thus links functionalism in economics as he thinks Veblen exemplified it, and its role in anthropology in the grasp of Bronislaw Malinowski (1884–1942) and A. R. Radcliffe-Brown (1886–1955), who articulated and formalized functionalism in the early twentieth century.

Criticisms of Veblen by Structural-Functional Theorists

What relationship does Veblen's social theory bear to the structural-functional paradigm established during the period from 1884 when he published his first article to his death in 1929? As a contemporary of the most influential theorists and practitioners of the first generation of structural-functionalists, Durkheim, Radcliffe-Brown, and Malinowski[6] among others, it might be instructive as a comparative study in the histo-

4. A. B. Wolfe, "Functional Economics," in *The Trend of Economics*, 447–75.
5. Kallen, "Functionalism," 525.
6. On the relationship between Veblen and Durkheim, see Rick Tilman, "Durkheim and Veblen on the Social Nature of Individualism," *Journal of Economic Issues* 36 (December 2002): 1104–10, and "Emile Durkheim and Thorstein Veblen on Epistemology, Religion and Social Order," *History of the Human Sciences* 15 (November 2002): 51–70. On Malinowski, see Michael W. Young, *Malinowski, Odyssey of an Anthropologist, 1884–1920* (New Haven, CT: Yale University Press, 2004).

ry of the human sciences to analyze what he had in common theoretically with the latter two. If space permitted it might also aid in the illumination of his social theory in a novel way.

Instead, while analyzing Veblen's corpus through the emerging structural-functional paradigm, the explicitly theoretical commentary on his work in that vein in the postwar era is briefly summarized. The last generation of its practitioners who were interested in his work were Talcott Parsons, Robert Merton, Arthur Davis, and Bernard Rosenberg. It should be kept in mind, however, that of these four men, only Merton attempted in any positive way to use Veblen's ideas to embellish structural-functional theory and his contribution has already been analyzed (see chapter 9). The other three used the theory to criticize Veblen's work and did not incorporate it into structural-functionalism as such.

In summation of structural-functional theory, it should be kept in mind that there are different versions of it; many social scientists question not only its theoretical validity, but its usefulness in forecasting future events, explaining the past, and analyzing the present. However, in view of the fact that some of the most detailed and explicit criticism of Veblen comes from scholars using the theory, whatever the truth of their claims about his work, it is important to recognize that it comes from the proponents of a particular theoretical viewpoint. For this reason alone, it is important to dissect their claims since this perspective had more influence among sociologists and anthropologists in the interwar and postwar eras than any other.

Talcott Parsons's Critique of Veblen

Talcott Parsons (1902–1979) attempted to undermine radical critiques of the American national power structure by labeling the normative aspects of them as "utopian." This was evident in his analysis of both C. Wright Mills and Veblen, whom he lumped together as having basically similar but fallacious views of the existing power structure.[7] Unfortunately, Parsons confused two theories that are substantially different by failing to distinguish Mills's power-elite theory from Veblen's ruling-class hypothesis. More important, however, he charged the two men with having unrealistic views of possible alternatives to the existing system of

---

7. Talcott Parsons, "Distribution of Power in American Society," in *C. Wright Mills and the Power Elite,* ed. G. W. Domhoff and Hoyt Ballard (Boston: Beacon Press, 1968), 84.

power, and further, with a highly selective treatment of the whole complex power problem.

Both were alleged to be guilty of exaggerating the importance of power by holding that it was only power that "really" determines what happens in a society. Parsons also maintained that Mills and Veblen were inclined to think of power as "presumptively illegitimate; if people exercise considerable power, it must be because they have somehow usurped it where they had no right and they intended to use it to the detriment of others."[8] Summarizing the case against Mills and Veblen, Parsons wrote:

> This is a philosophical and ethical background which is common both to utopian liberalism and socialism in our society and to a good deal of "capitalist" ideology. They have in common an underlying "individualism" of a certain type . . . Both individual and collective rights are alleged to be promoted only by minimizing the positive organization of social groups. Social organization as such is presumptively bad because, on a limited, short-run basis, it always and necessarily limits the freedom of the individual to do exactly what he may happen to want. The question of the deeper and longer-run dependence of the goals and capacities of individuals themselves on social organization is simply shoved into the background. From this point of view, both power in the individual enterprise and power in the largest society are presumptively evil in themselves, because they represent the primary visible focus of the capacity of somebody to see to it that somebody else acts or does not act in certain ways, whether at the moment he wants to or not.[9]

Parsons thus claimed that Mills and Veblen so distrust social organization of any sort that they can be lumped together with most of the other extremist critics, right and left, who want to minimize social organization so that individualism can flourish.

According to the Parsonian interpretation of the radical critique, all existing social restraints vanish in the "utopia" so that anarchy prevails. Parsons espoused an antiradical political viewpoint and refused in his own work to allow any serious normative consideration to forms of social organization that differed significantly from those already existing in the United States. The social property and egalitarian power system that Veblen and Mills saw as alternatives to the dominant form of corporate ownership and control could, therefore, be labeled "utopian" by Parsons.

Parsons saw little of value in any of Veblen's major contributions to modern social theory. For example, he disagreed with the emphasis Veb-

8. Ibid.
9. Ibid.

len allegedly placed on the role of technology in bringing about social change. As he put it:

> Some schools of thought, as of Veblen and Ogburn, give the former (technology) unquestioned primary. This is at least open to serious question since it is only in relatively highly developed stages of the patterning of functionally specialized roles that the most favorable situation for the functioning of scientific investigation and technological application is attained.[10]

Parsons suggested, rather, that the mobility of resources made possible through property and market relations and the institutions of personal freedom all greatly facilitate the influence of technology. In Parsons's principled pluralist explanation of social change, greater emphasis was placed on the role and value of the institutions of early capitalism, and Veblen was faulted for exaggerating the role of science and technology.[11]

Parsons was also highly critical of institutional economics, and, implicitly, Veblen, for repudiating the conceptual apparatus of orthodox economic theory without recognizing the possibilities of using its analytic tools even in a different economic system.[12] The denial of the legitimacy of analytical abstraction in economics was a serious error according to Parsons. And in a broader sense, the institutional movement was abortive and spread disillusionment because it "undoubtedly exaggerated the distance between the two disciplines. The combination (to us) of not very good sociology and a negative attitude toward economic and almost any other theory made this movement a poor entering wedge for exploring interdisciplinary relations on a theoretical level."[13]

Parsons found Veblen's distinction between business and industrial pursuits to be greatly exaggerated and destructive of the positive role of the business community.[14]

> Symptoms of disturbance appeared, e.g., the "technological" view of the destructive consequences of business . . . machinations as interfering with "efficiency," utopian exaggerations of the results to be obtained from abandoning "business" altogether and becoming purely "technological."[15]

10. Talcott Parsons, *Politics and Social Structure* (New York: Free Press, 1969), 87.
11. Ibid., 87–88.
12. Parsons, *The Structure of Social Action* (New York: Free Press, 1961), 125. Also, see Talcott Parsons and Neil J. Smelser, *Economy and Society* (Glencoe, NY: Free Press, 1956), 5–6.
13. Parsons and Smelser, *Economy and Society*, xviii.
14. Talcott Parsons, "General Theory in Sociology," in *Sociology Today*, vol. I, ed. Robert Merton et al. (New York: Harper & Row, 1965), 12–13.
15. Talcott Parsons, *Social Systems and the Evolution of Action Theory* (New York: Free Press, 1977).

Parsons argued that Veblen's application of the ceremonial aspect of the dichotomy to consumer behavior seriously distorted the significance of that behavior. He charged Veblen with believing that consumption under capitalism serves primarily a status function, and, is, therefore, a form of status emulation. Parsons wrote that:

> The very ready tendency to derogate such symbolism often takes the form immortalized by Veblen in the phrase "conspicuous consumption," with the allegation that people lived in comfortable and tasteful houses, or wore attractive clothes, in order, for instrumental motives, to enhance their prestige. This was then held to be a dishonorable motive with no "intrinsic" connection with the "real" functions of the unit.[16]

Parsons continued to the effect that "the aspect of the problem which needs to be noted here is that it arises wherever generalized media of interchange are involved in human action."[17] Parsons held that status emulation would occur wherever there is economic inequality and where money is used as a medium of exchange.[18] Parsons failed to note that status emulation is more intense in some societies than in others and that advanced capitalism is more effective, in part owing to mass advertising, than other kinds of societies in inducing such behavior. He thus assumed that status emulation on a massive scale is an inevitable feature of all industrial societies, about which little can be done.

Parsons believed that Veblen, in *The Theory of the Leisure Class:*

> Called attention to some of the relevant features of the role of women but did not relate it in this way to the functional equilibrium of the social structure. Moreover, what Veblen means by "conspicuous consumption" is only one aspect of the feminine role and one which is associated more with certain elements of malintegration than with the basic structure itself.[19]

Women's ostentatious display of goods, through which their men obtain vicarious status gratification, is not interpreted by Parsons as it was portrayed by Veblen, as "normal" behavior on the part of the leisure class and those who would emulate it. Instead, it is seen as evidence of "malintegration" with the basic social structure. Presumably, when women are more adequately integrated into the social structure, such behavior will

16. Ibid., 364.
17. Ibid.
18. Talcott Parsons, *The Social System* (New York: Free Press, 1963), 244–45.
19. Talcott Parsons, *Essays in Sociological Theory* (New York: Free Press, 1963), 80.

significantly diminish in intensity. Other than mere assertion, Parsons provided little evidence that this had occurred since Veblen's day, nor was there recognition in his analysis of the waste and predation such behavior signifies. Parsons also wrote that:

> High progressive taxation, both of incomes and of estates, and changes in the structure of the economy have "lopped off" the previous top stratum, where the symbols of conspicuous consumption were, in an earlier generation most lavishly displayed. A notable symbol of this is the recent fate of the Long Island estate of the J. P. Morgan family, which had to be sold at auction in default of payment of taxes. One wonders what Veblen would say were he writing today instead of at the height of the "gilded age."[20]

In keeping with the postwar tenor of American liberalism, Parsons argued that progressive taxation had significantly reduced conspicuous consumption by limiting the financial resources available for ostentatious display. Apparently, he believed this helped make Veblen's theory of status emulation obsolete.

Parsons indiscriminately lumped together different forms of emulatory behavior, thereby making them appear to be "normal" and generalized features of all societies. He did not adequately consider the waste and deprivation created by some forms of status emulation. Indeed, his systemic explanation of its functioning came perilously close to a justification or defense of it, although its most objectionable features were probably viewed by Parsons as having been modified by the tax, welfare, and regulatory system created by the New Deal and wartime policy.

Parsons believed that there exists no scientific or agreed-upon standard that would allow social scientists to condemn those practices Veblen described. The upshot of all this is that sociology of the Parsonian variety can brook no criticism of social action; what is, and what happens, exist and occur in some natural sense. Criticizing conspicuous consumption is as scientifically pointless as denouncing shifts of rock strata or the movements of planets.

Parsons was highly critical of Veblen's theory of instincts and habits. Yet nowhere did he attempt to go beyond the mere expression of these terms to draw out any further significance they might have. Veblen had cautioned against possible misinterpretation of his use of *instinct* and *habit*, and several observers since have pointed out that these terms are best understood in some other sense than the terms imply. Janice Harris, for ex-

20. Ibid.

ample, claimed that "Veblen's position on 'instinct' and 'habits' comes far closer to what Erich Fromm calls 'normative humanism' postulating a plausible relationship between basic drives and cultural determination, than to the tenets of biologistic determinism."[21] The best that Parsons could do was to label Veblen's system as quite "simple." Considered in the light of Parsons's own complex social theory, that may be true, but, nonetheless, to Veblen's credit.

Interestingly, Parsons recommended a fusion of aspects of neoclassical economics with the social theories of several of his favorite European thinkers.

> The older institutionalism has been essentially positivistic empiricism, unfortunate in its rejection of the solid achievements of the older economic theory and at best very one-sided in the factors put in its place. There is a great opportunity for a "new institutionalism" based on an enlightened and mutually respectful cooperation between the best and methodologically most sophisticated, of the orthodox economic theory, and the newer sociological theory of such men as Pareto, Durkheim, and Max Weber and their successors. In the understanding of concrete economic activities neither can get on without the other.[22]

Parsons simply dismissed Veblen and American institutionalism as irrelevant to the development of a new and more adequate institutional eco-

---

21. Janice Harris, "Thorstein Veblen's Social Theory: A Reappraisal" (Ph.D. diss., New School for Social Research, New York, 1956), 25. In this regard, Parsons wrote:

> But Veblen, underneath his empiricism, shared the positivistic bias of the thought of his time. When he went beyond merely pointing to the facts to fit them into a theoretical scheme, he took over the vague psychological concept of habit and then, conscious that it is impossible to derive a particular institutional structure from a general psychological mechanism alone, he resorted to very complicated combinations of his four "instincts" with each other and with particular environmental conditions and stages of social evolution to give some specific content to the concept. All this is within the circle of positivistic factors with which we have dealt. Veblen may be held up as the primary example of "positivistic institutionalism."

Parsons, "Institutionalism (1934)," 3, Talcott Parsons Collection, Nathan Pusey Library, Harvard University.

22. Ibid. The only positive comments Parsons made regarding any specific part of Veblen's social theory was in a draft of "Sociological Elements of Economic Thought," this portion of which was not included in the published version. Parsons wrote that "Veblen has some very important and clear insights into social phenomena. Above all, he saw and emphasized the historical relativity of economic activities. He also saw that they were related to a framework of factors, a "social structure," the main outline of which was independent of the individual ad hoc actions. From the point of view of freedom of adaptation to environment exigencies it was a restraining framework." Parsons, "Sociological Elements of Economic Thought" (Draft, Parsons Collection).

nomics. Few traces of Veblen were to permanently color his thinking and no residues of institutional economics were visibly embedded in his social theory.

Bernard Rosenberg's Critique of Veblen

Bernard Rosenberg (1923-1996), sociologist at the New School, paid little heed to Veblen's support for abolition of absentee ownership, his stress on the wasteful lifestyle of the upper classes, or his consistent praise of the egalitarianism of primitive societies. Perhaps he failed to come to terms with Veblen's egalitarianism because equality is incompatible with Rosenberg's own pessimistic vision of the inevitability of a permanently stratified bureaucratic society.

Rosenberg was aware of interpretations of Veblen that stressed his primitivism and his anarchism, but he did not recognize the extent to which these traits were connected with his egalitarianism and were the outgrowths of a mind that sanctioned no invidious distinctions among men. But functional distinctions based on the division of labor were, in Rosenberg's eyes, the basis of Veblen's theory of stratification, and inequalities of power, status, and income were their inevitable accompaniments. Although he believed that the technocracy of the 1930s was a distortion of Veblen's plan, the latter was no equalitarian with his eye on the creation of a classless society.[23]

Religion and Functionalism

Rosenberg portrayed Veblen as a philosopher descended from the enlightenment with all the antireligious bias that usually connotes. It is, of course, true that Veblen included the church when he spoke of "imbecile institutions." As Rosenberg put it:

It will be remembered that ritualists of every variety are berated by Veblen and that he regards religion itself as impermanent and on the wane. In this

---

23. See Bernard Rosenberg, "Veblen and Marx," *Social Research* 15 (March 1948): 88–117; "A Clarification of Some Veblenian Concepts," *American Journal of Economics and Sociology* 12 (January 1953): 179–87. "Thorstein Veblen: Portrait of the Intellectual as a Marginal Man," *Social Problems* 2 (January 1955): 181–87; *The Values of Veblen: A Critical Appraisal* (Washington, DC: Public Affairs Press, 1956); intro. to *Thorstein Veblen* (New York: Thomas Y. Crowell, 1963), 1–14.

role he is preeminently the Rationalist. Veblen always seems to be saying: unreason is rampant and it must be extirpated. The same cry resounds through so much late nineteenth century thought, notably as it comes down to us in psychopathological theory by way of Freud. Veblenism does not differ appreciably on this leap from Freudianism: for either the religious "illusion" should certainly be doomed with science making more and more headway.[24]

Rosenberg believed that Veblen, like Marx, was unable to adequately explain why and how religion has such a pervasive hold on the human mind. He attributed to him a belief that religion is a form of superstition resulting from the deceit and dupery of priests, a view that is incapable of explaining the continuing religiosity of the underlying population. Indeed, Rosenberg commented that "Veblen's functionalism has a number of blind spots of which religion is the most serious. It is affiliated scientifically with the early anthropology and philosophically with the Enlightenment."[25] The sociologist in Rosenberg criticized Veblen for failing to recognize "that from the very beginning society and religion were practically identical, or more correctly, that one was an expression of the other."[26] Rosenberg then showed his own hand, although admitting this to be a debatable point, when he commented that "religious organizations—like family structure—is *sui generic*, a completely neutral phenomenon."[27] This passage is reductionist in the sense that it portrays religious belief and practice in the same way that it does family structure. It is as though both were equally "natural" phenomena and permanently rooted in the human psyche and the social order. Rosenberg's own cultural relativism appears to be in conflict with his acceptance of the family and the church as permanent and integral parts of society.

Patriotism, War, and Peace

Rosenberg also evaluated Veblen's views on international politics and found them to be more realistic and subtle than those of Marx and his followers. He cryptically commented that "no devotee of Marx could, to begin with, have categorized patriotism as a part of the substructure, which is really what Veblen did."[28] But he then argued that:

24. Rosenberg, *The Values of Veblen*, 76.
25. Ibid., 49.
26. Ibid.
27. Ibid.
28. Rosenberg, "Veblen and Marx," 110.

Veblen was afflicted with ambivalence, and even in this book reverted occasionally to the other phase of his temperament. It is impossible to believe with any logical consistency that this was simply foisted upon the people by businessmen who profit from it and, at the same time, that it was as the result of patriotism which cannot be exorcised from the psyche. Yet Veblen alternately defended both points of view, and his gloom seems justifiable to us, irrespective of how soundly he has analyzed the total pattern.[29]

A different reading of Veblen based on a synthesis of his later writings would stress the latent convergence of the corporate profit motive with the patriotic sentiments of the underlying population. But Rosenberg argued in another context that Veblen could not make up his mind whether war was a cultural phenomenon or whether it was due to inherent impulses.[30] Consequently, Rosenberg believed Veblen was caught between believing that:

> On the one hand, patriotic devotion is falling into disuse, is little more than an archaic affection; on the other hand, it is a permanent and indissoluble part of human nature. To be sure, the first view appears more often and is stated more forcefully than the second. But together they reveal profound ambiguity in his thinking.[31]

When he treated Veblen's explanation of patriotism, war, and peace and the linkages between them, Rosenberg lay great stress on the ambivalence, indeed, inconsistency in his view of human nature. As regards the nature versus nurture controversy, Veblen, alas, could not make up his mind:

> As usual he is hard put to it to explain the persistence of such an untoward habit as patriotism. He does so in *The Nature of Peace*, by dismissing it as an instinct. This is the line of least resistance for Veblen at all times, the tautology that affords salvation. That it does violence to technological determinism is obvious. For clearly, some causal agent takes primacy over the state of the industrial arts. A quality of human nature exists, we learn—ubiquitous, intimate, and ineradicable—prior to technology. This view, however erroneous it may be, contradicts Veblen's customary hypothesis.[32]

Of course, if patriotism has indestructible roots in the human psyche, there is little hope for permanent peace and Veblen's pessimism is fully

29. Ibid.
30. Rosenberg, *The Values of Veblen*, 106.
31. Ibid.
32. Ibid., 104.

justified. On the other hand, if generic man is sufficiently plastic and malleable, then perpetual peace is at least conceivable. Rosenberg thus believed there is no clear or consistent portrayal of human nature in *The Nature of Peace*.

Instinct Theory and Human Nature

Rosenberg's interpretation of Veblen's view of human nature was inconsistent. At times he presented Veblen's views as instinctivist, but, contradictorily, he also represented them as stressing human plasticity and malleability.[33] Ultimately, however, despite his inconsistency (and Veblen's) he came down on the side of nurture in interpreting Veblen's ideological position in the nature versus nurture controversy:

> Instinct theory has bedeviled us for too long and Veblen may be numbered among those who have added to rather than detracted from the general confusion. Ordinarily instincts connote rigidity or inflexibility or the irreducible psychological and physiological elements in human behavior. This is the case with instinctivists like MacDougall; it is no longer the case with Veblen for whom human nature approaches something close to infinite malleability.[34]

Rosenberg believed that other liberal critics of Veblen, such as Talcott Parsons, attempted to discredit his work by focusing on his theory of instincts and, worse, by misrepresenting it. He commented thusly on Veblen's instinct theory starting with a quote from Veblen himself:

> "it is a concept of too lax and shifty a definition to meet the demands of exact biological science." So he debiologizes and sociologizes it. Where the biologists and the individual psychologists had ignored culture he plays it up; where they had slighted "habit" he gives it centrality; where their conception of man were inflexible his is plastic. From this it should be evident that the glib dismissal of Veblen as a theorist who subscribed to instinct psychology is altogether unfair.[35]

In spite of this defense of Veblen, Rosenberg did not find his instinct theory to be an adequate one for he argued that "four basic 'instincts' inhere

---

33. Rosenberg, "Veblen and Marx," 101, 108; also see *The Values of Veblen*, 46, and "A Clarification of Some Veblenian Concepts," 182.
34. Rosenberg, *The Values of Veblen*, 45.
35. Ibid.

in Veblen's schemata: the parental bent, the predatory bent, the bent of workmanship, and idle curiosity. There is vastly more to man than any composition of these instincts, and the first, third, and fourth are not as unqualifiedly beneficial as Veblen believes."[36] At this point the classical liberal residues in Rosenberg's thought manifested themselves as he questioned the value and consequences of altruism, proficiency of craftsmanship, and critical intelligence. By implication, does this also mean that the predatory bent, in the form of the sporting and pecuniary instincts, is more beneficial than Veblen supposed? Such an interpretation is consistent with Rosenberg's analysis.

Altruism and the Family

Rosenberg did not agree with Veblen's views concerning the family and the altruistic moral and behavioral traits that he alleged it supports. Veblen wrongly found nothing but virtue and sustenance in the family and viewed it as a nearly universal source of positive traits and other-regardingness that congealed in the "parental bent":

> It is supposed that some such polarity does exist, that collective satisfaction can only proceed from personal dissatisfaction, and that society should be served over the individual—conceding all these unverifiable assumptions, it is still not possible to support Veblen. The family as a social institution is neither the unmitigated evil that Marxists make it out to be nor the unmitigated good that it becomes in Veblenian psychology.[37]

Rosenberg believed that whether or not the family plays a positive role in society will vary greatly from one culture to another:

> A modicum of evidence about prehistorical times is enough to puncture the Noble Savage and reduce him to a kind of brutishness unfathomable to the Romantic mind. Infanticide is known to prevail in many primitive cultures—which does not argue very strongly for an innately beneficent parental bent. . . . If it is unrealistic to rhapsodize about primitive altruism in the family, then consider Western Culture with its oedipal affections? Relativistic norms govern the family structure in its irregular fluctuations from one gestalt to another; as comparative anthropology amply demonstrates, it may be constructive or destructive.[38]

36. Ibid., 46.
37. Rosenberg, "A Clarification of Some Veblenian Concepts," 183.
38. Ibid.

Rosenberg appeared to believe that there are some universal purposes and values that families should serve, but he was vague about what these might be. Consequently, it is difficult to tell what standards he used to measure the performance of the family and Veblen's attitudes toward it. Nevertheless, he criticized Veblen for uncritically sanctioning the family structure by assuming that it is the fountainhead from which altruism flows. Indeed, Rosenberg claimed that:

> The dedicatory spirit of parents for their children is culturally conditioned to a greater extent than Veblen imagined. He romanticizes primitive life, as we shall see from his anthropological precepts, and so it seems to him that preliterate peoples were selfless and devoted where civilized man with his narrow familism feels comparatively few parental obligations.[39]

Rosenberg failed to indicate to the reader what binding traits and values the family should protect and nurture. The extreme value-relativism *cum* moral agnosticism of modern secular liberalism had so infected his thinking that he was unwilling to articulate the very values he charged the family with betraying in certain cultures. On the one hand, he criticized Veblen's endorsement of the family because it does not foster certain values that Rosenberg prized. But, on the other hand, he was unwilling to adequately state these values. Is this additional evidence of the tendency of postwar liberals to obfuscate when analyzing and prioritizing values? If so, what does this moral posturing indicate about the structure of their own ethical values?

Rosenberg also criticized Veblen for his one-sided treatment of the family, which "obscures the reciprocity of a child-parent relationship":

> He would scorn the mother who neglected her son, but the one who cares for his mother is abetting conservatism and stagnation. Gerontocracies obstruct any innovation which would jeopardize their social supremacy. Veblen implies that reverence for elders is the cause of seniority, authenticity, tutelary oversight, class tabus, autocracy, chieftainship and aristocratic government; that his contention is specious can be shown by the ease with which other equally plausible causes could be summoned to explain the same effects.[40]

Unfortunately, Rosenberg did not tell what these other "equally plausible" causes might be. He also exaggerated the causal relations that Veb-

---

39. Rosenberg, *The Values of Veblen*, 47.
40. Ibid., 48.

len thought existed between reverence for elders and various undesirable social traits and processes.

Veblen's Attack on Culture: Role of Aesthetics and Athletics

Many writers of various ideological persuasions agreed with Rosenberg's criticisms of Veblen's narrow focus, which allegedly utilized rigid criteria of waste-avoidance to measure the role of aesthetics. It was *The Theory of the Leisure Class* that caught their eye, for it was in that work that Veblen most fully developed his functional aesthetic:

> Veblen gives himself away by referring to economic beauty which he inferentially equates with beauty itself; while chuckling over the aphorism that "cheap clothes make a cheap man" he practically says that cheap clothes make a fine man. The one is as manifestly wrongheaded an aesthetic as the other. If longer lasting goods made with a machine are scorned in favor of fragile handwrought goods, this indeed is laughable. But handwrought goods are not "intrinsically ugly" and that they are so marked by Veblen immeasurably weakens his excellent case against conventional taste.[41]

Rosenberg also found it objectionable that Veblen wanted to completely rationalize all other social processes. He complained that:

> Veblen defines conspicuous leisure as the *non-productive* consumption of time—i.e., as conspicuous waste. His excoriation of almost everything not purely economic blights the book. For there are areas which cannot sensibly be classified as wasteful or as economic; yet in *The Leisure Class* and elsewhere Veblen tends to ignore them.[42]

Veblen's attitude toward competitive sports and athletics was, of course, to focus on their waste and predation. His best-known comment in this regard from *The Theory of the Leisure Class* was that "the relation of football to physical culture is much the same as that of the bull-fight to agriculture."[43] Rosenberg complained that Veblen did not understand that sports and athletics provide physically wholesome escapes from boredom. However, he admitted that:

41. Ibid., 71.
42. Ibid.
43. Veblen, *The Theory of the Leisure Class*, 261.

Veblen would probably reply that this was all well and good for the boys and their surplus energy, but quite unbecoming for an adult who should work not to better his own physique but to perfect the body social. Veblen can see no critical difference between primary and secondary conflict; he does not speculate about a moral or an amoral equivalent of war? It is enough the exhibitions of prowess are irrational. Ergo, they may be written off as undesirable. The issue of athletics gives us a clue to *The Leisure Class* which from this angle, could be viewed as a brief in favor of total rationalization.[44]

Rosenberg's criticism was that Veblen favored "total rationalization" of society and this meant total elimination of all economic waste regardless of sources or consequences. But what if Veblen believed that only aesthetic and athletic endeavors which aimed at invidious comparisons be eliminated and that those activities that are noninvidiously meaningful to the common man be retained? Unfortunately, Rosenberg overlooked the latter possibility in his eagerness to convict Veblen of wanting complete rationalization of all social processes and institutions. For some forms of physical culture and aesthetic endeavor may be noninvidiously meaningful to the underlying population and can be spared in a drive toward total rationalization of social life. Unfortunately, Veblen did not advance beyond the functional aesthetics of his first book in which his theory of aesthetics is not fully developed. Indeed, he rarely turned his attention to aesthetics again, leaving the casual reader with the impression held by Rosenberg that waste avoidance and satire were the sum total of his contribution.

Arthur Davis: Order and Authority

Although Arthur Davis (1916-2001) was not a prominent American sociologist, his extensive writings on Veblen are of considerable interest and value. For there are important political similarities between his early interpretation of Veblen and that of Talcott Parsons in that both shared liberal ideological predilections that negatively influenced their view of our central figure.[45] Davis is also significant in this study because he is the

44. Rosenberg, *The Values of Veblen*, 76.
45. Davis's writings on Veblen from 1941 to 1948 were basically liberal in tone and included "Thorstein Veblen's Social Theory" (Ph.D. diss., Harvard University, 1941); "Veblen on the Decline of the Protestant Ethic," *Social Forces* 22 (March 1944): 282–86; "Veblen's Study of Modern Germany," *American Sociological Review* 9 (December 1944):

only scholar whose views underwent a pronounced ideological shift from liberalism to radicalism while he was writing on Veblen.[46] However, Davis's early criticisms of Veblen were rooted in a Parsonian model of the social order, with emphasis on an action theory of human behavior. The bias of this doctrinal stance was particularly evident in Davis's criticisms of Veblen for his failure to articulate an adequate theory of social order. Davis expressed this criticism in several ways: "Existing institutions appeared to Veblen as obstacles to Utopia . . . . A society without institutions such as Veblen sometimes leans toward, is a contradiction in terms. . . . Social life without organized institutions is tacitly assumed to be not only possible but ideal."[47] These criticisms of Veblen's attitude toward the problem of social order closely parallel Talcott Parsons's critique of both Veblen and C. Wright Mills which is not surprising since Davis wrote his doctoral dissertation under Parsons and shared most of his theoretical and ideological biases at this point in his career. Indeed, Davis succinctly summarized his and Parsons's case against Veblen when he commented that "Veblen's theory is highly consistent, but its main failing is the lack of an adequate concept of institutions. His explicit scheme leaves unsolved the problem of order."[48] Davis consistently found what he thought was a "latent strain of naïve philosophical anarchism in Veblen's thought"[49] that was rooted in his belief that contemporary technology and science are incompatible with all existing institutions. He put it this way:

> The three benevolent Veblenian instincts which would emerge as the decisive elements in action after the demise of current predatory institutions are

---

603–9; "Sociological Elements in Veblen's Economic Theory," *Journal of Political Economy* 53 (June 1945): 132–49; review of *The Freudian Psychology and Veblen Social Theory*, by Louis Schneider, *Social Forces* 27 (October 1948): 94–95. His later works on Veblen, produced in the mid-1950s, show strong evidence of his political radicalization. See "The Postwar Essays," *Monthly Review* 9 (July–August 1957): 91–98; and "Thorstein Veblen and the Culture of Capitalism," in *American Radicals: Some Problems and Personalities*, ed. Harvey Goldberg (New York: Monthly Review Press, 1957), 279–793. Also, see "Veblen, Thorstein," in *International Encyclopedia of the Social Sciences*, vol. 16, ed. David L. Sills (New York: Macmillan and Free Press, 1968), 303–8.

46. Davis's radicalization, however, apparently had its beginnings not in the 1950s but in the 1930s. The Depression weighed heavily on him and so did the Spanish Civil War. Indeed, he seriously considered going to Spain to join the Loyalists, but World War II temporarily diverted him from a radical path; the impact of the Cold War and the Chinese Revolution "reradicalized" him. See Arthur Davis, "Veblen Once More: A View from 1979," in *Thorstein Veblen's Social Theory* (New York: Arno Press, 1980). Davis held academic posts at Harvard, Union College, and Vermont before becoming a Canadian expatriate at the University of Calgary during the McCarthy era.

47. Davis, "Thorstein Veblen and the Culture of Capitalism," 291.

48. See Davis, "Thorstein Veblen Reconsidered," 52.

49. Ibid.

wholly inadequate as substitute institutional structures despite their implicit normative content. The problem of order in Veblen's future machine age therefore becomes exceedingly acute. On the institutionalization or integration of value patterns depends the very possibility of social order, organization, and stability. A society without institutions is inconceivable.[50]

Davis thus claimed that idle curiosity, proficiency of workmanship, and the parental bent, Veblen's three "instincts," "cannot begin to solve the inevitable and omnipresent problems of authority, property patterns, distributive standards, and punishment—reward patterns."[51] Yet Davis failed to explain how any viable social order could exist for long without those three values so esteemed by Veblen.

He complained that Veblen did not understand the need for or the methods by which fundamental social and cultural values were inculcated in the underlying population by basic socialization processes. Because he was so committed to the "generic ends of life," which emphasize "peaceful solidarity" and "maximum production," he could view contemporary institutions only as causing wasteful and predatory behavior. Apparently, at this point in his career, Davis believed that respect for authority, acceptance of the existing social order, and loyalty to the status quo must take priority over Veblen's normative goals. Davis was correct that Veblen failed to solve the problem of social order, but then its conventional resolution was of no particular concern to him.[52]

However, understanding the legitimization of social order and institutionalization of authority was a focus of Parsonian structural-functionalism as well as of Veblenian institutionalism. Even in Davis's later radical writings on Veblen, there was a focus on both processes. To illustrate, he wrote that:

> Workmanship may produce goods; and the parental bent, good will. But how are goods to be distributed? How are new members of society to be trained? How are rights and duties to be formulated and enforced? How to adjust to change? To rule-breakers? How are leaders to be chosen and trained? It is not that Veblen merely neglected to answer such questions: the point is that his theoretical system leaves no place for them at all.[53]

Davis believed that Veblen's views were ultimately rooted in his account of cultural evolution, starting with a primitive stage of savagery

---

50. Ibid.
51. Davis, "The Postwar Essays," 92, 98.
52. Davis, "Sociological Elements in Veblen's Economic Theory," 145; "Thorstein Veblen Reconsidered," 59; "Thorstein Veblen and the Culture of Capitalism," 282.
53. Davis, "Veblen on the Decline of the Protestant Ethic," 285.

that he would have liked to restore.[54] Unfortunately, this caused him to give no heed

> to such problems as allocation, the criteria of distribution, the organization of power. In short, his projected state of technological abundance had no provision for such indispensable institutions as property and authority. They are phenomena of cultural lag, which would not exist in a society governed by the sense of workmanship.[55]

Davis also criticized Veblen's failure to deal with the problem of social order and morality. As he put it:

> So broad a concept as the parental bent could never be an instinct. Its unadulterated altruism in part expresses Veblen's personal rejection of the highly competitive and self-interested values of contemporary society. Its formlessness springs from Veblen's lack of an adequate concept of a social system. Solidarity in the real world is not uniformly omnipresent but channelized in specific relationships—in patterns of moral obligation (rights and duties) which vary in content and extent in time and place.[56]

Veblen was thus criticized for his lack of specificity, which overlooks the fact that, in a generic evolutionary schema of the kind he constructed, specificity is not essential; indeed, it may even thwart a better understanding of the development of more adequate ideal-types, or "type forms" as Veblen called them.

Status Emulation and Emulatory Consumption

Davis was critical of the best known of Veblen's theoretical contributions, namely, his theory of status emulation. Interestingly, he believed that "in emphasizing competitive emulation, Veblen was universalizing a characteristic of modern American society by reading it back into antiquity. To the sociologist, history is a number of different social systems, more or less or not at all interrelated, depending on the particular case."[57] Veblen's history and his prehistory was thus infected with his view of the contemporary United States. He saw the past through a lens encrusted with the American present and, as a consequence, this past was permeated with emulatory consumption and invidious distinctions.

54. Davis, "Veblen's Study of Modern Germany," 606.
55. Ibid.
56. Davis, "Sociological Elements in Veblen's Economic Theory," 145.
57. Davis, "Review of the Freudian Psychology," 94.

Another related weakness Davis found was Veblen's "exclusive concern with the invidious effect of contemporary life [which] entirely overlooks the cohesive function of consumption habits in expressing the basic values of the group."[58] But the "cohesive function of consumption habits in expressing the basic values of the group" may be just as destructive of the generic ends of life as those coercively imposed from the top down. Davis's comments are of little value in helping determine whether the basic values of the group are authentically expressed or simply represent manipulated consensus.

It is also the case that the moral agnosticism characteristic of Davis's early liberal work on Veblen provided no standards by which to measure the proper development of the social order. Even in his later writings, Davis complained that "Veblen is one-sided about consumption motives. His streak of Tolstoyan austerity probably reflects his populist idealization of the countryman. But a neatly cropped lawn, for instance, is not necessarily wasteful; it can express a positive value as well as an invidious one."[59] However, Veblen carefully qualified his attack on emulatory consumption by pointing to *both* the emulatory and the functional, the latter signifying the utilitarian role of most commodities.

Davis gave a different stress to status emulation than most Veblen scholars who emphasize conspicuous consumption because, instead, he stressed conspicuous waste and conspicuous exemption from useful labor. As he put it:

> Leisure activities tended to divert time and resources from production, hence they ran counter to Veblen's idea of social welfare. He cited many examples: sports, war, religion, the study of Latin, gambling, idle wives, servants, government, anything above a fairly austere level of living. Such phenomena he presented as archaic survivals of earlier predatory traits.[60]

Perhaps Davis gave stress to these aspects of Veblen's thought because he rightly saw them as important to system maintenance as the more socially visible kinds of emulatory consumption, particularly in more traditional societies that have not yet reached the stage of mass production and consumption. In his early writing on Veblen, Davis also stressed the cultural-lag aspect of Veblen's theory, which he believed weakened it:

> His analysis is vitiated by the fact [that] he saw only the invidious aspect of the conspicuous consumption pattern and not its positive side. It is further

58. Davis, "Thorstein Veblen Reconsidered," 62.
59. Ibid., 59.
60. Davis, "Veblen on the Decline of the Protestant Ethic," 283.

weakened by his view that this pattern is an archaic survival, a culture-lag phenomenon long since obsolete from the standpoint of the sense of workmanship and technological efficiency.[61]

Davis again failed to note Veblen's view that many goods and services serve both an invidious and noninvidious role, that is, have both a functional or utilitarian significance as well as an emulatory one. Davis then went to considerable lengths to find merit in emulatory behavior:

> But his definition of conspicuous activity as waste is highly arbitrary. All activity that did not contribute directly to the production of material goods, all expenditure incurred on invidious grounds—Veblen styled wasteful. The test is whether the expenditure contributes to the "generically human ends of life." Exactly what these ends are Veblen nowhere specifies. . . . The unmistakable drift of his writings suggests that he considered all consumption notably above the level of subsistence as wasteful—a notion which scarcely comports with his ideal of technological abundance. Even his critical view of conspicuous display as a mere invidious competition for esteem may be questioned; surely the esteem of one's fellows is a "generically human end of life."[62]

Davis was still in his "liberal" stage when he wrote these lines in 1944, as the last sentence in the quotation makes abundantly clear. Nevertheless, he raised important questions about which values are most generically human and why this is so. Mostly, Veblen assumed that the prime values of life are critical intelligence, proficiency of workmanship, and altruism. But Davis was insistent on knowing why these ends of life are more important, for example, than the desire for esteem, even if it is achieved through invidious competition and conspicuous display. The issue may admit to no easy resolution, but in Veblen's view humanity must be seen as something other than atomized globules of desire, each trying to maximize their own utility.

Interestingly, Davis also criticized Veblen's theory of status emulation in an attempt to discredit Veblen's satire on women's dress:

> This "wasteful" aspect is most clearly shown in women's dress, according to Veblen. The skirt and the high heel sharply underline the nonproductive role of women, whose chief function is to symbolize the pecuniary status of their husbands. Here again, Veblen gives a one-sided picture of women's roles in our society. Differentiations in dress are primarily reflections of the segregation of sex roles in our society. Likewise, the rapid change in wom-

---

61. Davis, "Sociological Elements in Veblen's Economic Theory," 142,
62. Davis, "Veblen on the Decline of the Protestant Ethic," 284.

en's fashions is due more to the institutionalized definition of women's roles than to the seesaw struggle between the sense of workmanship and the norm of conspicuous waste which Veblen collaborates.[63]

Once again the theoretical influence of Parsonian structural-functionalism was evident in Davis's belief that even women's dress is system-sustaining; all social processes have a purpose or they would not exist. Apparently, it does not matter if emulatory behavior and invidious distinction adversely affect the well-being of the community.

Cultural Lag and Economic Self-Interest

In his last writings on Veblen, Davis commented that "Veblen made less of vested interest and more of culture lag in explaining contemporary institutions."[64] Davis believed this was unfortunate in that Veblen mistook economic self-interest in the form of exploitative and predatory behavior for archaic survivals. Davis believed that:

> Veblen clearly perceives the discrepancy between yesterday's theory and today's practice. But he one-sidedly attributes it to "culture lag." This throws the emphasis on inertia, ignorance and habit, at the cost of underestimating conscious exploitation of society by Big Business. Veblen isn't wrong, of course—only incomplete. His stress on culture lag neatly supplements the overemphasis common among Marxists on conscious exploitation.[65]

The theoretical consequence of Veblen's position was that "the concept of culture lag may give undue weight to factors of ignorance and drift, at the expense of vested interest rationality."[66] Davis believed Veblen was so preoccupied with cultural lag that he failed to develop an adequate conception of the imperialist state.[67] He did not sufficiently recognize that social customs "survive not only because of inertia and culture lag but also because certain groups profit from their survival."[68] It is interesting to note that in his later writings on Veblen, Davis criticized Veblen's use of the cultural-lag concept for its evasion of the realities of class interest and

---

63. Davis, "Veblen's Study of Modern Germany," 603.
64. Davis, "Thorstein Veblen Reconsidered," 71.
65. Ibid.
66. Davis, "Veblen on the Decline of the Protestant Ethic," 283.
67. Ibid., 285.
68. Ibid.

class conflict. Yet in his early Parsonian period, he attacked it for not leading Veblen to the truths of structural-functionalism! He commented that:

> Sociology considers a social phenomena not as survivals but as elements having definite functions, manifest or latent, in a social system. Survival as used by Veblen implies cultural lag—a conception which hinders more than it helps, because it obviates further analysis. The current sociological significance of social phenomena cannot often be explained by the study of their origins. Veblen's "genetic" method offers neither an adequate functional analysis nor a proof of causality. Indeed, it prevented him from seeing a social system as a functioning whole.[69]

The societal apologetics of Parsonian structural-functionalism are nowhere more evident than in Davis's insistence that "survivals" be accepted because they perform latent or manifest functions. The moral agnosticism of liberalism is also apparent in Davis's vacillation and, at times, refusal to judge whether the functions are positive or negative and can be contrasted with Veblen's emphasis on whether or not the generic ends of the community are being served. To Veblen, "survivals" were ceremonial in nature when they are destructive of the life process impersonally considered. To the Parsonian Davis, who lacked criteria by which to evaluate institutional growth and development, survivals, whether their functions be manifest or latent, were seen as equilibrating processes in the social system, and that ends the inquiry. Davis's interest in further investigation of the processes is, of course, a later story and is connected with his political radicalization.

In certain respects, Veblen's cultural-lag concept was utilized more intensively in his study *Imperial Germany and the Industrial Revolution* than elsewhere in his work. But this was not what attracted Davis's attention in his early writing on Veblen. Instead, once again showing the influence of Parsonian structural-functionalism, he stressed the role of value consensus and shared attitudes among commoners as the bulwark of support for the Hohenzollern regime; and it was this aspect of Veblen's analysis of Imperial Germany that impressed him. Davis wrote that "he took pains to emphasize the underlying meanings which the humans attached to their authoritarian behavior. He demonstrated the impossibility of making pure coercion as the essential cohesive factor in German social organization."[70] But if coercion was not important in coagulating the social

---

69. Davis, "Thorstein Veblen Reconsidered," 61.
70. Davis, "The Postwar Essays," 93.

and political fabric of the German community, what was its effective bonding agent? Davis answered that:

> Veblen argued that coercion is not the mainspring of the Dynastic State, as an American observer might assume. Instead, its intense patriotism is part of a common value pattern furnishing positive cohesion to the German state. Such coercion as exists is a symptom thereof rather than the essence. Unlike many political scientists of his day, Veblen looked beneath the external forms of government structures.[71]

Davis's early writings on Veblen thus manifested the influence of Parsonian structural-functionalism to such an extent that Veblen drew praise from Davis primarily when his analysis converged with it. However, Davis never quite laid his own ideological cards on the table. He came tantalizingly close to articulating the Parsonian theory as his frame of reference without explicitly endorsing it.[72] For example, he commented on "the lack of recognition of the absence in Veblen of a comprehensive conception of a functioning social system. The last is the basic weakness, shared by many institutionalists."[73] To this complaint against the institutionalists Davis added that:

> The last-mentioned type of category, having social referents, permitted further analysis. An adequate theory of change in social science must be stated primarily in terms of analytical social elements. Where, as in Veblen's case, the use of nonsocial and particularly normative elements have implicitly crept into such concepts as instinct and habit.[74]

On several occasions, Davis argued that Veblen's "evolutionary view of history, with its corollary of culture lag and his vein of Utopian anarchism ... prevented any consideration of a social system as a functioning whole."[75] But what is meant by the claim that Veblen was not able to explain the social system as a functioning whole? First, Davis meant that the fulfillment of Veblen's radical political vision could not be achieved within the framework of the existing social order. He claimed that Veblen believed that:

71. Davis, "Veblen, Thorstein," *International Encyclopedia of the Social Sciences,* 305.
72. Davis, "Thorstein Veblen and the Culture of Capitalism," 289.
73. Ibid., 285.
74. Davis, "Veblen on the Decline of the Protestant Ethic," 284.
75. Davis, "Veblen's Study of Modern Germany," 605.

Contemporary institutions were not competent media for conveying socialist attitudes. Actually, both the vagueness of the socialists and Veblen's own indefiniteness are due mainly to the negative character of their own sociological orientations. No specific institutional substitutes were offered because neither Veblen nor the socialists realized the necessity for such a program. In brief, they had only an incomplete idea of a functioning social system and its component elements.[76]

In short, Veblen's political hopes for the system were incompatible with a scientific (read Parsonian) understanding of it. Also, Davis believed his radical empiricism and his Darwinian bias made it impossible to develop his insights on a theoretical level.[77] Veblen's empiricism "accounts for the baffling vagueness" in his work[78] and the Darwinian elements sow confusion and inconsistency.

Related to these problems was Veblen's critical-genetic method, which Davis claimed could not adequately explain causation:

Temporal sequence established nothing about causality. Veblen's genetic method can produce only a descriptive history of socio-economic developments. Indeed, by focusing on an indiscriminate succession of concrete events, and by tending to overemphasize the causal role of one or a few elements, geneticism hinders the analytical consideration of a social system as a functioning whole.[79]

Of course, the opposite deficiency is often found in Parsonian theory where pluralistic conceptions of causation make causal factors and processes so diffuse and scattered that the main agents of social change cannot be identified.

Yet structural-functional models, since they focus on social stasis or equilibrium as a "normal" state of affairs, see an adaptive or integrative role for every social structure and process. The "objective" social scientist can make no statement of moral or evaluative preferences between them. For this reason Davis also found Veblen's attitudes toward religion to be inadequate. He wrote that:

Veblen implied that scientific knowledge, through the occupational discipline of industry, is increasingly becoming man's sole orientation to his

---

76. Ibid., 604. Also, see 608.
77. See Davis, "Sociological Elements in Veblen's Economic Theory," 146.
78. Ibid.
79. Ibid.

situation—nonscientific sources of orientation such as religion being the result of ignorance or error. This untenable view is to some extent characteristic of evolutionary and positivistic social thought.[80]

The clear implication of Davis's own position is that it is "unscientific" to
make judgments regarding the validity of religious beliefs, thus providing evidence once more of the epistemological underpinning of his brand
of liberalism.

The structural-functional elements that can be found in Veblen's social
theory do not a structural-functionalist of him make. But they do illuminate criticisms of his work and little attention has been paid to them. In
fact, with the exception of Robert Merton, other structural-functionalists
have strongly criticized Veblen because he ignores or downplays functional elements in the existing system! In short, Parsons, Davis, and
Rosenberg are very skeptical of his social theory because it does not incorporate the requisites of their paradigm in terms of explaining the positive role of social order and government, religion, sports, and politics. But
the focus of the rest of this chapter is on the functional nature of these and
other facets of the system in Veblen's analysis, although he mostly places
negative connotations on them. He does this by treating them as playing
fundamental roles to be sure, but as nevertheless exemplifying cultural
lag as composed of atavistic continuities. It is to these topics that I now
turn.

Veblen's social theory can be made part of the structural-functional paradigm as postulated by Malinowski's four basic requirements for a social
system (1944). These prerequisites were production and distribution, social control and regulation, education and socialization, and organization
and integration. If Veblen is treated as a structural-functionalist, the components of each of these four categories provide theoretical import for understanding him. More important, the categories listed below mesh to
form a system-maintaining and system-equilibrating social order because
the structural-functional approach, for better or worse, also assumes: (1)
the functional indispensability of norms and practices, that is, the existence of functional prerequisites and substitutes or alternatives to them;
(2) universal functionality, that is, the view that all persisting forms of culture are inevitably functional; and (3) the assumption of functional unity,
that is, the full integration of all societies. We will briefly consider how
Veblen's social theory relates to these latter three using textual exegesis,

---

80. Davis, "Veblen's Study of Modern Germany," 604.

anecdotal examples, and the casual historical empiricism of Veblen's own intellectual biography recently updated by revisionist scholarship.

Functional Indispensability, Functional Unity, and Universal Functionalism

What is functionally indispensable in Veblen's social theory and for that matter his ideal social order? In fact, the two are closely related in much of his analysis. Three illustrations might be given. They are adequate provision of "socially useful" goods properly distributed, maintenance of social order without unduly coercive authority and personal dominance, and perpetual peace among nations, ethnic groups, tribes or, at least, acceptable levels of friction and conflict within the human order national and international.

Much of Veblen's work suggests that the existing institutions, both as to structure and function, make the cost of functional unity high. One reason is that "cultural lag" plays a significant role in most of his writing and this signifies his belief that institutional and cultural life are misaligned with the development of science and technology. Nevertheless, what functional unity exists is promoted by social and cultural forces that Veblen clearly did not like, namely, emulatory consumption and nationalism-patriotism. But he does not deny their unifying effects, he only claims that their cost is excessive in that they lead to defeat or erosion of the generic ends of life.

What the structural-functional theory postulates is that atavistic continuities such as religious superstition, exploitation, coercion, honorific prowess, and invidious distinction play key roles in maintaining stasis and continuity; likewise, in Veblen's theory, nonemulatory behavior, that is, peacableness, honesty, goodwill, altruism, critical intelligence, and craftsmanship. In short, both "negative" and "positive" traits in the normative sense can perform essential functions in terms of system-maintenance. But, of course, Veblen is not content merely to make such observations; indeed, he often seeks and finds ways to satirize and mock or to justify values and behavior he dislikes. No serious reader of *The Theory of the Leisure Class* can ignore his value judgments in this regard so that it is not possible to entirely separate his social theorizing from his social criticism.

If structural-functionalism aims at explaining a system of self-maintaining stasis or even one rooted in dynamic equilibrium, it does so in part through normative acquiescence on the part of the theorist. Despite Veblen's "detachment" and putative "disinterest" in things as they are, he rarely acquiesced in the status quo. Instead, he favored an industrial republic based on an ungraded commonwealth of masterless humans (see tables 5 and 6).

So, insofar as structural-functionalist theorists have an inherent bias toward certain normative values and cultural/social practices and they converge with Veblen's, no problem arises. But to insist on the normative disaggregation of his strong preference for the generic ends of life and its associated values and practices in order to make Veblen a "structural-functionalist" defeats the analytics which separate structure from function; and it is this separation which ultimately erodes even the claims of so talented and eminent a theorist as Merton.

As I have already indicated, Veblen's interest in and knowledge of the data and literature of anthropology can probably be traced ultimately to his professor at Yale, William Graham Sumner and to the Anglo-American social scientists Edward B. Tylor, Herbert Spencer, and Lewis Henry Morgan, among other sources. His writing in an anthropological vein and his career also significantly overlap with the principal schools of early theory including evolutionism, diffusionism, historicism, and the structural-functionalism of his younger contemporaries, Bronislaw Malinowski and A. R. Radcliffe-Brown, respectively. This is not to suggest that there is marked theoretical congruence between Veblen and the latter two. Indeed, he probably had as much or more in common with Marx Engels as he did with the two British anthropologists. Yet, aside from his other ideational affinities, the question remains unanswered as to what extent and in what ways Veblen's work incorporates, wittingly or unwittingly, structural functionalism as it existed in his lifetime? A tentative answer to this question is found in the obiter dicta which follow.

Ruminations on Structural-Functionalism and Veblen

At first glance, it may appear that the criticism made by Marxists of structural-functionalism, and structural-functionalists of Marxism, might be relevant to the dialogue we have established in this chapter up to this point. Indeed, there is some validity in such a claim. To illustrate, compare Talcott Parsons's views on Veblen with what he said about Marx. About Veblen, Parsons wrote:

> The fact that a Veblen rather than a Weber gathers a school of ardent disciples around him bears witness to the great importance of factors other than

the sheer weight of evidence and analysis in the formation of "schools" of social thought.[81]

While few will question Parsons's novelty and brilliance as a theoretician, his attacks (and those of his students and disciples) on Veblen are one of the least impressive aspects of his legacy—perhaps because, like Parsons and Marx, they are very different kinds of thinkers and this is not simply due to the fact that they belonged to different generations.

To offer a few illustrations of how they were different is easy, for Veblen has no theory of equilibrium or stasis, several parts of the social systems he analyzes have no functional autonomy, institutions may be important to system-maintenance, that is, possess functional indispensability, but that does not mean there are no substitutes or alternatives to them; and functional and cultural universals may dominate many cultures, but that does not signify that the observer must acquiesce in them in their present form. It is to a brief analysis of each of these four points that I now turn.

Veblen's critics, of course, work from their own mostly centrist-liberal political-ideological slant that more than any conceptual-theoretical imperatives inspires and colors their interpretations of his work. But his implicit rebuttal of these critics starts with his view that neither equilibrium nor stasis exists in industrial society nor are they likely to evolve in the calculable future. Why? Briefly put, these societies are undergoing sustained scientific and technological change and their institutions and cultural superstructures are usually out of kilter with each other because they change at different speeds. To compound matters, cultural diffusion and absorption from other interacting systems makes cultural lag inevitable; for there is no effective way for the systems to seal themselves off from those which are more or less evolved. An external observer of Veblen's world would have to characterize it as always being in a state of disequilibrium and change.

Secondly, as to the view that different parts of any human society will exercise functional autonomy, Veblen's view is that they will not because they are powerfully integrated and restrained by the class structure, status bonds, emulatory behavior and ceremonial overlay which seeps into and fills what might otherwise be interstices or lacunae permitting functional autonomy. Power and control, that is, social hegemony thus congeal and lump instead of dispersing more or less evenly.

Structural-functional theorists allege functional indispensability, that

81. Talcott Parsons, *Essays in Sociological Theory*, 117.

is, the imperatives of religion, economic inequality, and patriotism as social necessities without which the social order cannot function harmoniously or efficiently. Veblen thought, however, that substitutes or alternatives such as secularism, equality, and neutralization of citizenship exist and could function effectively in their place. But the critics were alarmed by his negativity toward the functional dispensability of values and processes they held dear or felt were essential.

Some of these critics also believed that functional and cultural universals existed that were transcultural. In short, they insisted that social scientists acquiesce in these because of their vital role in primitive societies under study. But Veblen was not above mocking or satirizing them whether it was potlatch or satí or their equivalents in his own society.

Conclusion

This chapter is organized around Veblen's relationship to functionalism, criticisms of Veblen social theory by structural-functional theorists and evaluation of these criticisms by the author, and an effort to show the implications of his social theory for structural-functional analysis, including (a) the functional indispensability of norms and practices, that is, the existence of functional prerequisites as well as substitutes or alternatives to them; (b) universal functionality; the view that all persisting forms of culture are inevitably functional[82] and (c) the assumption of functional unity, that is, the full internal integration of all societies. It was shown that Veblen's social theory, while it is capable of being analyzed by each of these three prerequisites, can be used more effectively conversely to illuminate their political and ideological flavor. In fact, he focused on discrediting existing structures and functions and their system-maintaining propensities through elucidation of their actual roles.

Numerous examples have been drawn from textual exegesis of Veblen's writing to illustrate the premises of structural-functionalism. But the point is not to show that Veblen can best be labeled a "structural-functionalist." Rather, it is to show that although his work contains many

82. Although this analysis focuses on Veblen's sociology and anthropology, it is evident that the conventional economics of his time can also be understood using structural-functional theory as the main analytic device. See John Weeks, "Fundamental Economic Concepts and Their Application to Social Phenomena," in *The New Economic Anthropology*, ed. John Clammer (New York: St. Martin's Press, 1978), 23–24. That all cultural patterns, including the nuances of neoclassical economic theory, are "functionally indispensable" seems tautological, at best.

examples that fit the model, on the whole, it is misleading to place him in this theoretical genre because his theory lends itself more readily to the disequilibrium-conflict approach. Nevertheless, the use of a conventional structural-functional approach to analyze his thought illuminates his broader theory and also facilitates understanding of the conflictual nature of his thought on existing social orders. But it will provide little comfort to those in search of legitimization of social stasis particularly insofar as it rests on the existing structures and functions of religion,[83] patriotism, nationalism, private property, and emulatory consumption.

A theoretical appraisal of Veblen and structural-functionalism still awaits the theoretician who can locate and sanction the self-equilibrating and self-maintaining qualities of his social theory. Veblen is susceptible to treatment as a structural-functional theorist, but that would be at best a selective and, ultimately, inadequate way to interpret him. In any case, he is better understood as a theorist who uses functionalist premises to undermine the theory itself. What follows are several illustrations of this.

First, it is evident that he adopts a critical posture toward hegemonic institutions and classes in most of his work. He is critical, too, of the intellectuals and social scientists who attempt to justify and legitimize exploitation and power disparities as integral parts of the existing political and sociocultural order. In short, he may be interpreted as an opponent of those who support retention of the existing basic distribution of life chances through the institutions by which they are at present allocated. In fact, his work buttresses legitimation of claims for redistribution of both material goods and property/income. Thus his published work may accurately be interpreted as threatening capitalist property institutions as well as the existing system of hierarchy and social stratification.

Second, he challenges the structural-functional view of the social order

---

83. Arthur Vidich, too, argues that:

Veblen held to a radically secular, atheistic attitude and regarded religion as having no place in the world of learning. His secularism led him to dismiss the homiletics of ecclesiastical institutions and their officials as frauds. In his view a belief in other-worldly values was explained as a cultural lag, a failure of people's conceptions of the world to keep up with the institutional realities of their existence. He neither analyzed religious institutions nor noted that American Puritanism had placed a high value on learning and research. That Veblen failed to recognize the continuing relevance of religious values in contemporary secular institutions is a curious anomaly in his theoretical perspective. His conception of cultural lag states precisely that cultural residues from the past will continue to exert influence even after their time is past—that is, when they are no longer "serviceable." Veblen's radical secularism blinded him to his own theoretical insight.

Vidich, "The Legacy of Thorstein Veblen II," 608.

Table 5. Thorstein Veblen's Normative Proclivities

Positive Traits	Negative Traits
Secular (Scientific)	Religious (Animistic)
Egalitarian	Hierarchical
Efficient	Wasteful
Harmonious	Emulatory
Cooperative	Competitive

Sources: Veblen's published works

as based on shared values and spontaneity as leading to social stability. A conflict-free society based on mutuality of benefits where religious piety shores up social morality and serves as a diversion from recognition of the realities of power and privilege was not the meat of Veblen's dreams.

Parsons, Rosenberg, and Davis, collectively, indict Veblen for his neglect and/or misunderstanding of the system-maintaining, equilibrating functions of private property, particularly absentee ownership, religion, patriotism/nationalism, sports, gambling, and government. This list could, of course, be extended to include still other system-reinforcing traits and proclivities. But what is pertinent to grasping the main thrust of sociological criticism of Veblen by structural-functional theorists is their failure to recognize the political, cultural, and moral baggage they carry in tow with them in their critique of him. He was well aware of not only the *functional role* of each of these major aspects of the existing system in the United States, but of the *integrating* role they played in it. Most of his difference with these sociological critics rests on his view that either (1) they defeat fulfillment of the *generic ends* of life or (2) *alternatives* to, that is, more desirable *substitutes* for them might be found. Veblen was not optimistic about (2) at least in the short run. But he evidently felt that critics and outsiders should at least expose the underpinning of the system and this is, presumably, what his Parsonian-inspired sociological opponents denied he had accomplished. Veblen scholars will recognize the *functional role* each of the social beliefs and related practices play in his theory, the degree and ways in which these are *integrated* into and by the system, as well as the *alternatives* or *substitutes* which he suggests to modify or replace them. The structural-functional model can be used to indict, expose, and change existing systems, not merely analyze and legitimize them, or so Veblen's work suggests when the model is turned against its own proponents. His evolutionary naturalism provides an effective metatheoretical underpinning for such efforts.

Like Parsons, Veblen viewed "actors as socialized in ways that fulfill so-

Table 6. Contradictory Traits in Modern Social Theory: Structural Functionalism and Its Critics

Positive Traits	Negative Traits
Mutuality of Benefits	Exploitation
Shared Power	Power Disparities
Critical Posture toward Hegemonic Institutions	Class Support of Hegemonic Institutions and Classes
Emphasis on Problem of Changing the Social Order	Emphasis on Problem of Social Order and Its Maintenance
Shared Values, Spontaneity, Social Stability	Conflicting and Coerced Values, Social Disharmony

Sources: Thorstein Veblen, *The Theory of the Leisure Class* (New York: Macmillan, 1899), Alvin Gouldner, *The Coming Crisis of Western Sociology* (New York: McGraw-Hill, 1968).

ciety's need for value-consensus, role-playing skills, and motivational tendencies to conform to values and normative requirements of status position."[84] Unlike Parsons and his disciples, Veblen did not believe this would necessarily lead to either individual self-realization or collective emancipation. The result was as likely to be a stultifying social and individual inertia and the triumph of ceremonial pockets of inequity and irrationality. In short, Veblen did not view religion, patriotism, and status emulation not to mention sports and gambling as functional, that is, as mere expressions of the total structure of society with mostly integrative power. Rather he saw them as evidence of cultural lag with all that implies in his lexicon of satire and mockery.

84. Alexandra Maryanski and Jonathan Turner, *The Social Cage, Human Nature, and the Evolution of Society* (Stanford: Stanford University Press, 1992), 163.

Chapter 11

Evolutionary Naturalism: Its Progressive Enrichment

Introduction

Throughout this study the claim is made that Veblen's thought is held together by the strands of evolutionary naturalism. Nineteenth- and early-twentieth-century evolutionary science and philosophical naturalism, especially of the American variant, are major components of his thought (see tables 4, 5, and 6). This was evident by the early 1890s in his writing once he sloughed off the residues of German Idealism and Lutheran theology, which some have detected in his first published article in 1884. Suffice it to say that, when he began writing again in 1892 after a hiatus of eight years, his metaphysics and metatheoretical positions appear devoid of any supernatural postulates.

Be this as it may, our remaining task is to demonstrate the role that evolutionary naturalism plays in most of the topics around which his thought is organized in this book. Simply put, virtually every significant aspect of his thought including his social criticism is affected by his evolutionary naturalism. If at times he is made to sound even more like the middle-aged John Dewey than he really was, it is because Dewey's thought is better known to readers and will resonate more effectively in many ears. Indeed, Dewey and his fellow American naturalists were a most original and penetrating group of indigenous secular thinkers of their (and Veblen's) time. The intellectual movement(s) of which they were the catalysts is still rolling forward, converging with new, but not dissimilar currents and opening up novel lines of cultural and intellectual inquiry. It forms a main alternative to formal and informal religious belief and practice. In an age when fundamentalism and Jerry Falwell, the Roman pontiff and post-Vatican II Catholic teaching, and Mormonism provide one set of archaic alternatives, the presence and persistence of evolutionary naturalism is to be welcomed.

The values of science, technology, liberty, and equality underlie Veblen's belief that institutional and cultural retardants thwart and frustrate their realization, some more than others. These values are essentially the axiomatic core of naturalism and have shaped and been shaped by it. Yet their fulfillment in Veblen's paradigm depends on the undermining of their opposites such as predation, force, fraud, animism and patriotism as well as the reconstruction of non-invidious community through the use of instrumental knowledge. The degree of rigidity of existing institutional and cultural structures is a key variable in predicting stagnation or change as are the vigor and integrity of functional values and the efficacy of their thrust and flux. In the final analysis, it would be difficult to explain the directional changes Veblen wanted, and the forces facilitating or inhibiting these changes, without reference to his evolutionary naturalism.

Metaphysically and metatheoretically, politically and socially, culturally, aesthetically and morally, heuristically and historically, evolutionary naturalism was progressively enriched, or at least its horizons expanded by Veblen. Veblen scholars, naturally enough, have been prone to focus primarily on his institutional economics, his radical sociology, and his cultural criticism. This is appropriate, but there is more to his thought and it is my task in the rest of this chapter to summarize these as yet inadequately explored facets.

This study does not deal with the more technical aspects of Darwin and evolutionary theory, nor with its nuances and intricacies as it relates to philosophical naturalism as embedded in Veblen's thought. This has been done most recently by the Britisher Geoffrey M. Hodgson.[1] He deals not only with Veblen's Darwinism as it relates to his institutional economics,

---

1. I cannot resolve the hotly debated issue of Veblen's exact relationship to and use of Darwin. In any case, no serious Veblen scholar will deny the impact of Darwin on him or his use of evolutionary thought derived from Darwin and his disciples. Most recently, cf. Geoffrey M. Hodgson, "Darwin, Veblen and the Problem of Causality in Economics," *History and Philosophy Life Science* 23 (2001): 385–423; "Thorstein Veblen and Post-Darwinian Economics," *Cambridge Journal of Economics* 16, no. 3 (1992): 285–301; *Economy and Evolution: Bringing Life Back into Economics* (Cambridge, UK, and Ann Arbor, MI: Polity Press and University of Michigan Press, 1993); "The Approach of Institutional Economics," *Journal of Economic Literature* 36, no. 1 (1998a): 166–92; "On the Evolution of Thorstein Veblen's Evolutionary Economics," *Cambridge Journal of Economics* 22, no. 4 (1998b): 415–31; "Veblen's *Theory of the Leisure Class* and the Genesis of Evolutionary Economics," in *The Founding of Evolutionary Economics*, ed. Warren J. Samuels (London: Routledge, 1998), 170–200; *Evolution and Institutions: On Evolutionary Economics and the Evolution of Economics* (Cheltenham: Edward Elgar, 1999); "Is Social Evolution Lamarckian or Darwinian?" in *Darwinism and Evolutionary Economics*, ed. Laurent J. Nightingale (Cheltenham: Edward Elgar, 2001), 87–118;

but with the Darwinian aspects of his naturalism as well. His analysis of the metaphysics of Darwinism as natural selection, variation, adaptation, and the Lamarckian inheritance of acquired traits is accompanied by a discussion of the philosophic pivots of naturalism; namely intentionality, agency/volition, causation, emergence/novelty, teleology, individual and collective, and so forth. Readers who want to further investigate the more technical aspects of Veblen's evolutionary naturalism should consult Hodgson and his critics, some of whom are American institutional economists such as Bill Waller, Dale Bush, and Ann Jennings.[2] My own focus is not on the subtleties of evolutionary naturalism as such but on Veblen's enrichment, that is, social theoretical and critical additions to and embellishments of it. Also, it is not Veblen's economics in the narrower sense nor even his critique of neoclassicism, but other facets of his thought that I emphasize.

As a theorist, Veblen did not make the errors nor does his work contain the frailties of classical Marxism. To illustrate, he does not endorse the labor metaphysic, that is, the view that a wage-earning proletariat contains revolutionary potential and is the wave of a coming upheaval that will end the exploitation of man by man. He recognizes the deficiencies in the Marxist categories of social stratification such as the simple two-class model upon which Marx placed undue stress. Veblen avoids the error of assuming the omnipresent supremacy of economic causes in history and in explaining the mentality of classes, as well as the inadequacies of a rationalist psychological theory. Nor does he hold to the frailties of Marx's theory of power. Generally, too, he was cognizant of the "ambiguities and misjudgments about the psychological and political consequences of the development of the economic base."[3] However, he shares aspects of Marx's conception of the state as I have argued elsewhere. The reader may note that most of the above comes directly from C. Wright Mills's critique of Marx' and Engels's classical writings. In fact, every major point is taken from Mills, *The Marxists* (1962).[4]

It is simple enough to endorse most of Mills's critique of Marx since it closely parallels Veblen's in most respects. In any case, it has the advan-

2. Ann Jennings and William Waller, "Evolutionary Economics and Cultural Hermeneutics: Veblen, Cultural Relativism, and Blind Drift," *Journal of Economic Issues* 28, no. 4 (December 1994): 997–1030; "The Place of Biological Science in Veblen's Economics," *History of Political Economy* 30, no. 2 (1998): 189–217; Malcolm Rutherford, "Veblen's Evolutionary Programme: A Promise Unfulfilled," *Cambridge Journal of Economics* 22, no. 4 (July 1998): 463–77; see also Veblen, "Why Is Economics Not an Evolutionary Science?" and "The Evolution of the Scientific Point of View," in *The Place of Science in Modern Civilization.*
3. C. Wright Mills, *The Marxists* (New York: Dell Publishing Co., 1962), 129.
4. Ibid. So what separates Veblen's evolutionary naturalism from Marx' and En-

tage of being a brief but penetrating summation which is easy to translate back into the Veblenese from which much of it originally came. But since the Mills–Veblen critique of Marx is largely negative, it aids only modestly in explaining the meaning of the "progressive enrichment of evolutionary naturalism." If Veblen was a participant in this process, in what ways, with what consequences? Much of what he accomplished in this respect converged directly with the work of four other thinkers. The names of Americans John Dewey, John R. Commons, George H. Mead, and Charles Sanders Peirce come to mind and it is historically important to note that he knew them all personally. No mention is here made of the Britishers and Europeans who shared in this achievement since they will be dealt with in a subsequent study.

Specifically, what does the "progressive enrichment of evolutionary naturalism" mean as it is here used? It means first that its metatheoretical underpinnings came originally from Darwin and other evolutionary thinkers and from naturalist philosophers. Indeed, Veblen contributed little to the early stages of the Darwin debate or to the genesis of American naturalism. His contribution came at a later stage of development and is part of the transformation of evolutionary naturalism into social theory and criticism in the form of radical economics and sociology. It is in embellishing these two disciplines that Veblen progressively enriched them—"progressive" in the sense of cumulative enrichment and "progressive" in the politico-cultural sense of the values of the left-liberal side of the ideological spectrum. "Enrichment," too, in the sense of broadening and deepening the channels of economic and sociological inquiry. Also, in the social criticism a la social theory he aimed at every major institution in American life and at most of the weapons in the intellectual arsenal of the apologists, rationalizers, and defenders of these institutions.

Evolutionary Naturalism: Reiteration

Veblen's evolving evolutionary naturalism clearly set him off from the conventional cultural and behavioral traits of other Scandinavian Lutheran immigrants including most of his own family. He was exposed to the

---

gels's historical materialism in the larger philosophical sense? And how is this related to Veblen's more acute analysis of American capitalism and avoidance of the errors of classical Marxism in various ways? The answers to these two questions are important, but they lie outside the scope of this study. However, see Jaegwon Kim, *Physicalism or Something Near Enough* (Princeton, NJ: Princeton University Press, 2005), for a discussion of related philosophic, psychological, and physiological issues.

new ethnographic studies[5] in a way unlike the others, including his older brother Andrew. Andrew, for example, was familiar with Darwin's ideas, but he did not encounter the variety of cultures that Thorstein did because he was not as familiar with the data and literature of anthropology. Although an accomplished amateur historian and competent linguist with an earned doctorate in mathematics from Johns Hopkins, Andrew remained within Norwegian-American circles. This was in contrast to Thorstein's immersion in exotic ethnographic material, which, interestingly, represents a significant portion of his newly discovered library.[6]

Veblen's understanding and use of the data and literature of anthropology alone sets him off from the neoclassical economists of his time. Yet his grasp of primitive cultures and societies was not easily achieved, for by his own testimony, his study of anthropology was both a risky and arduous undertaking for a mere economist.[7] Although no adequate explanation as to why Veblen became so intrigued with anthropology nor any satisfactory analysis of his use of its data and literature exists, he showed a surprising detachment from the phenomena emphasized by contemporary economists because he was interested in calling attention to what they ignored. Because of his interest in the details of primitive culture, he delved into the contributions of anthropology, ethnology, and archeology in order to analyze the influences of the ruling classes on other institutions in obstructing the full development of the instincts of workmanship, parenthood, and idle curiosity. Veblen's knowledge of exotic cultures thus set him off from other economists and also enabled him to view his own Nor-

5. See Hugh J. Dawson, "E. B. Tylor's Theory of Survivals and Veblen's Social Criticism," *Journal of the History of Ideas* 54, no. 4 (July 1993): 489–504. Also, see John Laurent, "Charles Darwin on Human Nature," in *Evolutionary Economics and Human Nature*, ed. John Laurent (Cheltenham: Edward Elgar, 2003), 96–113.

6. See Russell Bartley and Sylvia Yoneda, "Thorstein Veblen on Washington Island: Traces of a Life," *International Journal of Politics, Culture and Society* 7 (Summer 1994): 589–614.

7.
As for the anthropological reading which I have [been] inveigled into, I do not know that it will be of much direct use; but it should be of some use in the sense of an acquaintance with mankind. Not that he is as viewed by the anthropologist is any more—perhaps he is less—human than we see him in everyday life and in ceremonial life; but the anthropological recovery should give a view of man in perspective and more in the generic than is ordinarily attained by the classical economist, and should give added breath and sobriety to the concept of "the economic man." My own keep of anthropological and ethnological lore is very meager and fragmentary, and it is somewhat presumptuous of me to offer advice, but it goes without saying that I shall want to try not to laugh at all questions that may occur to you, if you will give me a chance.

Thorstein Veblen to "my dear friend" [Sarah McLean Hardy], February 6, 1896, Dorfman Collection.

wegian-Lutheran cultural and familial life with a detachment denied to his many brothers and sisters.

Too, his vast store of learning included a substantial knowledge of the history of philosophy and religion, especially that of Western Civilization. His understanding of philosophy alone set him off biographically from the rest of his family and, indeed, from most other economists and sociologists. But it is philosophy with a particular twist because it blends evolutionary theory with naturalism, especially the American variant. Let me illustrate.

Veblen's evolutionary naturalism is related to his explanation of cultural lag in his own disciplines of economics and sociology. Evolutionary naturalism is both a derivative from and a catalyst for the growth of science and technology. Its interface with scientific inquiry and the proliferation of the machine process signifies that thinkers such as Veblen could only have developed the ideas they held in an industrial society where the ethos of science was rapidly gaining a foothold thus penetrating, although not enveloping, philosophy and social science. Not surprisingly, cultural lag would come to be used to explain the uneven permeation of human learning with the spirit of both technology and science. Nor is it surprising to note that the secularizing and egalitarianization of the byproducts of representative government would be to diminish racial, ethnic, class, and gender bias. That Veblen came to be a powerful articulator of these proclivities and should regret more than most their failure to adequately perform their historical and cultural mission is only too obvious. In short, cultural lag is a product of most of the same sociocultural vectors that produced evolutionary naturalism. The professional development of the social sciences, history, and modern philosophy as academic disciplines contains abundant evidence of this. Not surprisingly, Veblen and Durkheim in the social studies, even Turner in history, and Dewey in philosophy and psychology found their own fields infected with cultural lag and sought to remedy the damage it was doing.

Veblen's tongue-in-cheek claims to speak in "morally colorless" language and tone will deceive few. Nor will his realistic indictment of cultural lag which visibly manifests itself in his analysis of sports, religion, and gambling. Of course, as social phenomena the three are subject to naturalistic inquiry like any other phenomena. This means treating them in a matter-of-fact way, however make-believe, fashion, or sheer conjecture may have permeated or camouflaged them. In any case, evolutionary naturalism has its own criteria to make these judgments which provide standards by which to separate what is empirically demonstrable from what is make-believe. To illustrate, in sport, it enters Veblen's analysis through his separation of matter of fact from judgments based on invidious distinction, honorific prowess, and personal ferocity and fealty.

In religion, evolutionary naturalism enters Veblen's critique in several ways. First, it appears in metatheoretical form as epistemology, ontology, and axiology, that is, in terms of what can legitimately be regarded as true, real, and valuable. Second, it is visible at the level of functional consequences for both the individual and society in terms of the differences it makes in human lives. Third, it can be measured in terms of *denial* of fixed essences, eternal truths, and immutable cultural verities in favor of flux and process punctuated, perhaps, in a positive vein by morally and aesthetically consummatory experiences which are still historically and culturally specific.

In gambling, the relevance of evolutionary naturalism is evident in Veblen's indictment of the invocation of divine providence by the gambler participating in games of chance. What grounds could conceivably exist for the gambler to believe that deity or blind luck would favor him over his opponent? In any case, gambling is a waste of time and resources with a mischievous overlay of superstition which can only frustrate or defeat what Veblen takes to be the generic ends of life: critical intelligence, proficiency of workmanship, and altruism.

Veblen's ethics and aesthetics are also congruent with evolutionary naturalism and partly derived from it. In his view, value(s) including aesthetic value(s) are natural phenomena *neither* transcendental nor supernatural in origin or consequences. They are simply another facet of human experience perhaps with a consummatory phase to distinguish them from other kinds of endeavor. For Veblen never denied that aesthetic fulfillment involved a psychological or emotional pitch or peak as an organic part of its process. But he viewed this as induced by cultural experience and genetic endowment not as a partaking of the transnatural. Valuation, including aesthetic valuation, is thus a naturalistic psychosocial process accessible to the community, not merely a private state of mind. Nevertheless, aesthetics like other cultural and superstructural phenomena is susceptible to leisure-class exploitation and manipulation; and Veblen wanted to strip it of its honorific and emulatory functions in favor of a noninvidious process conducive to the edification of commoners.

Veblen's sociology of control, although it focuses on the social psychology of class and nation, is rooted in the explication of the production and transmission of subjugative, submissive, and emulatory types of behavior. Behavior is the product of the social and biological makeup of humanity and must be understood as an outgrowth of evolutionary naturalistic processes in its genesis and development. Consequently, insofar as class, hierarchy, gender, and race play a role in the sociology of control, Veblen insists that they be treated like any other natural phenomena when studied. For social control, itself, should not be understood as divinely preordained, metaphysically rationalized, or teleologically legitimated. It can only be explained through and by its empirical referents and he finds these in the biological template, social institutions, and cultural processes.[8] And as presently constituted these do not adequately lead to human emancipation or fulfillment of potential according to the generic ends of life.[9]

Veblen's "exceptionalism," namely, his efforts to deal with the claim that U.S. history and culture are different from those of Europe (and, obviously, from the rest of the world) is ambiguous. Although he comes down, in the final analysis, on the side of "non-exceptionalism," he was sensitive to the ways, social, economic, geographical, and cultural-political, in which America differed from its neighbors across the Atlantic. But, none of the myths and ideological legitimations such as "manifest destiny," "a city on a hill," and "making the world safe for democracy" appealed to him. It was doubtful if they could be derived from the historical methods and cultural techniques of evolutionary naturalism much less be sanctioned by such an approach. In Veblen's view, they remained in the realm of legal fiction, sacred myth, and manipulative illusion and provided rationales for the exploitation and subjugation of the underlying populace by elites and ruling classes. But I have dealt in more detail with American "exceptionalism" elsewhere.[10]

The revival of a Progressive tradition is perhaps in the minds of Veblen scholars insofar as they cherish social reconstruction along liberal-left doctrinal lines and many do. The Progressive's conviction that the government could become an instrument for reforming the United States was one not much shared by Veblen because he did not think the structures of the positive state were yet in place. But are the views of a man who praised

8. See Veblen, *The Place of Science in Modern Civilization*, 74–75.

9. Consider, for example, a Depression-era proposal for, and the role of, a new college named after Veblen to be located at Hightstown, New Jersey. "(a) Thorstein Veblen comes nearer than any other American to symbolizing the purpose of the college, although the name does not mean that the college will become a propagandist for his works. (b) His thinking in social and economic philosophy is indigenously American. (c) The name will arouse the interest and respect of educators and scholars of many shades of opinion. (d) The college name will help spread Veblen's fame, as he deserves. (e) Left-wing as well as liberal parties will respect the college for its name, while at the same time the name will prevent such party labels as communist being easily attached to the college, thus aiding to maintain its nonsectarian character." *Veblen College*, Oswald Veblen Collection, Library of Congress, Washington, DC.

10. See Rick Tilman, "Thorstein Veblen's Views on American Exceptionalism: An Interpretation," *Journal of Economic Issues* 39 (March 2005): 177–204.

the "discipline of the machine process" and believed it would help rid the mind of animism and superstition, and ground it in "opaque, impersonal cause and effect," adequate nutrition for contemporary reformers? The reader is entitled to make this judgment keeping in mind the cross-cutting cleavage of Veblen's analysis which pits invidious emulation, conspicuous consumption, and nationalism/patriotism against the ethos of science and machine technology, authentic representative government, and noninvidious behavior with the future up for grabs. Evolutionary naturalism, the proverbial scarlet thread which runs throughout this study, is the philosophic coagulant that underpins the study of all these. But his version of it exalts function not prescriptive status.

Veblen's sociology of knowledge, perhaps as much as any other facet of his thought, is derived from evolutionary naturalism. His value relativism and his view of the extent to which knowledge is socially constructed, neither of which is radical by postmodernist standards, placed him in an advantageous position to locate and analyze other schools of thought, intellectual traditions, and political ideologies. Since there is often no clear-cut or consistent demarcation between social criticism and theory in his work, he could identify moral, political, and teleological proclivities in others' work. Thus his sociology of knowledge rooted in evolutionary naturalism made it difficult for neoclassical economists to robe themselves in the garb of science and value neutrality without being unveiled by his discerning eye. Claims to a value-free, nonideological perspective were often grounds for mockery and satire since Veblen interpreted the social sciences through a counterfoil lens of evolutionary naturalism.

Veblen's sociology of knowledge is thus heavily evolutionary and naturalistic in that it relates the stages of social development to the increasing ability of humankind to grasp social and natural reality and articulate ideas and then theories about it. The social and cultural evolution of mankind has eliminated certain regressive ways of viewing this reality which are contaminated by animism and superstition, vested interest, and predatory impulse. Yet the possibility of the resurgence of these atavistic continuities remains.

As institutional economists have long argued, gender, ethnic, racial, and class bias all have their roots in treating equals unequally and this is contingent upon make-believe often with its tentacles in animism or related superstition. This view is derivative from Veblen and it gave him an advantage over early anthropologists in their interpretation of native peoples. Structural-functional theory in anthropology and sociology perpetrated its own myths regarding human behavior and applied these to Veblen's social theory and criticism. Veblen's evolutionary naturalism, itself derivative of the Darwinian and neo-Darwinian classics and the anthropological literature of his time, provides a philosophic backdrop for the critical analysis of such a theoretical overview. His metatheoretical stance, that is, his epistemology, ontology, and axiology, leads to a different view of the role of religion, sport, patriotism, and private property in market capitalism than that held by mainline sociologists including Talcott Parsons and his disciples.

Conclusion

Veblen was opposed to the assumptions of the politically conservative economists who belonged to the classical and what he was the first to label the "neoclassical" school. He did not agree with their confused conflation of the natural and the social sciences; their convictions regarding the immutability of economic laws; their deductive methodology; or their negativity toward the state as a tool for reform (but neither did he develop a positive theory of the state). Also, he strongly disagreed with their refusal to articulate economic policy on the basis of openly stated ethical and moral premises and social imperatives. Indeed, in place of their static view of economic institutions and practices he perceived process, flux, and dynamism as social and historical change. In place of the conservative endorsement of atomistic individualism he saw a functional relationship with the larger social organism. Veblen repudiated the rhetoric and the ideology of both natural law and natural rights, supported abandonment of the laissez-faire doctrine, satirically scrutinized the alleged identity of public and private interests, and demanded recognition of both the ethical parameters, genesis, and responsibilities of economics and economists. He held no static image of society, but, instead, looked to adjustment and adaptation as a process without end.

Beyond his attack on neoclassical economics was his effort to show that naturalism and social realism could be used effectively not only in the social studies proper but to undermine the false consciousness of religious belief and practice. He not only held a naturalistic and egalitarian view of humankind, and a dislike of supernaturalism and organized religion, but he strongly disliked the social injustice and economic inequities of his day. Yet the psychological individualism, social optimism, and social ameliorism of other political Progressives were not his either. Compassion for one's fellows were to be preferred to the pursuit of self-aggrandizement and economic success, to be sure, but this did not require naïveté and simple-minded utopian idealism.

Veblen's secular humanism is largely rooted in his evolutionary naturalism and related doctrine. As an antidote to formal religious belief and practice, it is reliant on the findings of modern science and its underpinnings, which are also found in naturalism. Although secular humanism is an outgrowth of evolutionary naturalism, Veblen did not join organizations or associations of such-minded people, nor did he actively seek to proselytize or promote such a viewpoint outside of his writings. Again, this seems to be an example of his detachment from social activism and political reconstruction at least in terms of direct personal involvement. Clearly, he preferred to work for radical change through his published work and it is this in addition to his satire and mockery for which he is remembered. The British philosopher Roy Bhaskar penetratingly comments that:

> So long as humanity survives and pays any attention to the condition of itself and its environment, there will always be a need for new, ever-deepening and more practically efficacious critical philosophy of and for the human sciences.[11]

It is to this end that this study of Veblen is directed since his evolutionary naturalism and its derivative, secular humanism, have long since become part of an indigenous critical theory which may achieve what Bhaskar recommends.

No Veblen scholar has briefly summarized the potential of Veblen's thought better than Max Lerner. Writing in 1935 he emphasized six traits. These included (1) the rigors and potentialities of the machine process, (2) the antithesis between business and industry, (3) the antisocial tendency of business enterprise, (4) legal and political institutions as the vesting of economic interests, (5) the compulsive force of idea patterns, and (6) the bankruptcy of leisure-class (business-class) values and of a culture dominated by them.[12] These traits cannot be adequately understood as Veblen intended without putting them in their cultural context and that also means understanding and using the tenets of evolutionary naturalism that undergird them. In his own time, Veblen was perturbed over the growing inequality of wealth and income, the constricted meaning of and opportunities available for leisure-time pursuits, the persistence of poverty, and the fixation on material goods for emulatory purposes. This also included the continued presence of war and the escalation of arms races

11. Roy Bhaskar, *The Possibility of Naturalism: A Philosophical Critique of the Contemporary Human Sciences* (New York: Routledge, 1998), 179.
12. Max Lerner, "What Is Usable in Veblen?" *New Republic* 83 (May 15, 1935): 7–10.

Table 7. Thorstein Veblen's Intellectual Pedigree: Positive and Negative Influences

Schools of Thought	Major Thinkers
Darwinian and Post-Darwinian Evolutionary Theory	Charles Darwin, T. H. Huxley, Alfred Russell Wallace, Hugo De Vries, G. Mendel, J. B. Lamarck
Cultural and Social Anthropology	L. H. Morgan, E. B. Tylor, Herbert Spencer, Franz Boas, Bronislaw Malinowski, W. G. Sumner
Social Darwinism: Reform and Conservative	E. A. Ross, Lester F. Ward, Herbert Spencer, W. G. Sumner
Racism and Racialism	William Z. Ripley, G. LaPouge, Count Gobineau, H. S. Chamberlain
Anglo-American Evolutionary Psychology	C. Lloyd Morgan, William James, William McDougal, Jacques Loeb
Psychoanalysis	Sigmund Freud, Carl Jung
Classical Socialism	Karl Marx, Friedrich Engels, Ferdinand La Salle
Revisionist Marxism	Karl Kautsky, Edward Bernstein, Antonio Labriola, Enrico Ferri, Henry Hyndman
American Socialism	Edward Bellamy, Eugene V. Debs, Norman Thomas
American Liberalism	Henry George, Woodrow Wilson, Henry Demarest Lloyd
British Liberalism	John Hobson, Victor Branford, Graham Wallas, Patrick Geddes, Lloyd George
Fabian Socialism	G. B. Shaw, H. G. Wells
Historians	Oswald Spengler, John F. Jameson, Charles A. Beard, Frederick Jackson Turner, Moses C. Tyler, James Harvey Robinson
Protestant Theologians	Martin Luther, Philip Melancthon, N. J. Grundtvig, John Calvin, Ulrich Zwingli

Table 7. (continued)

Schools of Thought	Major Thinkers
Classical Economists	J. E. Cairnes, Adam Smith, J. B. Say, Thomas Malthus, David Ricardo, J. S. Mill, Jeremy Bentham, John Rae, Richard Cantillon, Sismon de Simondi
Neoclassical Economists	Alfred Marshall, J. N. Keynes, W. S. Jevons, Carl Menger, E. von Böhm-Bawerk, Vilfredo Pareto, J. B. Clark, Frank Taussig, Irving Fisher, J. L. Laughlin, Knut Wicksell
German Historical School	Karl Knies, Adolf Wagner, Gustav Schmoller, Wilhelm Roscher
Reform Economics	J. M. Keynes, J. R. Commons, John Hobson, Stuart Chase, Wesley Mitchell, Herbert Davenport, Richard Ely, Walton Hamilton
Economic History	William Ashley, John Urrie, Werner Sombart
Revolutionary Marxism	Rosa Luxembourg, V. I. Lenin, Leon Trotsky
Literature: Writers and Poets	Rudyard Kipling, Edgar Lee Masters, Sinclair Lewis, William Shakespeare, Charles Dickens, Maxwell Anderson, Henrik Ibsen, Geoffrey Chaucer, John Milton, Theodore Dreiser, Hamlin Garland, William Dean Howells, Mark Twain, Jack London, Robert M. Lovett, Hans O. Storm, Jonathan Swift, George Borrow
Sociology	Albion Small, Ferdinand Tönnies, G. Tardé, W. I. Thomas, Henry Maine, W. H. Mallock, Emile Durkheim
Classical Thinkers	Plato, Aristotle, Aquinas, Augustine, Calvin, Rousseau, Burke, Montesquieu, Newton, Galileo, Machiavelli, Cervantes, Homer, Thucydides

Table 7. (continued)

Schools of Thought	Major Thinkers
Social Uplift and Philanthropy	Jane Addams, Andrew Carnegie
Anarchists and Anarcho-Syndicalists	Peter Kropotkin, Enrico Malatesta, Bill Haywood, Daniel De Leon
Journalists	George Creel, Robert Duffus, H. L. Mencken, George Soule
Populists	Jacob Coxey, James Weaver, W. J. Bryan, Ignatius Donnelly, W. H. Harvey
Technocracy	Guido Marx, Howard Scott, Morris L. Cooke
Labor Unionism	Samuel Gompers, Eugene V. Debs
Feminism	J. S. Mill, Harriet Martineau, Charlotte Perkins Gilman
Art	B. J. Nordfeldt, John Ruskin, William Morris, O. L. Triggs
Social Criticism	H. L. Mencken, Lewis Mumford, Stuart Chase, Benjamin Franklin, Felix Adler
Biography	Ida M. Tarbell
American Pragmatism	John Dewey, C. S. Peirce, William James, George H. Mead
German Idealism	Immanuel Kant, G. W. F. Hegel, Noah Porter, George S. Morris
British Empiricism	Thomas Hobbes, John Locke, David Hume, J. S. Mill
Scottish "Common Sense" Realism—Philosophy	Thomas Reid, William Hamilton

Sources: Joseph Dorfman, *Thorstein Veblen and His America* (New York: Viking Press, 1934); Rick Tilman, *Thorstein Veblen and His Critics, 1891-1963* (Princeton: Princeton University Press, 1992); *The Intellectual Legacy of Thorstein Veblen: Unresolved Issues* (Westport, CT: Greenwood Press, 1996); Becky Veblen, *Becky's Biography* (Long Version), Veblen Collection Archives, University Library, Carleton College, Northfield, MN.

**Table 8. The Meaning of Evolutionary Naturalism—
Thorstein Veblen's Version**

Religion

1. Unless and until adequate empirical evidence is offered to the contrary only the natural world can be said to exist. This signifies (a) God does not exist, (b) humans do not have souls, (c) there is no afterlife.

2. Religion, nevertheless, may provide important moral, aesthetic, and psychological insights into the human condition; but these insights must be experiential, that is, subject to pragmatic and empirical tests; not accepted on an a priori, intuitive, or revelatory basis.

Relationship between Natural and Social Sciences

1. Naturalism presupposes an external objective natural and social order which exists in part independently of investigation.

2. The interpretative qualities of the social sciences are largely neglected in mainstream economics. At the same time, the Darwinian and post-Darwinian (scientific) character of heterodox economics dictates a culturally informed version of naturalism.

3. Symmetry exists between the natural and the social sciences. But the methods and techniques used to study the two vary depending on the subject(s) of inquiry. Methodological pluralism is prescribed and requires quantitative-statistical, constructed types, comparative historical and exegetic methods, among others.

Epistemology, Ontology, Axiology in Social Science Praxis

1. No strong dualism exists between the positive/normative, fact/value, theory/practice distinctions because these break down as social science merges with its subject matter.

2. The study of society interprets subjects that already embody interpretation.

3. Values are intrinsic to theory and practice and are not introduced from outside as "basic" value judgments.

4. The ontology of evolutionary naturalism avoids any reductionism that denies the role of collective entities or, conversely, of individuals. Individual agency and social structure are both allowed as influences on behavior, with no reduction of social action to one or the other.

5. An interactionist position is thus a logical emergent from a rejection of both individualistic and structural reductionism.

6. Satire and mockery are useful antidotes to codified, formalized perspectives including the above.

Sources: Thorstein Veblen's published works, John Ryder, editor, *American Philosophical Naturalism in the Twentieth Century* (Amherst, NY: Prometheus Books, 1994), Yervant H. Krikorian, editor, *Naturalism and the Human Spirit* (New York and London: Columbia University Press, 1944). W. J. Samuels, "The Self-Referentiability of Thorstein Veblen's Theory of the Preconceptions of Economic Science," *Journal of Economic Issues* 24 (September 1990): 695-718. William A. Jackson, "Naturalism in Economics," *Journal of Economic Issues* 29 (September 1995): 761-80.

Table 9

Archaic Knowledge	Evolutionary Naturalism
Quasi-Spiritual Preconceptions: Animism	Opaque Materialistic Interpretation: Inanimate
Teleology	Causal Sequence
Spiritual Bearing	Scientific Animus
Astrology-Alchemy	Dispassionate Apprehension of Facts
Theological and Metaphysical	Empirical-Narrative Generalizations
Natural Law	Mechanical Continuity—Causal Efficacy
Superstition	Material Facts
Archaic	Up-to-date
Honorific Bearing	Equality of Status
Ceremonial Nature	Rationality and Efficiency as Prescription
Hierarchical Gradation	Egalitarian Collectivism
Invidious Comparison	Functional Relevance
Personal Fealty and Discretionary Mastery	Democratic Leadership
What one "is"	What one "does"

Source: Thorstein Veblen, *The Place of Science in Modern Civilization and Other Essays* (New York: Viking Press, 1930).

and the encroachments of imperialism abroad; gender, racial, ethnic inequality; and the persistence of religious superstition—in short, the continued ascendancy of what he regarded as "imbecile institutions," which only a pervasive evolutionary naturalism and its concomitants could checkmate.

Few competent critics of our industrial and financial order can deny the mass support that exists for business domination and control of our economic and political life. As others have suggested, business approval tends to be an essential condition and structural limitation of any moves toward favoring the generic interests of the American community as a whole. Veblen earlier and more effectively than most identified and analyzed the archaic traits that made this so essential to understanding the dominance of business culture and its pervasiveness in our social existence. America has produced no better discernment of the enduring power and malignity of archaism in the four hundred years of its intellectual history than its indictment in his writing.

To understand why this is so, his life and work must themselves be treated as evolutionary naturalist phenomena. To Veblen, the human drama and situation falls entirely within nature. The generic uniformities and variations of nature are the main determinants of the shape and mold of the natural-social order. There is no way to isolate humanity from its natural existence—whether motivated by class or individual self-interest, emotion, routine, or authority—that is, the constraints and opportunities of nature. Any common measures of value among humans as well as exotic deviations from the norm are no more or less than artifacts of the natural-social order. It is only natural that nature should manifest various tendencies and that humans should adapt themselves as one part of nature to other parts of it. Homo sapiens, in Veblen's view, is thus the biological template and the repository of prior experience both social and individual within the natural order.

Veblen attacks as "ceremonial" what appears throughout the social order in the form of deposits of irrationality and inequity which he would replace with a more rational and just existence. Underlying the latter are the forces of progressive change found in the ethos and practice of science, the secularism and egalitarianism secured by the advance of technology as well as humanistic ends; in short, the upsurge of the "generic ends of life," impersonally considered.

Yet lurking in the shadows are the persistence of atavistic continuities and imbecile institutions whose regressive resurgence in the forms of war, exploitation, waste, and superstition still threaten humankind. Veblen was no Pollyanna who believed we have reached the land of milk and honey where we can live together peaceably under conditions of voluntary simplicity and economic abundance. But, perhaps, such a tranquil order has now appeared on the historical scene and political agenda as a distinct possibility whose ultimate realization is worth striving for. In any case, as an evolutionary naturalist Veblen knew he had no choice but to act within the world as it was; and in order to change it he had to adapt himself as one part of nature to nature's other parts.

Bibliography

On most topics concerning Veblen, the reader may consult J. L. Simich and Rick Tilman, *Thorstein Veblen: A Reference Guide* (Boston: G. K. Hall, 1985), which has a listing and abstracting of literature concerning Veblen in various languages from 1891 to 1982. The reader is also referred to Solidelle and Felicity Wasser, "Veblen References, 1983–1996," in *Research in the History of Economic Thought and Methodology*, Archival Supplement 9 (Kidlington, UK: JAI Press, 1999), 275–315, which updates Simich and Tilman.

The most important archival collections in researching this study were the Thorstein Veblen Collections at Carleton College, Northfield, Minnesota, and the Wisconsin State Historical Society, Madison, Wisconsin; Joseph Dorfman Collection, Rare Book and Manuscripts Room, Butler Library, Columbia University, New York City; John Dewey Papers, Center for Dewey Studies and Morris Library, Southern Illinois University, Carbondale, Illinois; the C. Wright Mills and Clarence E. Ayres Collections, Center for American History, University of Texas, Austin; and the David Starr Jordan Collection, Stanford University Archives. The following collections were also useful.

Other Collections

Franz Boas Collection, American Philosophical Society, Philadelphia.

John P. Diggins, Correspondence and interviews with Becky Veblen and Ann Sims, 1981, Palo Alto, California.

William Rainey Harper Papers, Regenstein Library, University of Chicago, Chicago, Illinois.

Albert Ross Hill Papers, Joint Collection, University of Missouri Western

Historical Manuscript Collection, State Historical Society of Missouri Manuscripts, Columbia, Missouri.

Jacques Loeb Papers, Library of Congress, Washington, D.C.

Isador Lubin Papers, Franklin D. Roosevelt Memorial Library, Hyde Park, New York.

James Mavor Collection, Fisher Rare Book and Manuscript Library, University of Toronto, Toronto, Canada.

George H. Mead Collection, Regenstein Library, University of Chicago, Chicago, Illinois.

Talcott Parsons Collection, Harvard University Libraries, Cambridge, Massachusetts.

Noah Porter Collection, Sterling Library, Yale University, New Haven, Connecticut.

Norwegian-American Historical Association Archives, St. Olaf College, Northfield, Minnesota.

President's Office Papers (1892–1966), Joint Collection, University of Missouri Western Historical Manuscript Collection, State Historical Society of Missouri, Columbia, Missouri.

W. G. Sumner Papers, Sterling Library, Yale University, New Haven, Connecticut.

F. J. Turner Collection, Huntington Library, San Marino, California.

Andrew Veblen and Veblen Family Manuscript Biographies Collection, Minnesota State Historical Society, St. Paul, Minnesota.

Oswald Veblen Collection, Library of Congress, Washington, D.C.

Jacob Warshaw Papers, Joint Collection, University of Missouri Western Historical Manuscript Collection, State Historical Society of Missouri, Columbia, Missouri.

Articles

Adorno, Theodor. "Veblen's Attack on Culture," *Studies in Philosophy and Social Science* 9, no. 3 (1941): 389–413.

Arrow, Kenneth. "Thorstein Veblen as an Economic Theorist," *American Economist* 19, no. 1 (1975): 5–9.

Atherton, Lewis E. "The Midwestern Country Town: Myth and Reality," *Agricultural History* 26 (July 1952): 73–80.

Bartley, Russell, and Sylvia. "Thorstein Veblen on Washington Island: Traces of a Life," *International Journal of Politics and Society* 7 (Summer 1994): 589–614.

————. "Stigmatizing Thorstein Veblen: A Study in the Confection of Academic Reputations," *International Journal of Politics, Culture and Society* 14 (Winter 2000): 363–400.

Boyles, Michael, and Rick Tilman. "Thorstein Veblen, Edward O. Wilson and Sociobiology: An Interpretation," *Journal of Economic Issues* 27 (December 1993): 1195–1218.

Brette, Olivier. "Thorstein Veblen's Theory of Institutional Change: Beyond Technological Determinism," *European Journal of the History of Economic Thought* 10, no. 3 (Autumn 2003): 455–77.

Brinkmann, Svend. "Psychology as a Moral Science: Aspects of John Dewey's Psychology," *History of the Human Sciences* 17, no. 1 (2004): 1–28.

Burger, John S., and Mary J. Deegan. "George Herbert Mead on Internationalism, Democracy and War," *Wisconsin Sociologist* 18 (Spring–Summer 1981): 72–83.

Campbell, Colin. "Conspicuous Confusion? A Critique of Veblen's Theory of Conspicuous Consumption," *Sociological Theory* 13, no. 1 (March 1995): 37–47.

Chamberlain, J. E. "Oscar Wilde and the Importance of Doing Nothing," *Hudson Review* 25 (Summer 1972): 194–218.

Coats, A. W. "The Influence of Veblen's Methodology," *Journal of Political Economy* 62 (December 1954): 529–37.

Davis, Arthur. "Sociological Elements in Veblen's Economic Theory," *Journal of Political Economy* 53 (June 1945): 132–49.

————. "The Postwar Essays," *Monthly Review* 9 (July–August 1957): 91–98.

————. "Thorstein Veblen and the Culture of Capitalism," in *American Radicals: Some Problems and Personalities*, ed. Harvey Goldberg (New York: Monthly Review Press, 1957), 279–93.

————. "Thorstein Veblen Reconsidered," *Science and Society* 21 (Winter 1957): 52–85.

————. "Veblen, Thorstein," in *International Encyclopedia of the Social Sciences*, vol. 16, ed. David L. Sills (New York: Macmillan and Free Press, 1968), 303–8.

————. "Veblen on the Decline of the Protestant Ethic," *Social Forces* 22 (March 1944): 282–86.

————. "Veblen's Study of Modern Germany," *American Sociological Review* 9 (December 1944): 603–9.

Dawson, Hugh. "E. B. Tylor's Theory of Survivals and Veblen's Social Criticism," *Journal of the History of Ideas* 54 (July 1993): 489–504.

Dewey, Ernest W., and David L. Miller. "Veblen's Naturalism Versus Marxian Materialism," *Southwestern Social Science Quarterly* 35 (September 1954): 165–74.

Dewey, John. "Context and Thought," *University of California Publications in Philosophy* 12, no. 3 (1921): 203–24.

———. "Psychology and Social Practice," *Psychological Review* 7, no. 2 (1900): 105–24.

———. "The Need for Social Psychology," in *The Middle Works*, Vol. 10: 1916–1917, ed. J. A. Boydston (Carbondale: Southern Illinois University Press, 1980), 53–63.

———. "The Reflex Arc Concept in Psychology," *Psychological Review* 3, no. 4 (1896): 357–70.

Diggins, John P. "Animism and the Origins of Alienation: The Anthropological Perspective of Thorstein Veblen," *History and Theory* 16 (May 1977): 113–36.

Dimaggio, Paul, and Michael Useem. "Social Class and Arts Consumption: The Origins and Consequences of Class Differences in Exposure to the Arts in America," *Theory and Society* 5 (March 1978): 141–62.

Dugger, William. "Do Genes Hold Culture on a Leash?" *Social Science Quarterly* 62 (June 1981): 243–46.

———. "Sociobiology for Social Scientists: A Critical Introduction to E. O. Wilson's Evolutionary Paradigm," *Social Science Quarterly* 62 (June 1981): 221–33.

Eby, Clare. "Boundaries Lost: Thorstein Veblen: The Higher Learning in America and the Conspicuous Spouse," in *Prospects*, vol. 26, ed. Jack Salzman (Cambridge: Cambridge University Press, 2002), 251–93.

Edgell, Stephen. "Thorstein Veblen's Theory of Evolutionary Change," *American Journal of Economics and Sociology* 34 (July 1975): 267–80.

Edgell, Stephen, and Jules Townshend. "Marx and Veblen on Nature, History and Capitalism: Viva La Difference," *Journal of Economic Issues* 27 (September 1993): 721–40.

Edgell, Stephen, and Rick Tilman. "The Intellectual Antecedents of Thorstein Veblen: A Reappraisal," *Journal of Economic Issues* 23 (December 1989): 1003–26.

Feuer, Lewis. "Thorstein Veblen: The Metaphysics of the Interned Immigrant," *American Quarterly* 5 (Summer 1953): 99–112.

Fischler, Claude. "Interview with Edward O. Wilson," *Le Monde* (February 24, 1980): 15.

Gibbs, Jack. "Control as Sociology's Central Notion," *Social Science Journal* 27 (January 1990): 1–27.

Gilman, Nils. "Thorstein Veblen's Neglected Feminism," *Journal of Economic Issues* 33 (September 1999): 689–712.

Glade, William P. "The Theory of Cultural Lag and the Veblenian Contribution," *American Journal of Sociology and Economics* 11 (July 1952): 427–37.

Goldberg, Harvey. "Thorstein Veblen Reconsidered," *Science and Society* 21 (Winter 1957): 52–85.

Gould, Stephen J. "The Confusion over Evolution," *New York Review of Books* 39, no. 19 (November 19, 1992).

Hamilton, Walton. "The Institutional Approach to Economic Theory," *American Economic Review,* Suppl. 9 (March 1919): 312.

Hare, Peter. "The American Naturalist Tradition," *Free Inquiry* 16, no. 1 (Winter 1995–1996): 38.

Herskovits, Melville J. "The Significance of Thorstein Veblen for Anthropology," *American Anthropologist* 38 (April–June 1936): 351–52.

Hodgson, Geoffrey M. "Darwin, Veblen and the Problem of Causality in Economics," *History and Philosophy of Life Science* 23 (2001): 385–423.

———. "On the Evolution of Thorstein Veblen's Evolutionary Economics," *Cambridge Journal of Economics* 22, no. 4 (1998): 415–31.

———. "Social Darwinism in Anglophone Academic Journals: A Contribution to the History of the Term," *Journal of Historical Sociology* 17 (December 2004): 429–63.

———. "The Approach of Institutional Economics," *Journal of Economic Literature* 36, no. 1 (1998): 166–92.

———. "Thorstein Veblen and Post-Darwinian Economics," *Cambridge Journal of Economics* 16 (December 1992): 295–301.

Jennings, Ann, and William Waller. "Evolutionary Economics and Cultural Hermeneutics: Veblen, Cultural Relativism, and Blind Drift," *Journal of Economic Issues* 28 (December 1994): 997–1030.

———. "The Place of Biological Science in Veblen's Economics," *History of Political Economy* 30, no. 2 (1998): 189–217.

Kallen, Horace M. "Functionalism," *Encyclopedia of the Social Sciences,* ed. E. R. A. Seligman and Alvin Johnson, vol. 6 (New York: Macmillan Company, 1931), 523–26.

Kapp, K. William. "In Defense of Institutional Economics," *Swedish Journal of Economics* 70, no. 1 (1968): 1–18.

Kimmelman, Michael. "Spinning American Myths without Sentimentality," *New York Times* (June 21, 1996): B1, B10.

Laumann, Edward O., and James S. House. "Living Room Styles and Social Attributes: The Patterning of Material Artifacts in a Modern Urban Community," *Sociology and Social Research* 54 (April 1970): 321–42.

Leibenstein, Harvey. "Bandwagon, Snob and Veblen Effects in the Theory of Consumer's Demand," *Quarterly Journal of Economics* 64 (May 1950): 183–207.

Lerner, Max. "What Is Usable in Veblen?" *New Republic* 83 (May 15, 1935): 7–10.

Levine, Donald N. "Rationality and Freedom: Weber and Beyond," *Sociological Inquiry* 51, no. 1 (1981): 5–25.

Loader, Colin, and Rick Tilman. "Thorstein Veblen's Analysis of German Intellectualism," *American Journal of Economics and Sociology* 54 (1995): 339–56.

Lunsden, Charles, and Edward O. Wilson. "Genes, Mind and Ideology," *The Sciences* 21 (November 1981): 6–9.

———. "The Relation between Biological and Cultural Evolution," *Journal of Social and Biological Structures* 8 (1985): 343–59.

Manasse, Ernest M. "Moral Principles and Alternatives in Max Weber and John Dewey," I and II, *Journal of Philosophy* 41 (January 20 and February 3, 1944): 29–48, 57–68.

Mattson, Vernon, and Rick Tilman. "Thorstein Veblen, Frederick Jackson Turner and the American Experience," *Journal of Economic Issues* 21 (March 1987): 219–35.

Mayhew, Anne. "Contrasting Origins of the Two Institutionalisms: The Social Science Context," *Review of Political Economy* 1 (November 1989): 319–33.

Mead, George H. "Review of *An Inquiry into the Nature of Peace* by Thorstein Veblen," *Journal of Political Economy* 26 (June 1918): 752–62.

Mestrovic, Stjephan G., and Russell and Sylvia Bartley. "III. An Exchange on Veblen," *International Journal of Politics, Culture and Society* 16 (Fall 2002): 153–63.

Metzger, Walter P. "Ideology and the Intellectual: A Study of Thorstein Veblen," *Philosophy of Science* 16 (April 1949): 125–33.

Mukerji, Chandra. "Artwork: Collection and Contemporary Culture," *American Journal of Sociology* 84 (September 1978): 348–65.

Murphree, Idus. "Darwinism in Thorstein Veblen's Economics," *Social Research* 26 (June 1959): 311–24.

O'Hara, P. A., and H. J. Sherman. "Veblen and Sweezy on Monopoly Capital, Crises, Conflict and the State," *Journal of Economic Issues* 38 (December 2004): 969–88.

Parsons, Talcott. "General Theory in Sociology," in *Sociology Today*, vol. I, ed. Robert Merton, Leonard Broom, and Leonard S. Cottrell Jr. (New York: Harper and Row, 1965), 12–13.

Pfeifer, Edward J. "The Genesis of American Neo-Lamarckism," *Isis* 56, no. 2 (184) (1965): 156–67.

Rawls, A. W. "Durkheim's Epistemology: The Neglected Argument," *American Journal of Sociology* 102 (September 1996): 430–82.

Rojek, Chris. "Veblen, Leisure and Human Need," *Leisure Studies* 14 (April 1995): 73–86.

Rosenberg, Bernard. "A Clarification of Some Veblenian Concepts," *American Journal of Economics and Sociology* 12 (January 1953): 179–87.

———. "Thorstein Veblen: Portrait of the Intellectual as a Marginal Man," *Social Problems* 2 (January 1955): 181–87.

———. "Veblen and Marx," *Social Research* 15 (March 1948): 88–117.

Ruse, Michael, and Edward O. Wilson. "Moral Philosophy as Applied Science," *Philosophy* 61 (1986): 173–99.

Rutherford, Malcolm. "Thorstein Veblen and the Processes of Institutional Change," *History of Political Economy* 16, no. 3 (Fall 1984): 331–48.

———. "Veblen's Evolutionary Programme: A Promise Unfulfilled," *Cambridge Journal of Economics* 22, no. 4 (July 1998): 463–77.

Samuels, Warren J. "The Self-Referentiability of Thorstein Veblen's Theory of the Preconceptions of Economic Science," *Journal of Economic Issues* 24, no. 3 (September 1990): 695–718.

Schneider, Louis. "The Postwar Essays," *Monthly Review* 9 (July–August 1957): 91–98.

———. "Thorstein Veblen and the Culture of Capitalism," *American Radicals: Some Problems and Personalities,* ed. Harvey Goldberg (New York: Monthly Review Press, 1957).

Sheehan, Michael, and Rick Tilman. "A Clarification of the Concept of 'Instrumental Valuation' in Neoinstitutional Economics," *Journal of Economic Issues* 26 (March 1992): 731–44.

Shields, Currin V. "The American Tradition of Empirical Collectivism," *American Political Science Review* 46 (March 1952): 104–20.

Sowell, Thomas. "Veblen, Thorstein," in *The New Palgrave: A Dictionary of Economics,* vol. 4, ed. John Eatwell, Murray Milgate, and Peter Newman (New York: Stockton Press, 1987), 799–800.

Stabile, Donald R. "Thorstein Veblen and His Socialist Contemporaries: A Critical Comparison," *Journal of Economic Issues* 16 (March 1982): 1–28.

Steiner, Robert L., and Joseph Weiss. "Veblen Revised in the Light of Counter-Snobbery," *Journal of Aesthetics and Art Criticism* 9 (March 1951): 263–68.

Stocking, George W. "Lamarckianism in American Social Science, 1890–1915," *Journal of the History of Ideas* 23 (1962): 239–56.

Tilman, Rick. "Colin Campbell on Thorstein Veblen on Conspicuous Consumption," *Journal of Economic Issues* 39 (March 2006): 97–112.

———. "Control as Sociology's Central Notion: An Appraisal," *Social Science Journal* 27 (January 1990): 35–40.

———. "Durkheim and Veblen on the Social Nature of Individualism," *Journal of Economic Issues* 36 (December 2002): 1104–10.

———. "Emile Durkheim and Thorstein Veblen on Epistemology, Religion and Social Order," *History of the Human Sciences* 15 (November 2002): 51–70.

———. "Institutional Economics, Instrumentalist Political Theory, and the American Tradition of Empirical Collectivism," *Journal of Economic Issues* 35 (March 2001): 117–38.

———. "Some Recent Interpretations of Veblen's Theory of Institutional Change," *Journal of Economic Issues* 21 (June 1987): 683–90.

———. "The Aesthetics of Thorstein Veblen Revisited," *Cultural Dynamics* 10 (November 1998): 325–40.

———. "The Frankfurt School and the Problem of Social Rationality in Thorstein Veblen," *History of the Human Sciences* 12 (February 1999): 91–109.

———. "Theoretical Parallels in George H. Mead and Thorstein Veblen," *Social Science Journal* 29 (July 1992): 241–58.

———. "Thorstein Veblen and Western Thought Fin de Siécle: A Recent Interpretation," *Journal of Economic Issues* 35 (March 2001): 117–38.

———. "Thorstein Veblen: Incrementalist and Utopian," *American Journal of Economics and Sociology* 32 (April 1973): 155–69.

———. "Thorstein Veblen's Views on American Exceptionalism: An Interpretation," *Journal of Economic Issues* 39 (March 2005): 177–204.

Tilman, Rick, ed. *The Intellectual Legacy of Thorstein Veblen*, vol. 1 (Cheltenham, UK: Edward Elgar Publishing, 2003).

Tool, Marc. "A Social Value Theory," *Journal of Economic Issues* 11 (December 1977): 823–46.

Toulouse, Teresa. "Veblen and His Reader: Rhetoric and Intention in 'The Theory of the Leisure Class,'" *Centennial Review* 29 (Spring 1985): 249–67.

Veblen, Thorstein. "Christian Morals and the Competitive System," *Ethics* 20 (January 1910): 168–85.

———. "Fisher's Capital and Income," *Political Science Quarterly* 23, no. 1 (March 1908): 112–18.

———. "Gustav Schmoller's Economics," *Quarterly Journal of Economics* 16, no. 1 (November 1901): 69–93.

———. "Kant's Critique of Judgment," *Journal of Speculative Philosophy* (July 1884): 260–74.

———. "On the Nature of Capital, I: The Productivity of Capital Goods," *Quarterly Journal of Economics* 22 (August 1908): 517–42.

———. "Professor Clark's Economics," *Quarterly Journal of Economics* 22, no. 2 (February 1908): 147–95.

———. "The Food Supply and the Price of Wheat," *Journal of Political Economy* 2 (June 1893): 365–79.

———. "The Price of Wheat since 1867," *Journal of Political Economy* 1 (December 1892): 68–103.

———. "The Preconceptions of Economic Science I," *Quarterly Journal of Economics* 13, no. 2 (January 1899): 121–50.

———. "The Preconceptions of Economic Science II," *Quarterly Journal of Economics* 13, no. 4 (July 1899): 396–426.

———. "The Preconceptions of Economic Science III," *Quarterly Journal of Economics* 14, no. 2 (February 1900): 240–69.

———. "The Socialist Economics of Karl Marx and His Followers: I. The Theories of Karl Marx," *Quarterly Journal of Economics* 20, no. 4 (August 1906): 575–95.

———. "The Socialist Economics of Karl Marx and His Followers: II. The Later Marxism," *Quarterly Journal of Economics* 21, no. 2 (February 1907): 299–322.

———. "Why Is Economics Not an Evolutionary Science?" *Quarterly Journal of Economics* 12, no. 4 (July 1898): 373–97.

Weatherly, Ulysses G. "Review of Thorstein Veblen: The Instinct of Workmanship and the State of the Industrial Arts," *American Economic Review* 4 (December 1914): 860–61.

Webb, James, and Thomas De Gregori. "Notes and Communications," *Journal of Economic Issues* 37 (December 2003): 1161–74.

Weed, Frank J. "Thorstein Veblen's Sociology of Knowledge," *Review of Social Theory* 1 (September 1972): 1–11.

Wells, D. Collin. "Review of *The Theory of the Leisure Class*," *Sewanee Review* 7 (July 1899): 373.

Books

Alcock, John. *Animal Behavior* (Sunderland: Sinauer Associates, Inc., 1989).

Alexander, Thomas M. *John Dewey's Theory of Art, Experience and Nature: The Horizons of Feeling* (Albany: SUNY Press, 1987).

Ashley, David, and David M. Orenstein. *Sociological Theory Classical Statements,* 5th ed. (Boston: Allyn and Bacon, 2001).

Ayres, Clarence E. "Gospel of Technology," in *American Philosophy Today and Tomorrow,* ed. Horace Kallen and Sidney Hook (New York: Books for Libraries Press, 1968), 25–44.

———. *The Theory of Economic Progress* (Chapel Hill: University of North Carolina Press, 1944).

Ayres, Clarence E., Joseph Dorfman, and R. A. Gordon. *Institutional Economics: Veblen, Commons, and Mitchell Reconsidered* (Berkeley: University of California Press, 1964), 45–62.

Banta, Martha. *Taylored Lives* (Chicago: University of Chicago Press, 1993).

Bevir, Mark. *The Logic of the History of Ideas* (Cambridge: Cambridge University Press, 1999).

Bhaskar, Roy. *The Possibility of Naturalism: A Philosophical Critique of the Contemporary Human Sciences,* 3d ed. (London and New York: Routledge, 1998).

Boorstin, Daniel. *The Genius of American Politics* (Chicago: University of Chicago Press, 1962).

Boris, Eileen. *Art and Labor: Ruskin, Morris and the Craftsman Ideal in America* (Philadelphia: Temple University Press, 1986).

Bourdieu, Pierre. *Distinction: A Social Critique of the Judgment of Taste,* trans. Richard Nice (Cambridge: Harvard University Press, 1984).

Bowler, Peter. *The Non-Darwinian Revolution: Reinterpreting a Historical Myth* (Baltimore: Johns Hopkins University Press, 1988).

Bowles, Samuel, and Herbert Gintis. *Democracy and Capitalism* (New York: Basic Books, 1986).

Broda, Phillip. "Veblen and Commons on Private Property: An Institutionalist Discussion around a Capitalist Foundation," *Evolution and Path Dependence in Economic Ideas Past and Present,* ed. Pierre Garrouste and Stavros Ioannides (Cheltenham, UK: Edward Elgar Publishing, 2001).

Burkhardt, Richard W. "The Zoological Philosophy of J. B. Lamarck," intro. *Zoological Philosophy* (Chicago: University of Chicago Press, 1984), iv—lxvi.

Cady, Edwin. *The Big Game: College Sports and American Life* (Bloomington: Indiana University Press, 1976).

Campbell, James. *Understanding John Dewey, Nature, and Cooperative Intelligence* (La Salle, IL: Open Court Publishers, 1995).

Carter, Paul. *Revolt against Destiny: An Intellectual History of the United States* (New York: Columbia University Press, 1989).

Chandler, Charles. "Institutionalism and Education: An Inquiry into the

Implications of the Philosophy of Thorstein Veblen" (Ph.D. diss., Michigan State University, 1959).

Clayton, Bruce. *Forgotten Prophet: The Life of Randolph Bourne* (Baton Rouge: Louisiana State University Press, 1984).

Commager, Henry Steele. *The American Mind* (New Haven, CT: Yale University Press, 1950).

Curti, Merle. "The Section and the Frontier in American History: The Methodological Concepts of Frederick Jackson Turner," in *Methods in Social Science: A Case Book,* ed. Stuart A. Rice (Chicago: University of Chicago Press, 1931), 357.

Darwin, Charles. *The Descent of Man and Selection in Relation to Sex,* vol. 1 (London: John Murray, 1871).

———. *Metaphysics, Materialism and the Evolution of Mind* (Chicago: University of Chicago Press, 1980).

———. *The Descent of Man and Selection in Relation to Sex,* in *Great Books of the Western World* (Chicago: University of Chicago Press, 1982).

———. *The Origin of Species,* ed. J. W. Burrow (Middlesex: Penguin Books, 1968).

Daugert, Stanley. *The Philosophy of Thorstein Veblen* (New York: King's Crown Press, 1950).

Davis, Arthur. "Thorstein Veblen's Social Theory" (Ph.D. diss., Harvard University, 1941).

———. "Veblen Once More: A View from 1979," in *Thorstein Veblen's Social Theory* (New York: Arno Press, 1980).

Degler, Carl. *American Social Thought* (New York: Oxford University Press, 1991).

Dennes, William R. "The Categories of Naturalism," in Krikorian, ed., *Naturalism and the Human Spirit,* 270–94.

Dewey, John. *A Common Faith* (New Haven, CT: Yale University Press, 1934).

———. "Antinaturalism in Extremis," in Krikorian, ed., *Naturalism and the Human Spirit,* 1–16.

———. *Democracy and Education* (New York: Macmillan Company, 1916) [1966].

———. *Experience and Nature* (Chicago: Open Court Publishing Company, 1925) [1958].

———. *Experience and Education* (New York: Macmillan Company, 1938).

———. *Human Nature and Conduct: An Introduction to Social Psychology* (New York: Modern Library, 1922).

———. *Individualism, Old and New* (New York: Minton, Balch and Company, 1930).

———. *Logic: The Theory of Inquiry* (New York: Henry Holt and Company, 1938).

———. *Reconstruction in Philosophy* (New York: Henry Holt and Company, 1920).

———. *The Later Works,* vol. 3, ed. Jo Ann Boydston (Carbondale: Southern Illinois University Press, 1984).

———. "The Need for Social Psychology," in *The Middle Works,* Vol. 10: 1916–1917, ed. J. A. Boydston (Carbondale: Southern Illinois University Press, 1980).

———. *The Public and Its Problems: An Essay in Political Inquiry* (Chicago: Gateway Books, 1927).

———. "The Reflex Arc Concept in Psychology," in *Philosophy, Psychology and Social Practice,* ed. Joseph Ratner (New York: Capricorn Books, 1956).

———. *Theory of Valuation* (Chicago: University of Chicago Press, 1939).

Dewey, John, and J. H. Tufts. *Ethics* (New York: Henry Holt, 1910).

Diggins, John P. *The Bard of Savagery: Thorstein Veblen and Modern Social Theory* (New York: Seabury Press, 1978).

———. *Thorstein Veblen, Theorist of the Leisure Class* (Princeton, NJ: Princeton University Press, 1998).

Dobriansky, Lev. *Veblenism: A New Critique* (Washington, DC: Public Affairs Press, 1957).

Dorfman, Joseph. "New Light on Veblen," in *Essays, Reviews and Reports* (New York: Augustus M. Kelley, 1973).

———. *Thorstein Veblen and His America* (New York: Viking Press, 1934) [1966].

Dowd, Douglas. "The Strengths and Weaknesses of Veblen," in *Veblen's Century: A Collective Portrait,* ed. I. L. Horowitz (New Brunswick, NJ: Transaction Publishers, 2002), 17–40.

Dowd, Douglas, ed. *Thorstein Veblen: A Critical Reappraisal* (Ithaca, NY: Cornell University Press, 1958).

Dugger, William, ed. *Radical Institutionalism: Contemporary Voices* (Westport, CT: Greenwood Press, 1989).

Duncan, Hugh D. *Culture and Democracy* (New Brunswick, NJ: Transaction Publishers, 1989).

Durkheim, Emile. *Durkheim on Religion,* ed. and trans. W. F. Dickering (London: Routledge-Kegan Paul, 1975).

———. *The Elementary Forms of the Religious Life,* intro. K. E. Fields (New York: Free Press, 1995).

Edel, Abraham. *Ethical Theory and Social Change: The Evolution of John Dewey's Ethics, 1908–1932* (New Brunswick, NJ: Transaction Publishers, 2001).

———. "Naturalism and Ethical Theory," in Krikorian, ed., *Naturalism and the Human Spirit*, 65–95.

Edgell, Stephen. *Veblen in Perspective: His Life and Thought* (Armonk, NY, and London: Myron E. Sharpe, 2001).

Eff, Ellis Anthon. "Veblen's Paradox of Development and Manufacturing Productivity in U.S. Metropolitan Areas, 1958–1977" (Ph.D. diss., University of Texas, Austin).

Friedrich, Carl H. *Inevitable Peace* (New York: Greenwood Press, 1969).

Fuchs, Victor. *How We Live* (Cambridge: Harvard University Press, 1983).

Gambs, John. *Beyond Supply and Demand: A Reappraisal of Institutional Economics* (New York: Columbia University Press, 1946).

Goldberg, Harvey, ed. "Thorstein Veblen and the Culture of Capitalism," in *American Radicals: Some Problems and Personalities* (New York: Monthly Review Press, 1957), 279–93.

Goldstene, Paul. *The Bittersweet Century: Speculations on Modern Science and American Democracy* (Novato: Chandler and Sharp Publishers, 1989).

Gould, Stephen J. *Ever since Darwin: Reflections in Natural History* (Middlesex: Penguin Books, 1980).

Gouldner, Alvin W. *For Sociology: Renewal and Critique in Sociology Today* (New York: Basic Books, 1973).

Gramsci, Antonio. *Selections from the Prison Notebooks*, ed. and trans. Quintin Hoare and Geoffrey Nowell Smith (London: Lawrence and Wishart, 1971).

Griffin, Robert A. *Thorstein Veblen: Seer of American Socialism* (Hampden, CT: Advocate Press, 1982).

Gruber, Carol S. *Mars and Minerva: World War I and the Uses of the Higher Learning in America* (Baton Rouge: Louisiana State University Press, 1975).

Gruchy, Allan G. *Modern Economic Thought: The American Contribution* (New York: Prentice Hall, 1947).

Gunnell, John. *Political Theory: Tradition and Interpretation* (Cambridge, MA: Winthrop Publishers, Inc., 1979).

Hare, Peter. "Problems and Prospects in the Ethics of Belief," in *Pragmatic Naturalism and Realism* ed. John R. Shook (Buffalo: Prometheus Books, 2003), 239–61.

———. "Classical Pragmatism, Recent Naturalistic Theories of Representation, and Pragmatic Realism," in *The Role of Pragmatics in Contemporary Philosophy*, ed. P. Weingartner, et al. (Vienna: Hölder-Pickler-Tempsky, 1998), 58–65.

———. "Dewey, Analytic Epistemology, and Biology," in *Dewey, Pragmatism, and Economic Methodology*, ed. Elias Khalil (London and New York: Routledge, 2004), 144–52.

Harris, Janice. "Thorstein Veblen's Social Theory: A Reappraisal" (Ph.D. diss., New School for Social Research, New York, 1956).

Harris, Marvin. *Cultural Materialism: The Struggle for a Science of Culture* (New York: Random House, 1979).

Harvey, David. *The Condition of Post-Modernity* (Oxford: Basil Blackwell, 1990).

Haselberg, Peter von. *Functionalismus und Irrationalitat: Studien über Thorstein Veblen's "Theory of the Leisure Class* (Frankfurt, AM: Europaische Verlagsanstalt, 1962).

Herman, Sondra. *Eleven against War: Studies in American Internationalist Thought, 1898–1921* (Stanford, CA: Hoover Institution Press, 1969).

Hoberman, John M. *Sport and Political Ideology* (Austin: University of Texas Press, 1984).

Hodgson, Geoffrey M. *Economy and Evolution: Bringing Life Back into Economics* (Cambridge, UK, and Ann Arbor: Polity Press and University of Michigan Press, 1993).

———. *Evolution and Institutions: On Evolutionary Economics and the Evolution of Economics* (Cheltenham, UK: Edward Elgar Publishing, 1999).

———. "Is Social Evolution Lamarckian or Darwinian?" in *Darwinism and Evolutionary Economics,* ed. Laurent J. Nightingale (Cheltenham, UK: Edward Elgar Publishing, 2001), 87–118.

———. *Reconstructing Institutional Economics: Evolution, Agency and Structure in American Institutionalism* (London: Routledge, 2004).

———. *The Evolution of Institutional Economics: Agency, Structure and Darwinism in American Institutionalism* (London and New York: Routledge, 2004).

———. "Veblen's *Theory of the Leisure Class* and the Genesis of Evolutionary Economics," in *The Founding of Evolutionary Economics,* ed. Warren J. Samuels (London: Routledge, 1998), 170–200.

Hook, Sidney. *Pragmatism and the Tragic Sense of Life* (New York: Basic Books, 1974).

Horowitz, Irving Louis. *War and Peace in Contemporary Social and Philosophical Theory,* 2d ed. (New York: Humanities Press, 1973).

Howard, Michael. *The Causes of War and Other Issues* (London: Unwin Paperbacks, 1983).

Hull, David L. "Lamarck among the Anglos," *Zoological Philosophy* (Chicago: University of Chicago Press, 1984), iv–lxvi.

Jacoby, Russell. *The End of Utopia Politics and Culture in an Age of Apathy* (New York: Basic Books, 1999).

James, William. *Pragmatism and Four Essays from "The Meaning of Truth"* (Cleveland: Meridian Books, 1969).

—. *The Principles of Psychology* (New York: Henry Holt and Company, 1890).

Jensen, Robert. *Marketing Modernism in Fin-de-Siécle Europe* (Princeton, NJ: Princeton University Press, 1994).

Joas, Hans. *G. H. Mead: A Contemporary Re-Examination of His Thought* (Cambridge, MA: MIT Press, 1985).

Jones, Susan Stedman. *Durkheim Reconsidered* (Cambridge: Polity Press, 2001).

Jordan, John M. *Machine Age Ideology: Social Engineering and American Liberalism, 1911–1939* (Chapel Hill: University of North Carolina Press, 1994).

Jorgensen, Elizabeth and Henry. *Thorstein Veblen: Victorian Fire Brand* (Armonk, NY: Myron E. Sharpe, 1999).

Khalil, Elias, ed. "Dewey, Analytic Epistemology, and Biology," in *Dewey, Pragmatism, and Economic Methodology* (London and New York: Routledge, 2004).

Kim, Jaegwon. *Physicalism, or Something Near Enough* (Princeton, NJ: Princeton University Press, 2005).

Krikorian, Yervant H. "A Naturalistic View of Mind," in *Naturalism and the Human Spirit* (New York: Columbia University Press, 1944), 242–69.

Kuhn, Thomas S. *The Structure of Scientific Revolutions*, 2d ed., enlarged (Chicago: University of Chicago Press, 1970).

Lamar, Howard R. "Frederick Jackson Turner," in *Pastmasters*, ed. Marcus Cunliffe and Robin Winks (New York: Harper and Row, 1969), 74–109.

Lamarck, J. B. Intro. essays by Richard W. Burkhardt, "The Zoological Philosophy of J. B. Lamarck," and David L. Hull, "Lamarck among the Anglos," *Zoological Philosophy* (Chicago: University of Chicago Press, 1984).

Lamprecht, Sterling P. "Naturalism and Religion," in Krikorian, ed., *Naturalism and the Human Spirit*, 17–39.

Laurent, John. "Charles Darwin on Human Nature," in *Evolutionary Economics and Human Nature*, ed. John Laurent, preface by Geoffrey M. Hodgson (Cheltenham, UK: Edward Elgar Publishing, 2003), 96–113.

Lavine, Thelma. Intro. to John Dewey and Arthur Bentley, *Knowing and the Known* (Boston: Beacon Press, 1949).

—. "Naturalism and the Sociological Analysis of Knowledge," in *Naturalism and the Human Spirit*, ed. Yervant H. Krikorian (New York and London: Columbia University Press, 1944), 183–209.

Lerner, Max. Intro. to *The Portable Veblen* (New York: Viking Press, 1948).

Long, Norton. Foreword to Eugene J. Meehan, *Value Judgment and Social Science: Structures and Processes* (Homewood, IL: Dorsey Press, 1969).

Lopreato, Joseph. *Human Nature and Biocultural Evolution* (Boston: Allen and Unwin, 1984).

Lubin, Isador. "Recollections of Veblen," in *Thorstein Veblen*, ed. Carlton C. Qualey (New York: Columbia University Press, 1968), 131–48.

Lukes, Stephen. *Essays in Social Theory* (New York: Columbia University Press, 1977).

Lunsden, Charles, and Edward O. Wilson. *Genes, Mind and Culture: The Co-Evolutionary Process* (Cambridge: Harvard University Press, 1981).

MacIver, Robert. *The Web of Government*, rev. ed. (New York: Free Press, 1965).

Mannheim, Karl. *Man and Society in an Age of Reconstruction* (New York: Harcourt, Brace and Co., 1940).

———. *Sociology as Political Education*, ed. David Kettler and Colin Loader (New Brunswick, NJ: Transaction Publishers, 2001).

Margolis, Howard. *Paradigms and Barriers: How Habits of Mind Govern Scientific Beliefs* (Chicago: University of Chicago Press, 1993).

Maryanski, Alexandra, and Jonathan Turner. *The Social Cage: Human Nature and the Evolution of Society* (Stanford: Stanford University Press, 1992).

Maxwell, Mary, ed. *The Sociobiological Imagination* (Albany: SUNY Press, 1991).

McCormick, Ken. *Veblen in Plain English* (Youngstown, NY: Cambria Press, 2006).

McWilliams, Wilson Carey. *The Idea of Fraternity in America* (Berkeley: University of California Press, 1973).

Meehan, Eugene. *Contemporary Political Thought: A Critical Study* (Homewood, IL: Dorsey Press, 1967).

Merton, Robert K. *Science, Technology and Society in Seventeenth-Century England* (New York: Howard Fertig, 1970).

———. *Social Theory and Social Structure* (New York: Free Press, 1956).

Mestrovic, Stjephan G. *The Barbarian Temperament toward a Postmodern Critical Theory* (London and New York: Routledge, 1993).

Mills, C. Wright. *Power, Politics and People*, ed. I. L. Horowitz (New York: Oxford University Press, 1963).

———. *Sociology and Pragmatism* (New York: Oxford University Press, 1966).

———. *The Marxists* (New York: Dell Publishing Co., 1962).

Mitchell, Wesley, ed., with intro., *What Veblen Taught: Selected Writings of Thorstein Veblen* (New York: Augustus M. Kelley, 1964).

Morgan, C. Lloyd. *Habit and Instinct* (London: Edward Arnold, 1896).

Mouhammed, Adil H. *An Introduction to Thorstein Veblen's Economic Theory* (Lewiston, NY: Edwin Mellen Press, 2003).

Mumford, Lewis. *The Condition of Man* (New York: Harcourt, Brace and Company, 1944).

Murphey, Murray G. "Thorstein Veblen: Instinctive Values and Evolutionary Science," in *Values and Value Theory in Twentieth-Century America: Essays in Honor of Elizabeth Flower,* ed. Murray G. Murphey and Ivar Berg (Philadelphia: Temple University Press, 1988), 122–45.

Murphy, Robert F. *The Dialectics of Social Life* (New York: Basic Books, 1971).

Myers, Frances. *The Warfare of Democratic Ideals* (Yellow Springs, Ohio: Antioch Press, 1956).

Nightingale, Laurent J., ed. *Darwinism and Evolutionary Economics* (Cheltenham, UK: Edward Elgar Publishing, 2001).

Nisbet, Robert. *History of the Idea of Progress* (New York: Basic Books, 1980).

———. *The Sociology of Emile Durkheim* (New York: Oxford University Press, 1974).

Novick, Peter. *That Noble Dream: The "Objectivity Question" and the American Historical Profession* (New York: Cambridge University Press, 1988).

Odegaard, Holton. *The Politics of Truth* (Tuscaloosa: University of Alabama Press, 1971).

Ogburn, William F. *On Culture and Social Change,* ed. with intro. O. D. Duncan (Chicago: University of Chicago Press, 1964).

———. *Social Change* (New York: Viking Press, 1928).

Parker, Richard. *John Kenneth Galbraith: His Life, His Politics, His Economics* (New York: Farrar, Straus and Giroux, 2005).

Parsons, Talcott. "Distribution of Power in American Society," in *C. Wright Mills and the Power Elite,* ed. G. W. Domhoff and Hoyt Ballard (Boston: Beacon Press, 1968), 60–87.

———. *Essays in Sociological Theory* (New York: Free Press, 1963).

———. "General Theory in Sociology," *Sociology Today,* vol. 1, ed. Robert Merton, Leonard Broom, and Leonard S. Cottrell Jr. (New York: Harper and Row, 1965).

———. *Politics and Social Structure* (New York: Free Press, 1969).

———. *Social Systems and the Evolution of Action Theory* (New York: Free Press, 1977).

———. *The Social System* (New York: Free Press, 1963).

———. *The Structure of Social Action* (New York: Free Press, 1961).

Parsons, Talcott, and Neil J. Smelser. *Economy and Society* (Glencoe, NY: Free Press, 1956).

Patsouras, Louis. *Thorstein Veblen and the American Way of Life* (Montreal: Black Rose Books, 2004).

Peel, J. D. Y. *Herbert Spencer: The Evolution of a Sociologist* (London: Heinemann, 1971).

Plotkin, Sidney. "Illiberal Habits: Veblen's Theory of Power in Pre-Capitalist Cultures" (Presented at ITVA meetings, New School, May 2002).

Plotkin, Sidney, and Rick Tilman. "Thorstein Veblen and the American Tradition of Empirical Collectivism," in *The Political Thought of Thorstein Veblen* (Forthcoming).

Raines, J. Patrick, and Charles Leathers. *The Economic Institutions of Higher Education: Economic Theories of University Behavior* (Cheltenham, UK: Edward Elgar Publishing, 2003).

Randall, John Herman. "Epilogue: The Nature of Naturalism," in Krikorian, ed., *Naturalism and the Human Spirit*, 354–82.

Rasmussen, Charles, and Rick Tilman. *Jacques Loeb: His Science and Social Activism and Their Philosophical Foundations* (Philadelphia: American Philosophical Society, 1998).

Riesman, David. *Thorstein Veblen,* with new intro. by Stjephan G. Mestrovic (New Brunswick, NJ: Transaction Publishers, 1995).

———. *Thorstein Veblen: A Critical Interpretation* (New York: Charles Scribner's Sons, 1953).

Rorty, Richard. *Philosophy and the Mirror of Nature* (Princeton, NJ: Princeton University Press, 1979).

Rosenberg, Bernard. *The Values of Veblen: A Critical Appraisal* (Washington, DC: Public Affairs Press, 1956).

———. Intro. to *Thorstein Veblen* (New York: Thomas Y. Crowell, 1963), 1–14.

Ross, Dorothy. *The Origins of American Social Science* (Cambridge: Cambridge University Press, 1991).

Runes, Dogobert D., ed. *Dictionary of Philosophy* (Totowa, NJ: Littlefield, Adams and Co., 1968).

Russett, Cynthia. "Thorstein Veblen: Darwinian, Skeptic, Moralist," in *Darwin in America: The Intellectual Response, 1865–1912* (San Francisco: W. H. Freeman & Co., 1976), 147–71.

Rutherford, Malcolm. *Institutions in Economics: The Old and the New Institutionalism* (Cambridge: Cambridge University Press, 1994).

Ryan, Frank X., ed. *Social Darwinism and Its Critics* (Bristol, UK: Thoemmes Press, 2001).

Ryder, John, ed. *American Philosophic Naturalism in the Twentieth Century* (Amherst, NY: Prometheus Books, 1994).

Samuels, Warren J., ed. *The Founding of Evolutionary Economics* (London: Routledge, 1998).

Schneider, Herbert W. "The Unnatural," in Krikorian, ed., *Naturalism and the Human Spirit*, 121–32.

Searle, John. *How to Derive "Ought" from "Is" in the Is-Ought Question*, ed. W. D. Hudson (London: Macmillan, 1969), 120–34.

Shannon, Christopher. *Conspicuous Criticism, Tradition, the Individual, and Culture in American Social Thought, From Veblen to Mills* (Baltimore: Johns Hopkins University Press, 1996).

Shook, John R., ed. *Pragmatic Naturalism and Realism* (Buffalo, NY: Prometheus Books, 2003).

Smith, Benjamin. "The Political Theory of Institutional Economics" (Ph.D. diss., University of Texas, 1969).

Sowell, Thomas. *"Veblen, Thorstein,"* in *The New Palgrave: A Dictionary of Economies*, vol. 4, ed. John Eatwell, Murray Milgate, and Peter Newman (New York: Stockton Press, 1987).

Spengler, J. J. *Weltwirtschaftliches Archiv: Zeitschrift des Institus fue Weltwirtschaft and der Universitat Kiel* 82, no. 1 (1959): 35–67.

Spindler, Michael. *Veblen: Modern America Revolutionary Iconoclast* (London: Pluto Press, 2002).

Stark, Werner. *The Sociology of Knowledge* (London: Routledge and Kegan Paul, 1958).

Stocking, Jr., George W. *Anthropology at Chicago* (Chicago: University of Chicago Press, 1979).

Strong, Edward W. "The Materials of Historical Knowledge," in Krikorian, ed., *Naturalism and the Human Spirit*, 154–82.

Sweezy, Paul, and Paul Baran. *Monopoly Capital* (New York: Monthly Review Press, 1966).

Tillich, Paul. *The Interpretation of History* (New York: Charles Scribner's Sons, 1959).

Tilman, Rick. *C. Wright Mills: A Native Radical and the American Intellectual Tradition* (University Park: Penn State University Press, 1984).

———. "Introduction" to Frank X. Ryan, ed., *Social Darwinism and Its Critics* (Bristol, UK: Thoemmes Press, 2001).

———. *The Intellectual Legacy of Thorstein Veblen: Unresolved Issues* (Westport, CT: Greenwood Press 1996).

———. *Thorstein Veblen and His Critics, 1891–1963: Conservative, Liberal and Radical Perspectives* (Princeton, NJ: Princeton University Press, 1991).

———. *Thorstein Veblen, John Dewey, C. Wright Mills, and the Generic Ends of Life* (Lanham, MD: Rowman & Littlefield, 2004).

Tilman, Rick, ed. *The Legacy of Thorstein Veblen*, vol. 1 (Cheltenham, UK: Edward Elgar Publishing, 2003).

Tilman, Rick, and Robert Griffin. "The Aesthetics of Thorstein Veblen Revised," *Cultural Dynamics* 10 (November 1998): 325–40.

Tilman, Rick, and Cyrill Pasterk. *The Intellectual Legacy of Thorstein Veblen: Unresolved Issues* (Westport, CT: Greenwood Press, 1966), ch. 5.

Tilman, Rick, and Ruth Porter Tilman. "Veblen's Leisure Class Theory and Legalized Gambling: Jackpot Realism," in *Thorstein Veblen in the Twenty-First Century*, ed. Doug Brown (Cheltenham, U.K.: Edward Elgar Publishing, 1998): 104–16.

Timasheff, Nicholas S. *Sociological Theory: Its Nature and Growth* (New York: Random House, 1964).

Tiryakian, Edward. "Emile Durkheim," in *A History of Sociological Analysis*, ed. Tom Bottomore and Robert Nisbet (New York: Basic Books, 1978), 187–236.

Tool, Marc. *The Discretionary Economy: A Normative Theory of Political Economy* (Santa Monica, CA: Goodyear Publishing Company, 1979).

Toulouse, Teresa. "Veblen and His Reader: Rhetoric and Intention in 'The Theory of the Leisure Class,'" *Centennial Review* 29 (Spring 1985): 249–67.

Turner, F. J. *Rise of the New West* (New York: Harper and Brothers, 1906).

———. *Sections in American History* (New York: Henry Holt, 1932).

———. *The Frontier in American History* (New York: Henry Holt, 1920).

———. "The Significance of the Frontier in American History," in *The Turner Thesis concerning the Role of the Frontier in American History*, rev. ed., ed. with intro. George Rogers Taylor (Boston: D. C. Heath and Company, 1956), 1–18.

———. *The United States, 1830–1850* (New York: Henry Holt, 1935).

Veblen, Thorstein. *Absentee Ownership and Business Enterprise in Recent Times* (New York: B. W. Huebsch, 1923).

———. *Absentee Ownership and Business Enterprise in Modern Times*, intro. Robert Lekachman (Boston: Beacon Press, 1967).

———. *Absentee Ownership* (New Brunswick, NJ: Transaction Publishers, 1997).

———. *An Inquiry into the Nature of Peace and the Terms of Its Perpetuation* (New York: Ben Huebsch, 1917).

———. *Essays in Our Changing Order*, ed. Leon Ardzrooni (New York: Augustus M. Kelley, 1964).

———. *Imperial Germany and the Industrial Revolution* (New York: B. W. Huebsch, 1915).

———. *Imperial Germany and the Industrial Revolution* (New York: Augustus M. Kelley, 1964).

———. "Mr. Cummings' Strictures on 'The Theory of the Leisure Class,'" in *Essays in Our Changing Order,* ed. Leon Ardzrooni (New York: Augustus M. Kelley, 1964).

———. *The Engineers and the Price System* (New York: Augustus M. Kelley, 1965).

———. *The Higher Learning in America* (1918) (New York: Augustus M. Kelley, 1965).

———. *The Higher Learning in America: A Memorandum on the Conduct of Universities by Business,* intro. David Riesman (Stanford: Stanford University Press, 1954).

———. "The Intellectual Pre-eminence of Jews in Modern Europe," in *Essays in Our Changing Order,* ed. Leon Ardzrooni (New York: Augustus M. Kelley, 1964).

———. *The Instinct of Workmanship* (New York: B. W. Huebsch, 1918).

———. *The Laxdaela Saga,* translated from the Icelandic with intro. (New York: B. W. Huebsch, 1925).

———. "The Place of Science in Modern Civilization," in *The Place of Science in Modern Civilization and Other Essays* (New York: Viking Press, 1930).

———. *The Place of Science in Modern Civilization and Other Essays* (New York: Transaction Publishers, 1991).

———. *The Theory of Business Enterprise* (New York: Mentor Books, 1958).

———. *The Theory of Business Enterprise* (New York: Augustus M. Kelley, 1975).

———. *The Theory of the Leisure Class* (New York: Macmillan Company, 1899).

———. *The Theory of the Leisure Class,* with intro. by C. Wright Mills (New York: New American Library, 1953).

———. *The Theory of the Leisure Class* (London: Unwin Books, 1970).

———. *The Theory of the Leisure Class* (New York: Augustus M. Kelley, 1965).

———. *The Vested Interests and the Common Man* (New York: Capricorn Books, 1969).

———. *The Vested Interests and the State of the Industrial Arts* (New York: B. W. Huebsch, 1919).

———. "Why Is Economics Not an Evolutionary Science?" and "The Evolution of the Scientific Point of View," in *The Place of Science in Modern Civilization and Other Essays* (New Brunswick, NJ, and London: Transaction Publishers, 1998).

Vivas, Eliseo. "A Natural History of the Aesthetic Transaction," in Krikorian, ed., *Naturalism and the Human Spirit,* 96–120.

Vollmer, Gerhard. "How Is It That We can Know This World?" in *Dar-*

winism and Philosophy, ed. Victorio Höslet and Christian Illis (Notre Dame: University of Notre Dame Press, 2005), 259–74.

Weber, Max. *The Protestant Ethic and the Spirit of Capitalism,* trans. Talcott Parsons (New York: Scribner's, 1930).

———. *From Max Weber,* trans., ed., and with an introduction by Hans H. Gerth and C. Wright Mills (New York: Oxford University Press, 1946).

———. *The Religion of China,* trans., ed. Hans H. Gerth, with an introduction by C. K. Yang (New York: The Free Press, 1964).

———. *The Theory of Social and Economic Organization,* ed. with an introduction by Talcott Parsons (New York: The Free Press, 1967).

———. *Ancient Judaism,* trans., ed. Hans H. Gerth and Don Martindale (New York: The Free Press, 1967).

———. *Economy and Society,* ed. Guenther Roth and Claus Wittich (New York: Bedminster Press, 1968).

Weeks, John. "Fundamental Economic Concepts and Their Application to Social Phenomena," *The New Economic Anthropology,* ed. John Clammer (New York: St. Martin's Press, 1978), 21–30.

Weingartner, Paul, Gerhard Schurz, and Georg Dorn, eds. "Classical Pragmatism, Recent Naturalistic Theories of Representation, and Pragmatic Realism," *The Role of Pragmatics in Contemporary Philosophy* (Vienna: Hölder-Pickler-Tempsky, 1998).

White, Morton. *Social Thought in America: The Revolt against Formalism* (Boston: Beacon Press, 1949).

Wilde, Oscar. "The Soul of Man under Socialism" (1891), *De Profundis and Other Writings,* intro. Hesketh Pearson (New York: Penguin Books, 1986), 17–53.

Wilson, Edward O. *On Human Nature* (Cambridge: Harvard University Press, 1978).

———. *Nature Revealed: Selected Writings, 1949–2006* (Baltimore: Johns Hopkins University Press, 2006).

Wolfe, A. B. "Functional Economics," *The Trend of Economics,* ed. Rexford G. Tugwell (New York: Alfred E. Knopf, 1924), 447–75.

Wolff, Kurt H. "Cultural Lag," in Julius Gould and William Kolb, eds., *A Dictionary of the Social Sciences* (New York: Free Press, 1964).

Ylvisaker, Erling. *Eminent Pioneers: Norwegian American Pioneer Sketches* (Freeport, NY: Books for Libraries Press, 1970).

Young, Michael W. *Malinowski: Odyssey of an Anthropologist, 1884–1920* (New Haven, CT: Yale University Press, 2004).

Index

Absentee ownership, 86–87, 89, 93, 182
Absentee Ownership and Business Enter-
prise in Recent Times (Veblen), 35, 86–
87; on religion, 119, 123
Adaptation, 201; aesthetics and, 148,
155; to demands of environment, 15–
17, 93, 107, 191, 204; Lamarck's theories of, 191, 255; natural selection *vs.*,
216; social, 195, 303; in sociobiology,
202–8
Addictions, 113, 131–32
Adler, Felix, 36n39
Adorno, Theodor, 115–16
Aesthetics: class and, 142–43, 148–50,
152; definitions of, 156–57; deterioration of, 143–44; experience of, 147–
49; of pure and applied art, 149–52;
Rosenberg's criticism of Veblen on,
275–76; utilitarianism and, 146–48;
Veblen's, 5, 141, 144–47, 152–59; of
workmanship, 142–43, 149–51
Agency, 26, 40
Agnosticism, Veblen's, 2. *See also* Religion
Altruism, 171–72; definitions of, 209–10;
in generic ends of life, 197; patriotism
and, 178; propensity toward, 174;
Rosenberg's criticism of Veblen on,
273–75; in social Darwinism, 194; in
sociobiology, 195
Anarchy, 264, 277
Animism, 110, 302; persistence of, 33–
35, 106; in religion, sports, and gambling, 5, 133, 137; Veblen rejecting, 70,
95, 97–98, 119
Anthropology, 259, 262; of religion, 127–

29; Veblen's use of, xiii, 7, 10, 259–60,
298, 302–3
"Anthropology and Ethics" (Dewey),
106
Apperceptive mass, 50
Apriorism, 18–19
Art. *See* Aesthetics
Arts and crafts societies, 149–52
Atavistic continuities, 69n55; persistence
of, xiii, 310; resurgence of, xiii, xvi, 67,
186–87, 261; in structural functional-
ism, 287. *See also* Force and fraud;
Superstition
Austria, 177
Authoritarianism, 35; intercollegiate
sport and, 112; Veblen's expectation
of, 57, 78
Ayres, Clarence, 207n36, 235; as empiri-
cal collectivist, 56–57; on technology,
63, 77; on Veblen, 27, 55–56, 68

Bartley, Russell, 7
Beard, Charles A., 105
Becker, Gary, 221
Behavior, 192, 199, 216, 287; causes of,
205–7, 212–13, 300–301; influences
on, 85, 205–6, 215; learned, 206, 209;
myths of, 302–3; sociobiology used to
justify, 204–5
Behavioral change, 57; difficulty of, 78,
254–55; knowledge problem and, 79–
80; sociobiology on, 202–3; through
eras of human history, 203–4
Beliefs, truth of, 95
Bell, Daniel, 12
Benedict, Ruth, 199

333

Permissions

Naturalism and the Human Spirit, Yervant H. Krikorian, Editor. Columbia University Press, 1944.

Talcott Parsons Collection, Harvard University Archives.

Association for Evolutionary Economics, Rick Tilman, "Karl Mannheim, Max Weber and the Problem of Social Rationality in Thorstein Veblen," *Journal of Economics Issues* 38 (March 2004): 155–72; and Michael Boyles, "Thorstein Veblen, Edward O. Wilson and Sociobiology: An Interpretation," *Journal of Economic Issues* 27 (December 1993): 1195–1218; and Stephen Edgell, "The Intellectual Antecedents of Thorstein Veblen: A Reappraisal," *Journal of Economic Issues* 23 (December 1989): 1003–26, all reprinted by special permission of the copyright holder, the Association for Evolutionary Economics.

"The Aesthetics of Thorstein Veblen Revisited," by Robert Griffin and Rick Tilman, in *Cultural Dynamics* 10 (November 1998): 325–40, reprinted by permission of Sage Publications, Ltd.

The Western Social Science Association, "Control as Sociology's Central Notion: An Appraisal," *Social Science Journal* 27 (January 1990): 35–40.

Rick Tilman and Ruth Porter Tilman, "Veblen's Leisure Class Theory and Legalized Gambling: Jackpot Realism," in Doug Brown, editor, *Thorstein Veblen in the Twenty-first Century,* Edward Elgar Publishing, 1998: 104–16.

Eric Hilleman, Archivist, Carleton College Archives, Gould Library, Thorstein Veblen Collection.

Esther Baran, Becky Veblen Meyers correspondence with John P. Diggins.

Joseph Dorfman Papers, Rare Book and Manuscript Library, Columbia University.

Rick Tilman, *Thorstein Veblen and His Critics, 1891–1963,* Princeton University Press, 1992.